GALEN ON THE BRAIN

STUDIES IN ANCIENT MEDICINE

EDITED BY

JOHN SCARBOROUGH

PHILIP J. VAN DER EIJK
ANN HANSON
NANCY SIRAISI

VOLUME 26

GALEN ON THE BRAIN

Anatomical Knowledge and Physiological Speculation in the Second Century AD

BY

JULIUS ROCCA

BRILL
LEIDEN · BOSTON
2003

Cover Illustration

Galen in Roman dress forms part of a striking stained-glass triptych in the Old Medical School, University of Sydney, Australia. With kind permission of Professor Jonathan Stone, Department of Anatomy, and Professor Anne Sefton, Department of Physiology, University of Sydney.

This book is printed on acid-free paper.

Die Deutsche Bibliothek - CIP-Einheitsaufnahme

Rocca, Julius:

Galen on the brain
Anatomical knowledge and physiological speculation in the second century AD/ by Julius Rocca. – Leiden ; Boston : Brill, 2003
(Studies in ancient medicine ; Vol. 26)
ISBN 90-04-12512-4

Library of Congress Cataloging-in-Publication Data

The Library of Congress Cataloging-in-Publication Data is also available

ISSN 0925-1421
ISBN 90 04 12512 4

PRINTED IN THE NETHERLANDS

To my mother, Joan Alice
and
to the memory of her parents,
Emma Jane Crowley and Thomas Samuel Bull

CONTENTS

PART TWO

THE ARCHITECTURE OF THE BRAIN

PART THREE

A PHYSIOLOGY OF THE BRAIN

ACKNOWLEDGMENTS

This work was made possible by a two year Wellcome Trust Post-doctoral Research Fellowship held in the Faculty of Classics, University of Cambridge, 1997–1999.

A book should stand on its own merits, but it is appropriate to sketch the circumstances of its creation. My first qualification is medical. An abiding interest in the history of my profession, especially its Greek origins, brought me to pursue, also at the University of Sydney, a BA and an MA in philosophy, followed by a Ph.D. which examined the concept of *pneuma* in Galen's physiology. The generous encouragement of my supervisor, Professor Paul Crittenden of the School of Philosophy, enabled my associate supervisor, Harold Tarrant, then Senior Lecturer in Classics at the University of Sydney and now Professor of Classics at the University of Newcastle, Australia, to encourage and develop my interest in ancient philosophy and propelled me in two significant directions, Galen and Cambridge. Professor Tarrant kindly read through earlier draft versions of this text and has made many incisive and constructive suggestions.

In Cambridge I was privileged to work with Professor Sir Geoffrey Lloyd, first as a graduate student in 1992–93, and then as a Wellcome Trust Research Fellow. He raised the idea that an interdisciplinary approach to Galen's handling of the brain would prove interesting. Professor Vivian Nutton encouraged me to apply for a Wellcome Fellowship, and has throughout strongly supported the interdisciplinary nature of this work. Professor Tore Frängsmyr, Professor of History of Science at Uppsala University, kindly agreed to serve as a referee in my Wellcome application and has been an enthusiastic sounding board to my ideas. I should like to thank Professor Paul Cartledge and the Faculty Board of Classics, Cambridge, for the generous use of their facilities. I am also grateful to Professor David Sedley for his comments on part of this work, and to Professor Heinrich von Staden for his encouragement.

I also wish to thank Professor William Brown, Master of Darwin College, Cambridge, for allowing me to stay in such a pleasant and stimulating environment whilst I revised the manuscript.

The dissections for this work would not have been possible without the help of the Department of Veterinary Anatomy, University of Uppsala. Professor Stig Drevemo listened to my requests and placed himself, his technical support staff and the facilities of his department at my disposal. I must also thank Dr Ian Parkin, Department of Anatomy, University of Cambridge, for taking the time to listen to my interpretation of Galenic brain anatomy and helping to verify some points of detail.

I owe a great deal to Professors Philip van der Eijk and Ann Hanson, whose encouragement and scrupulously thorough and fair criticisms and suggestions helped guide this book to publication. Their comments, together with those of the two anonymous readers, have greatly improved the text and saved me from several embarrassments. I am also grateful to the Brill editorial team, especially Michiel Klein Swormink and Ms Gera van Bedaf, for their patience and professionalism. All remaining errors of fact and interpretation remain mine alone.

I owe so much to my wife Hélena, who has been more than patient through the various ups and downs of this work's progress. My mother has been the lodestar for much of my life, and in all the sea changes of my career paths has provided counsel, support and encouragement It is to her and to the memory of her parents that I humbly dedicate this book.

Stockholm
October, 2002

ABBREVIATIONS

AGM	*Archiv für Geschichte der Medizin.* This journal is now *Sudhoffs Archiv*, and is abbreviated *ZWG* (*Zeitschrift für Wissen-schaftsgeschichte*). Older issues are abbreviated as *AGM.*
ANRW	Haase, W., Temporini, H. (eds.), *Aufstieg und Niedergang der römischen Welt*, Berlin, 1972–.
BHM	*Bulletin of the History of Medicine.*
CAG	*Commentaria in Aristotelem Graeca.*
CAH²	Bowman, A.K., Garnsey, P., Rathbone, D. (eds.), *The Cambridge Ancient History. Volume XI*, Second Edition, Cambridge, 2000.
CHCL	Easterling, P.E., Knox, B.M.W. (eds.), *The Cambridge History of Classical Literature. Volume I part 4. The Hellenistic Period and the Empire*, Cambridge, 1989.
CMG	*Corpus Medicorum Graecorum.* Ed. Academiae Berolinensis, Hauniensis, Lipsiensis, Leipzig-Berlin, 1947–.
CML	*Corpus Medicorum Latinorum.* Ed. Consilio et auctoritate instituti Puschmanniani Lipsiensis, Leipzig-Berlin, 1947–
Deichgräber	Deichgräber, K., *Die griechische Empirikerschule*, Berlin-Zurich, Second Edition, 1965.
DG	Diels, H. (ed.), *Doxographi Graeci.* Berlin, Leipzig, 1879. (reprinted: Berlin, 1958).
DK	Diels, H., Kranz, W. (eds.), *Die Fragmente der Vorsokratiker*, 3 vols., Berlin, 1961^{10}.
DNP	Cancik, H., Schneider, H. (eds.), *Der Neue Pauly*, Stuttgart-Weimar, 1998–.
Herophilus	Von Staden, H., *Herophilus. The Art of Medicine in Early Alexandria*, Cambridge, 1989.
HGP	Guthrie, W.K.C., *History of Greek Philosophy*, vols. I–VI, Cambridge, 1962–81.
J.Hist.Med.	*Journal of the History of Medicine and Allied Sciences.*
K	Kühn, C.G. (ed.), *Claudii Galeni Opera omnia*, vols. 1–20, Lipsiae, 1821–1833 (reprinted: Hildesheim, 1964–65; Oxford, 1976).
KRS	Kirk, G.S., Raven, J.E., Schofield, M., *The Presocratic Philosophers*, Second Edition, Cambridge, 1991.
L	Littré, E., *Œuvres complètes d'Hippocrate.* 10 vols., Paris, 1839–61.
L-MS	*Claudius Ptolemaeus. On the Kriterion and Hegemonikon*, ed. Liverpool-Manchester Seminar on Ancient Greek philosophy (L-MS), in P. Huby and G. Neal (eds.), *The Criterion of Truth*, pp. 179–230, Liverpool, 1989.
L-S	Long, A.A., Sedley, D.N., *The Hellenistic Philosophers.* 2 vols., Cambridge, 1989.

LSJ	Liddell, H.G., Scott, R., Jones, H.S., *A Greek-English Lexicon*, Oxford, 1996^9 (with revised supplement).
Med.Hist.	*Medical History*.
OCD	Hornblower, S., Spawforth, A., (eds.), *The Oxford Classical Dictionary*, Third Edition, Oxford, 1996.
ODB	Kazhdan, A.P., Talbot, A-M., Cutler, A., Gregory, T.E., Sevcenko. W.P. (eds.), *The Oxford Dictionary of Byzantium*, 3 vols., Oxford, 1991.
RE	Wissowa, G. et al. (eds.). *Paulys Real-Encyclopädie der classichen Alterumswissenschaft*, Stuttgart-Munich, 1893–.
SM	Marquardt, J., Müller, I., Helmreich, G. (eds.). *Claudii Galeni Pergameni Scripta minora*, 3 vols., Leipzig, 1884, 1891, 1893.
Steckerl	Steckerl, F., *The Fragments of Praxagoras and His School*, Leiden, 1958.
SVF	Von Arnim, H. (ed.), *Stoicorum Veterum Fragmenta*, 4 vols., Leipzig, 1903–1924.
vdE	Van der Eijk, P., *Diocles of Carystus*. A Collection of the Fragments with Translation and Commentary. Volume one: Text and Translation, Leiden, 2000.
ZWG	*Zeitschrift für Wissenschaftsgeschichte*. See *AGM*, above.

Hippocratic texts are cited by volume and page number of the Littré edition (L), as well as by editor, page and line number of the most recent critical editions in the *Corpus Medicorum Graecorum* (*CMG*) or in the *Collection des Universitiés de France* (Budé). Works of the Galenic Corpus are cited in accordance with the critical editions of the *CMG* or the Teubner series (where these exist) as well as by volume and page number from Kühn's edition, where they exist (K). Thus *De praenotione* p. 94,12–15 Nutton; XIV. 625 K refers to the page and line numbers of the CMG edition (V 8, 1), and volume XIV, p. 625 of Kühn's edition. There are as yet no standard abbreviations for Galen's works. LSJ list only a few, which are used here. For all other texts, the abbreviations used are based on Hankinson (1991c), and for convenience are placed beside each work in the *Index of Sources and List of Editions*. Other Greek medical authors are cited in the text according to the most recent critical editions (again where these exist). Thus "Fr. 1 vdE", refers to the first fragment in the edition of *Diocles of Carystus* by P. Van der Eijk, 2000. "T 1 *Herophilus*" means the first text in *Herophilus of Alexandria* by H. Von Staden, 1989. The Latin medical writer Aulus Cornelius Celsus is cited by page number and line number of the CML edition by Marx and by page number of the Budé edition by Serbat (*praefatio*). Plato and Aristotle are cited according to the *Oxford Classical Texts*. The critical edition of Roselli is used for the Pseudo-Aristotelian work *De spiritu*.

Unless otherwise indicated, all other abbreviations in this work are used according to Liddell, Scott, Jones, with the 1996 revised supplement (LSJ, above), P.G.W. Glare (ed.), *Oxford Latin Dictionary*, Oxford, 1982, and S. Hornblower and A. Spawforth, (eds.), *The Oxford Classical Dictionary*, Third Edition, Oxford, 1996 (*OCD*, above).

ILLUSTRATIONS

Fig. 1. Base of the ox brain. Scale not given. From: Sisson, S., *The Anatomy of the Domestic Animals*, Philadelphia-London, 1930, p. 836.

Fig. 2. Midline section through the ox brain. Scale not given. From Sisson, 1930, p. 838.

Fig. 3. Right side of the ox brain. Scale not given. From: Sisson, 1930, p. 837.

Fig. 4. Retiform plexus in the calf. Scale not given. From: Solly, S., *The Human Brain: its Structure, Physiology and Diseases*, London, 1847, p. 136.

Fig. 5. Photograph of the ox brain from behind. The upper cranial bones have been removed to show the rear of the cerebrum, the cerebellum, and the upper part of the spinal cord (Scale in cm. Dissection and photograph by the author).

a. The left cerebral hemisphere, meningeal coverings intact.
b. Cerebellum covered by meninges.
c. Meningeal coverings split to reveal surface of right cerebral hemisphere.
d. Top of spinal cord.

Fig. 6. Photograph of the ox brain from behind. The upper cranial bones and the meningeal coverings have been removed. The cerebellum has been removed to expose the floor of the fourth ventricle (Scale in cm. Dissection and photograph by the author).

a. Left cerebral hemisphere reflected to show inner surface.
b. Pia mater.
c. Part of anterior medullary velum dissected from the cerebellum (and slightly displaced upwards).
d. Cut edge of vermis of cerebellum.
e. Centre of floor of fourth ventricle showing characteristic midline groove. The approximate shape of the calamus scriptorius can also be discerned.
f. Reflected part of tentorium cerebelli remaining after removal of cerebellum.
g. Right cerebral hemisphere overlying corpus callosum.
h. Part of choroid plexus in middle (third) ventricle.

i. Cut edge of dura mater.
j. Pineal.
k. Right gloutia.
l. Branchium conjunctivum covering right testis.
(k + l = corpora quadrigemina).
m. Opening of aqueduct ("duct of Erasistratus") leading into fourth ventricle.
n. Cut edge of inferior cerebellar peduncle.
o. Occipital condyle.

Fig. 7. Photograph of the inner aspect of the base of the ox skull. The brain has been completely removed and part of the dura covering the skull base has been excised (Scale in cm. Dissection and photograph by the author).
a. Cut edge of skull.
b. Needle holding part of left retiform plexus which has been displaced from the cavernous sinus and teased out.
c. Remainder of left retiform plexus in situ.
d. Cut edge of right optic nerve.
e. Pituitary (cut).
f. Cut edge of dura mater overlying right cavernous sinus.
g. Right retiform plexus *in situ.*
h. Cut edge of middle cranial fossa.

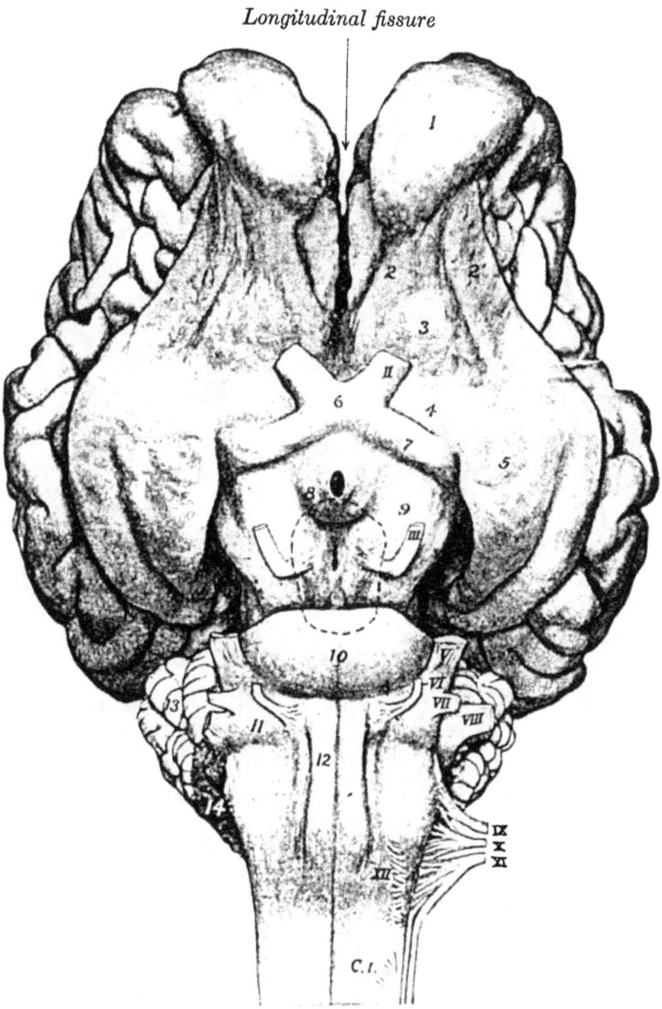

1, Olfactory bulb; *2,2'*, olfactory striae; *S*, trigonum olfactorium; *4*, fossa lateralis; *5*, piriform lobe; *6*, optic chiasma; *7*, optic tract; *8*, tuber cinereum; *9*, cerebral peduncle; *10*, pons; *11*, corpus trapezoideum; *12*, pyramid; *13*, cerebellum; *14*, chorioid plexus of fourth ventricle; *C.I.*, first cervical nerve roots. The stumps of the cranial nerves are designated by Roman numerals. The hypophysis has been removed and its contour is indicated by dotted line. The central black area is the infundibular recess of the third ventricle opened up when the infundibulum is torn off.

Fig. 1. Base of the ox brain.

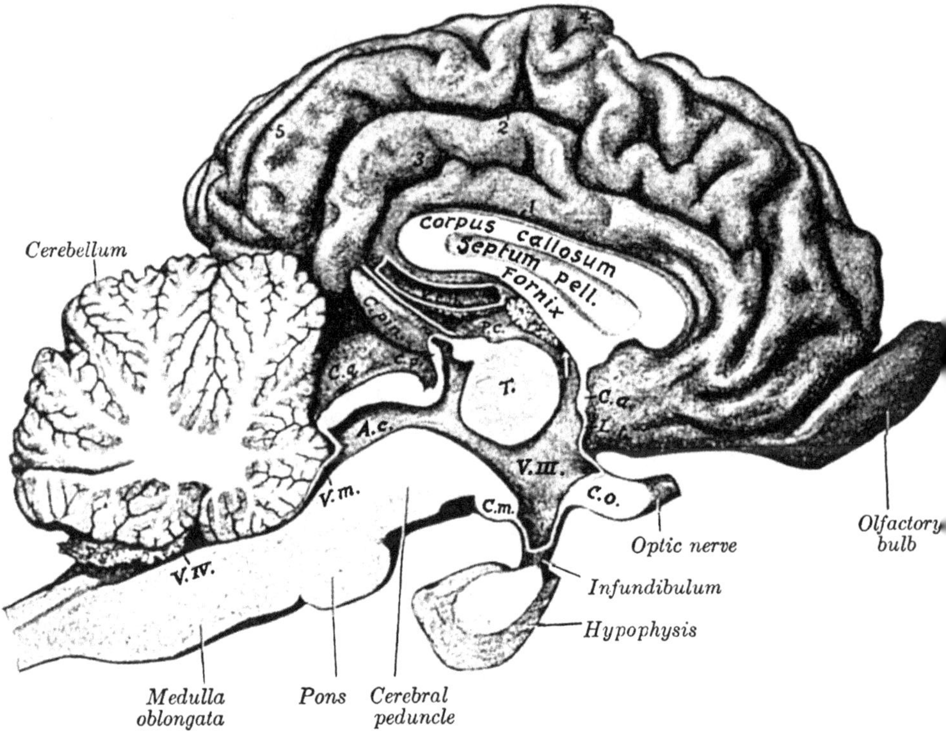

V.IV. Fourth ventricle; *P.c.*, choroid plexuses; *V.m.*, anterior medullary velum; *A.c.*, cerebral aqueduc *C.q.*, corpora quadrigemina; *C.p.*, posterior commissure; *C.pin.*, pineal body; *V.c.i.*, great cerebral vei *T*, thalamus; *V.III.*, third ventricle (arrow points to interventricular foramen); *C.a.*, anteri commissure; *L.t.*, lamina terminalis; *C.o.*, chiasma opticum. Fissures: *1*, callosal; *2*, calloso-marginal; splenial; *4*, transverse; *5*, entomarginal.

Fig. 2. Midline section through the ox brain.

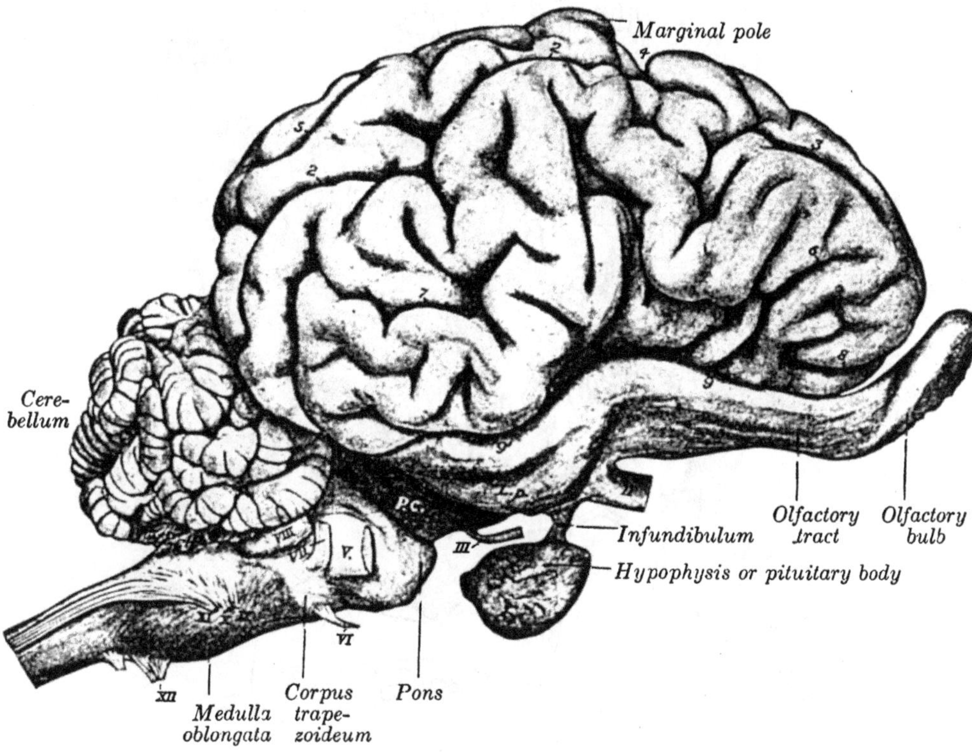

Fissures: *1,* Lateral; *2,2,* suprasylvian; *3,* coronal; *4,* transverse; *5,* ectomarginal; *6,* diagonal; *7,* posterior ectosylvian; *8,* presylvian; *9, 9',* rhinal (anterior and posterior); *P.ch.,* chorioid plexus of fourth ventricle; *P.c.,* cerebral peduncle; *L.p.,* piriform lobe; *I,* insula. Stumps of cranial nerves are designated by Roman numerals.

Fig. 3. Right side of the ox brain.

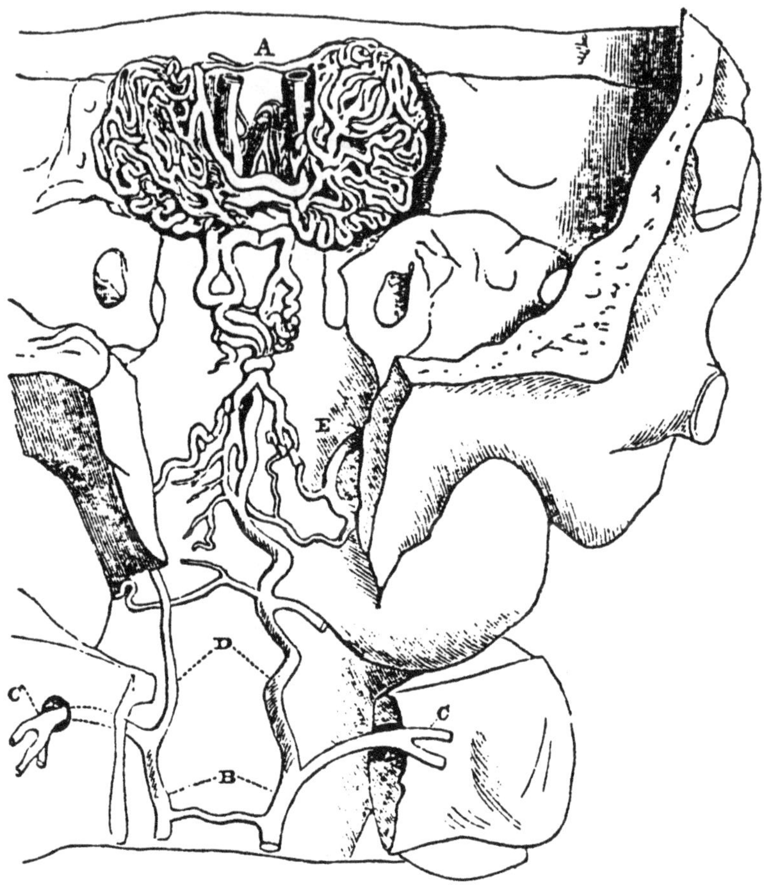

Rete mirabile of the calf. (Rapp.) The cavity of the skull and the spinal canal are opened: the brain, spinal cord, and dura mater have been removed. *A.* The cerebral carotid arteries arising from the rete mirabile. *B.* The vertebral arteries in the spinal canal. *C C.* The vertebral arteries passing through the foramen in the atlas. *D.* Branches of the vertebral arteries passing forwards under the dura mater to anastomose with the arteria condyloidea, forming thus a plexus which is connected to the rete mirabile. *E.* Arteria condyloidea.

Fig. 4. Retiform plexus in the calf.

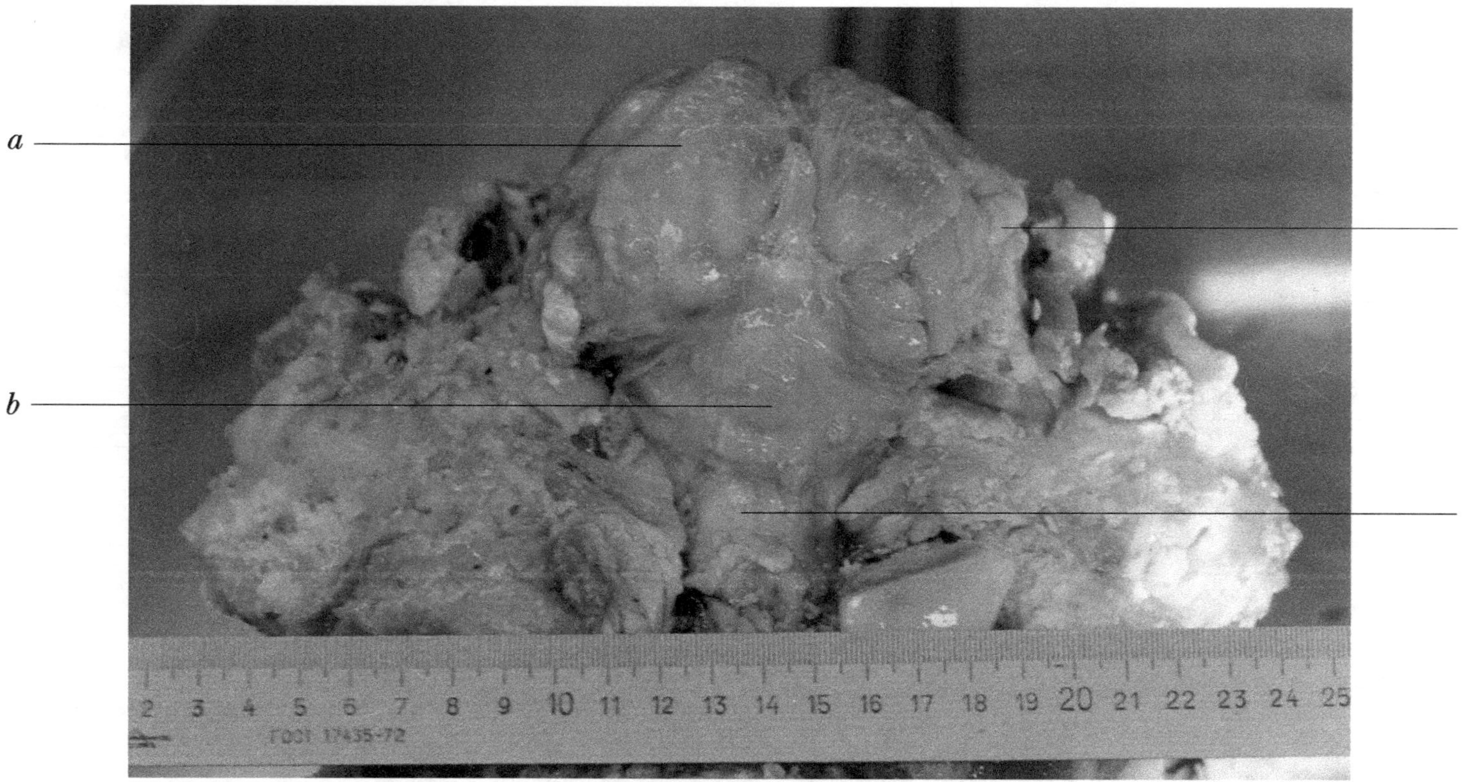

Fig. 5. The ox brain *in situ* from behind.

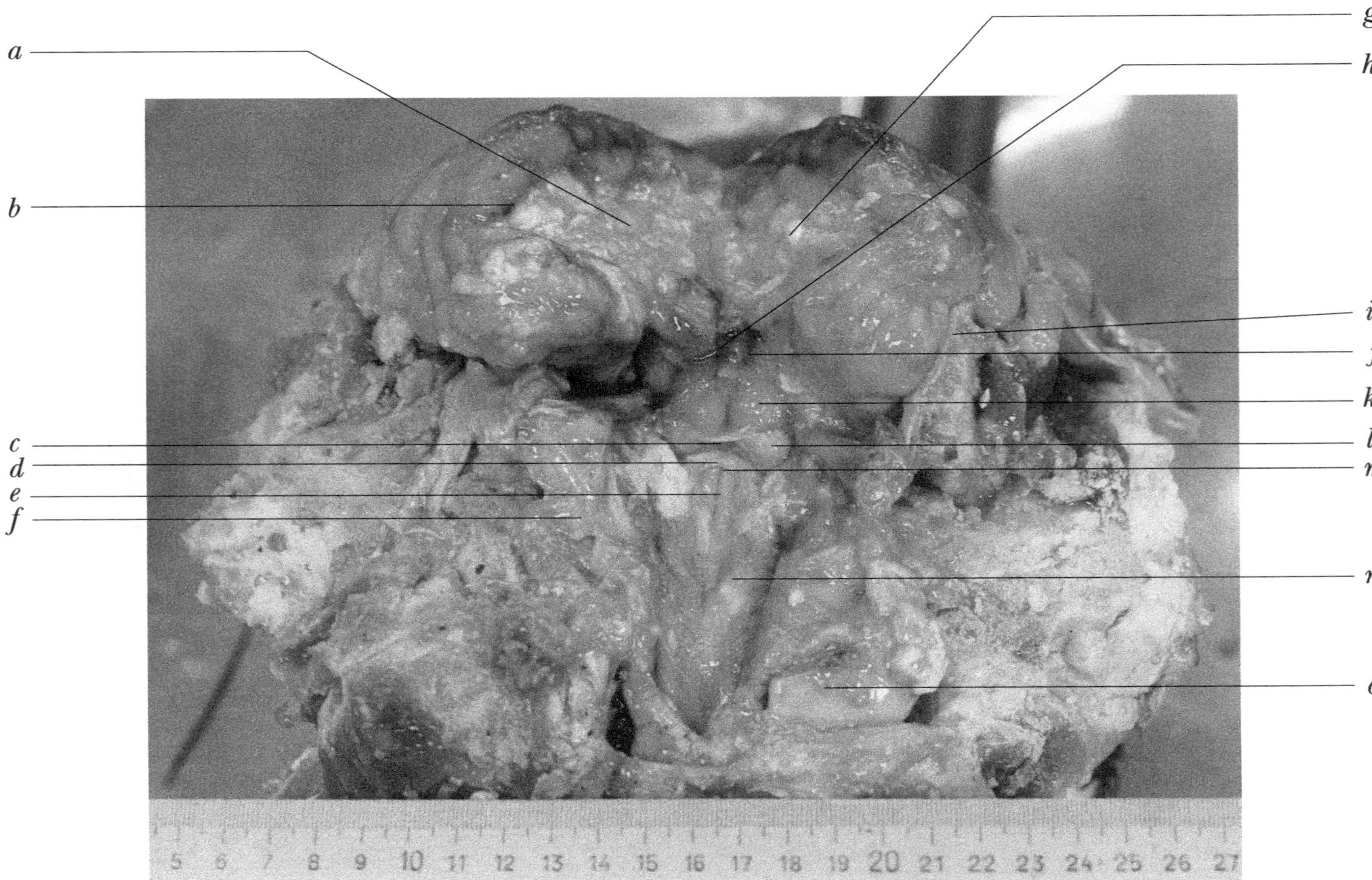

Fig. 6. The ox brain from behind. The cerebellum has been removed.

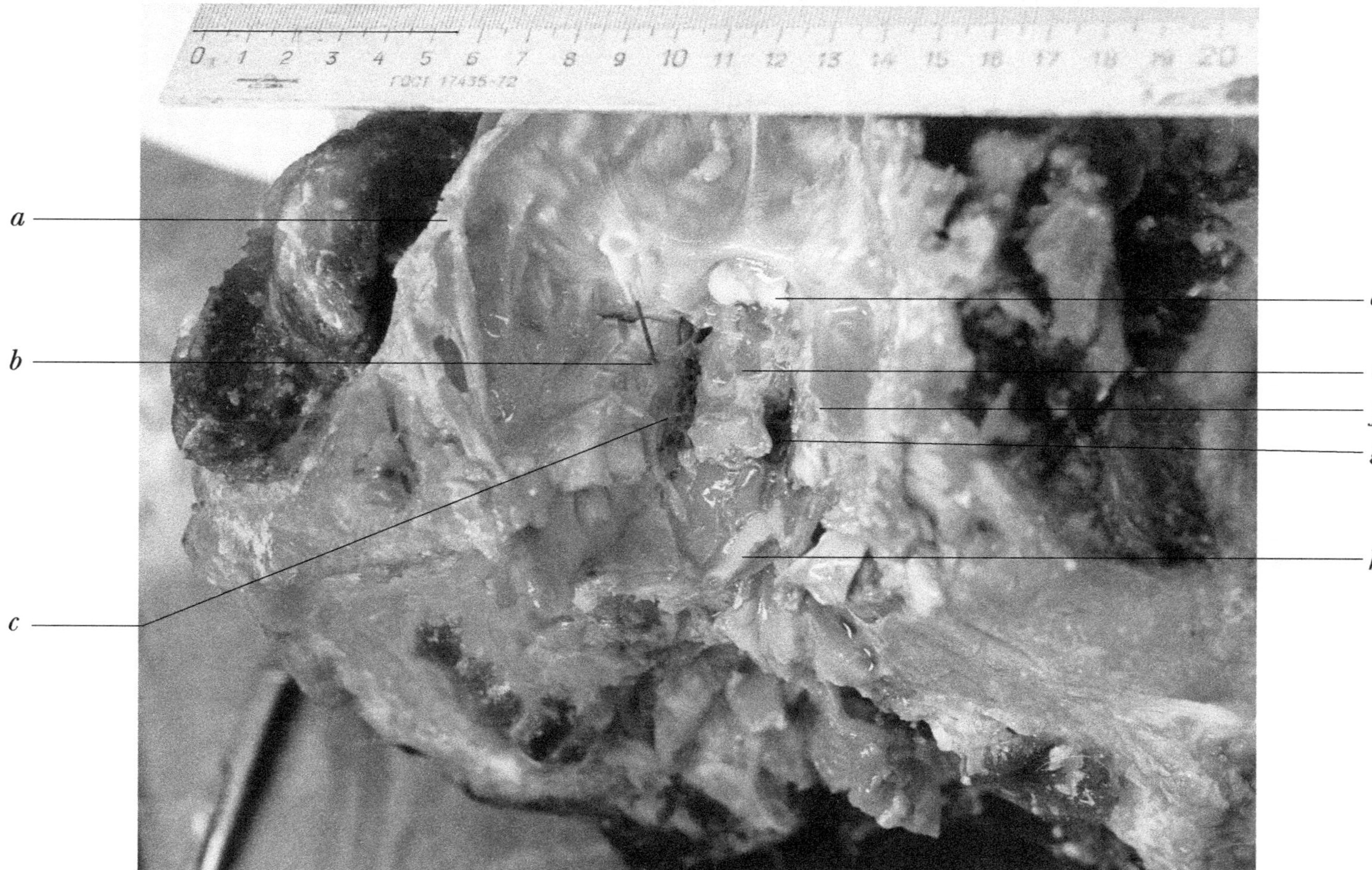

Fig. 7. The retiform plexus in the ox.

NOTE TO THE READER

Unless indicated in the text or the footnotes, all translations from the Greek are my own. Where Galenic works exist only in Arabic versions, the relevant critical editions have been used. Thus, citations from *De optimo medico cognoscendo* and *De partibus artis medicativae* are from the editions and translations of Iskandar and Lyons respectively. The second half of Galen's *De anatomicis administrationibus* (Books IX,6–XV), exists only in the Arabic version of Hunayn Ibn Ishaq. All citations are from Duckworth's translation (1962), which is largely based on Simon's critical German edition (1906). All citations have been compared with Simon's edition as well as that of Garofalo (1991). Any relevant discrepancies between these three editions are clearly stated in the footnotes. For the extant Greek part of Book IX of *De anatomicis administrationibus*, Kühn's edition is used. In places where I. Garofalo's edition (*Anatomicarum administrationum libri qui supersunt novem: earundem interpretatio Arabica Hunaino Isaaci filio ascripta, tomus alter libros V–IX continens*, Naples, 2000) suggests improvements to the Greek, these have been incorporated and are so indicated in the footnotes.

Although the Greek for "brain" is ἐγκέφαλος" (*encephalos*), it was decided that for reasons of convention the neuter, *encephalon*, would be employed throughout this work. This is the form in which it appears in modern medical texts and is cited in this way in the Oxford English Dictionary.

All brains dissected in the course of research for this book were from young adult cattle, killed in accordance with the ethical protocols of the Department of Veterinary Anatomy, University of Uppsala, Sweden. The heads were immediately frozen and thawed prior to the cranial vault being opened. No vivisectional procedures whatsoever were performed. Galen's descriptions of his vivisection experiments are critically examined within the context of his own theoretical and practical paradigms.

Earlier versions of parts of chapters 2, 4, 5 and 6 have appeared as the following:

"Galen and the ventricular system", *Journal of the History of the Neurosciences*, 6 (3), 1997, 227–239.

"A note on the term *fornix*", *Journal of the History of the Neurosciences*, 7 (3), 1998, 243.

"Galen and Greek neuroscience (notes towards a preliminary survey)", *Early Science and Medicine*, 3 (3), 1998, 216–240.

"Galen and Greek pneuma theory. The limitations of physiological explanation", in Boudouris, K.J. (ed.), *Philosophy and Medicine. II. Studies in Greek Philosophy, 29*, Athens, 1998, 171–197.

"The brain beyond Kühn: reflections on *Anatomical procedures*, book IX", in Nutton V. (ed.), *The Unknown Galen: Galen beyond Kühn.* BICS Supplement 77, 2002, 87–100.

PREFACE

Galen stands in no want of praise. By a combination of talent and relentless self-promotion, he is unrivalled as the most successful exponent of medicine in antiquity. His force of personality, undoubted skill and ability, together with an often wearying verbal prolixity, succeeded in influencing Western medicine for almost 1500 years. Galen's reputation as an anatomist and experimental physiologist is formidable. Consider one example. His efforts to determine the function of the nerves of the voice place him, together with Herophilus and Erasistratus, among the chief figures of ancient medical science. However, whilst Galen's anatomical achievements have merited notice, a significant part of his physiology, especially that pertaining to the brain, has sometimes been regarded as *jejune*. The purpose of this book is to provide a comprehensive study of the ways in which Galen sought to establish the brain as the regent part or *hegemonikon* of the body, utilising both a rigorous anatomical epistemology and a sophisticated (but perforce limited) set of physiological arguments.

In 1841, C. Daremberg published his doctoral dissertation, *Exposition des connaissances de Galien sur l'anatomie, la physiologie et la pathologie du système nerveux*. Although more exegetic than analytic, it was nevertheless the first systematic treatment of Galen's handling of the brain and nerves. In 1899, J. Soury produced a general history of neurology from Greek antiquity to the end of the nineteenth century, *Le système nerveux central. Structure et fonctions. Histoire critique de théories et des doctrines*. This work, although written in the context of its times, nevertheless contains much useful source material. In 1936, A. Souques' *Étapes de la neurologie dans l'Antiquité grecque (d'Homère à Galien)*, provided the first detailed examination of Galen's work in this field since that of Daremberg, although like Soury, Souques approached the subject with a positivist mindset. The only other author who has attempted to deal with Galen's writings on the brain and nerves in anything like a comprehensive way is R. Siegel, whose 1973 work unfortunately is marred by an anachronistic handling of the material, to say nothing of an excessive reliance on Kühn's Latin translations. Siegel presents Galen's extensive writings on the brain in an abridged form, with an inevitable loss of distinction and contextual relevance.

The last twenty five years have witnessed a recrudescence in Galenic scholarship. Since 1979, six international colloquia have devoted themselves to investigating the many-faceted parts of his *oeuvre.* His physiological output has not passed unnoticed, as A. Debru's impressive 1996 work on Galen's respiratory physiology attests. T. Tieleman's 1996 study of Books II and III of Galen's *De placitis Hippocratis et Platonis* has provided useful insights into Galen's handling of the site of the controlling centre or *hegemonikon* of the body. The purpose of this work is to provide a full-scale examination of Galen's description of the structure and function of the brain. This monograph offers for the first time in English a detailed, integrated and critical account of Galen's anatomy and physiology of the brain, set within the cultural and intellectual interstices of his time. Such a study is important because Galen's account of the brain is arguably one of the best examples of the apogee of Greek anatomical science, and in its own right is an intellectual achievement whose scope was not matched until Vesalius.

Any examination of the manner in which Galen investigates the brain must first place him within the context of his time. Whilst the details of Galen's life are familiar to historians of ancient medicine, a General Introduction provides the non-specialist with a summary of the circumstances that led to Galen's establishment in Rome. This sketch includes those medical sects who denied the importance of anatomical science, for Galen's anatomical and physiological writings must be considered in part as a reaction to these physicians.

Chapter 1 orients the reader to the relevant medical and philosophical past in which the study of the brain took place. Galen's anatomical understanding of the brain is significantly dependent, to an almost irrecoverable extent, on key earlier sources. Galen's greatest debt is to those innovative Alexandrian physicians of the third century BC, Herophilus and Erasistratus. Together with Galen, they may be placed on one side of the hegemonic polemic. Against what can be referred to as their encephalocentric hypothesis, which maintains that the brain is the controlling principle of the body, is the cardiocentric thesis, which accorded the heart hegemonic status. This latter concept received strong support from, among others, Stoicism, Aristotle and the Peripatetic tradition, the physicians Diocles of Carystus, Praxagoras of Cos, the author of the Hippocratic text *Diseases I*, and Athenaeus of Attalia, founder of the Pneumatist medical sect. It should be noted from the outset that the terms "encephalocentrist" and "cardiocentrist" are modern expressions which are employed throughout this work as shorthand labels to represent the two principal hegemonic views.

Chapter 2 examines the materials and methods which Galen employs to legitimate his hegemonic argumentation: his appropriation of an Aristotelian empirical research methodology; his use of *pneuma* in his physiology of the brain; the ways in which he used animals as anatomical and experimental subjects. This study then proceeds to decipher the Galenic framework of brain dissection and experimentation by a thorough analysis of Galen's investigation of the brain, commencing, in Chapter 3, with a discussion of Galen's techniques of its dissection. Apart from examining Galen's record on this subject, the most effective way to comprehend his agenda is to perform these dissections along the lines he has provided. This has rarely been attempted in a systematic way, and although some questions concerning Galen's dissection of the anatomy of the brain can never be resolved entirely, this study concludes that Galen's dissections are scrupulously undertaken, his descriptions meticulous, and the results consistent with his empirical methodology.

For Galen, the focal point of his anatomy and physiology of the brain consists of a set of cavities deep within the brain substance, the ventricles. Chapter 4 examines the anatomy of the Galenic ventricular system. It is crucial to Galen's notion of the brain as *hegemonikon* of the rational soul that it functions in terms of a set of ventricles which communicate with the spinal cord and the nerves. Exactly when the ventricles were first examined with a view to apportioning them a physiological role is by no means clear, although Aristotle was the first to record that the brain possessed a cavity. Given the physiological importance of the ventricles for Galen, their anatomy was examined by him in great detail, and his account requires careful evaluation before the question of his physiology of the brain can be considered. Ventricular anatomy is the high water mark of Galen's anatomy of the brain.

Galen's physiology of the brain forms the basis of Chapters 5 and 6. Galen made extensive use of animal vivisection, and it must be assessed on its merits and on what it provides. Underlying Galen's vivisectional experiments is the legacy of Herophilus and Erasistratus, whose pioneering work on the brain involved the use of vivisected animals (as well as human beings). Chapter 5 examines the achievements and limitations of Galen's experimental physiology upon the ventricles. In noting the effects of pressure and incision at various places on the ventricles of living animals, Galen also attempted to catalogue clinical examples of damage to the brain, the condition of such patients being interpreted by him strictly in ventricular terms. It must be said however, that his recording of such cases was not systematic and was more opportunistic than anything else. Although Galen's work on the anatomy of the brain is a remarkable empirical accomplishment, the physiology behind it is necessarily speculative.

The examination of Galen's physiology of the brain is concluded in chapter 6, which begins by examining two vascular structures which for Galen serve to elaborate *psychic pneuma*, his chosen physiological agent. These consist of the *retiform plexus*, a network of small arteries in the base of the brain (also referred to by Galen as a "marvellous net" and known to later anatomists as the *rete mirabile*), and the *choroid plexuses*, a complex of veins and arteries within the ventricles. The function of these vascular networks is only partly elucidated by Galen's recourse to an ingenious set of arterial ligation experiments, the results of which leave him with the difficult task of accommodating them to the demands of a pneuma-based physiology. Whilst this affirms once more the conjectural nature of such a physiological theory (and, *inter alia*, of all types of physiological speculation until the nineteenth century), it is to Galen's credit that he does not entirely gloss these difficulties. Galen remains cognisant of the limitations of pneuma as a physiological agent and its inability to provide a complete solution to the question of brain and nerve function.

On several occasions, Galen comes very close to elaborating contemporary neurological concepts. The notion of the brain as *hegemonikon* is the most obvious example. This monograph does not seek to portray Galen as an illustrious ancestor in the history of the science of the brain, although seeing him in this way is understandable. This study provides an examination and analysis of Galen's descriptive, experimental and speculative examination of the brain based on the philosophical and medical standards of his time. Viewed by these criteria, Galen's work can be seen for what it is—a record of considerable achievement in ancient science.

Ὡς δ' ἄν τις κάλλιστα θεάσαιτο τὰ διὰ τῆς ἀνατομῆς ἐν ἐγκεφάλῳ . . . φαινόμενα, τεθνεῶτός τε καὶ ζῶντος ἔτι τοῦ ζῴου, διὰ τοῦδε τοῦ λόγου δηλωθήσεται.

How best to observe the things that become manifest within the brain through dissection . . . in the dead as well as in the living animal, will be explained in this book.

De anatomicis administrationibus II. 707 K.

GENERAL INTRODUCTION: GALEN AND HIS TIME

Indeed, by high summer I had made many praiseworthy predictions and cures in cases involving leading Romans: my reputation stood high among all, as you know, and great was the name of Galen.

De praenotione.[1]

In the Autumn of 162 AD, during the first year of the reign of the co-emperors Marcus Aurelius Antoninus and Lucius Aurelius Verus, a Greek from Mysia in Asia Minor arrived in Rome with the intention of constructing the foundations of a remarkable career.[2] Galen of Pergamum would rise to become an Imperial Physician under Marcus Aurelius and several of his successors, dying, it is now reckoned, towards the end of the reign of Caracalla, sometime after 210 AD.[3] Apart from an absence of three years, from 166 to 169, Rome was Galen's world. It provided him with a stable and productive working environment. It was where he carried out his public, and later private anatomical demonstrations, his physiological experiments and composed the bulk of his writings. Yet for all the advantages Rome would give him, Galen's was a Greek mentality, committed to a medical ideology which for him was part of a Hippocratic tradition.[4] The following sketches Galen's background and outlines some of those factors which contributed to his position as an anatomist.[5]

[1] P. 94,12–15 Nutton; XIV. 625 K (tr. Nutton, 1979, 95).

[2] Galen introduces *De anatomicis administrationibus* by stating that he wrote on anatomical demonstration shortly after coming to Rome at the beginning of the Principate of Marcus Aurelius (II. 215 K). Ilberg, 1889, 1892, 1896, 1897, remains fundamental for the chronology of Galen's works. On Galen's life and achievements see Debru, 1997b; Moraux, 1985b; Nutton, 1970, 1972b, 1973, 1984a, b, 1987a, 1990, 1993a, b, 1995a, 2000; Singer, 1997a; Temkin, 1973.

[3] Galen's traditional death date has been given in the *Suda* lexicon as 199 AD. However, as Nutton, 1984b and 1995a, has persuasively argued, this date was based partly on a misunderstanding of an Arabic source which reckoned as *seventy* the years of Galen's medical practice, not his lifespan. If this figure is added to the age at which Galen began his medical studies, a revised date for his death is then about 210–216 AD.

[4] A tradition assiduously cultivated and augmented by Galen himself. Cf. Lloyd, 1991a.

[5] Although Galen is often the only source of information regarding his anatomical

Galen was born in Pergamum in 129 AD. His father Nicon was a prominent citizen and successful architect.[6] Initially, his son was slated for a career in philosophy and from the age of fifteen undertook an education that included teaching from representatives of the four chief philosophical schools, Stoic, Aristotelian, Academic and Epicurean.[7] When Galen was seventeen, his father added medicine to his son's curriculum.[8] Nicon did so as the result of persuasion from Asclepius in a dream (Pergamum housed a renowned temple to Asclepius).[9] Someone doing the bidding of a god in this way was taken as a genuine form of religious experience.[10] His father's death (149/150) left Galen financially secure and able to continue his exceptional medical education, described as "the longest on record."[11] In Pergamum, Galen's first medical teacher was Satyrus, the author,

and physiological claims, this does not mean that everything he states must be discounted because of a lack of cross references from other authors. Cf. Scarborough, 1981. Nutton, 1984b, adduces three pieces of contemporary evidence for Galen's high reputation as a physician and philosopher.

[6] On Galen's references to his father's background and training see *Bon. Mal. Suc.* pp. 392,21–393,3 Helmreich; VI. 755–6 K.

[7] Cf. *Aff. Dig.* p. 28,9–19 De Boer; V. 41–42 K; *Lib. Propr.* p. 116,12–26 Mueller; XIX. 39–40 K; *Ord. Lib. Propr.* pp. 88,7–89,4 Mueller; XIX. 59–60 K.

[8] Galen makes clear that medicine and philosophy were read concurrently when he states that at Smyrna he studied both under Pelops the physician and Albinus the Platonist (*Lib. Propr.* p. 97,6–11 Mueller; XIX. 16 K). On Albinus, pupil of the Middle Platonist Gaius, see Dillon, 1993, ix–xi; Whittaker, 1990.

[9] *Ord. Libr. Propr.* p. 88,15–17 Mueller; XIX. 59 K. Cf. *Meth. Med.* X. 609 K; *Praecog.* pp. 76,29–78,2 Nutton; XIV. 608 K. On the cult of Asclepius, the fundamental study is Edelstein and Edelstein, 1945. See also Temkin, 1991, chapter 7. For the general background to dreams in antiquity, see Burkert, 1996, 35–6; Dodds, 1965, 39–45; Swain, 1996, 254–297. According to Galen, the intervention of Asclepius persuaded Marcus Aurelius not to take him on the German campaign (*Libr. Propr.* p. 99,8–13 Mueller; XIX. 18–19 K; cf. Swain, 1996, 375), although in *De praenotione* Galen says he persuaded the Emperor to allow him to stay in Rome (p. 118,25–27 Nutton; XIV. 650 K). In any case, the decision left Galen with ample time to pursue his studies (*Praecogn.* pp. 118,33–120,2 Nutton; XIV. 650 K).

[10] Cf. Kudlien, 1981. Galen's devotion to Asclepius was strengthened by the medical advice imparted to him by the god to cure his potentially fatal sub-diaphragmatic abscess (*Lib. Propr.* p. 99,9–11 Mueller; XIX. 19 K). In *De curandi ratione per venae sectionem* XI. 314–315 K, Galen describes its successful treatment by venesection (Cf. Nutton, 1990, 253–4; Oberhelman, 1983). The guidance of a dream persuaded Galen to be thorough in his handling of the anatomy and physiology of the eye (*UP* II, pp. 92,23–93,10 Helmreich; III. 812–813 K). Galen also used dreams for polemical purposes, for example, against the Empiricists, who, according to him, employed dreams uncritically. Cf. Kudlien, 1981, 120–122.

[11] Nutton, 1995b, 61. Galen admits his inherited wealth saved him from the everyday travails of most doctors who had no choice but to charge fees (*Meth. Med.* X. 561 K). Cf. Drabkin, 1944, 337.

together with the physicians Pelops, Numisianus, Quintus and Lycus, of several Hippocratic commentaries.[12] Satyrus was taught by Quintus, whom Galen refers to as a man possessed with the "greatest skill in anatomy" (ἀνατομικώτατος).[13] Galen could afford to travel throughout the Mediterranean in search of the best teachers. In Smyrna, Galen attended lectures by Pelops, who like Satyrus was a pupil of Quintus.[14] Another influence on Galen was Lycus, also a pupil of Quintus.[15] Galen visited Corinth and several other places to glean information from Numisianus or his pupils.[16] Galen would later present these

[12] *AA* II. 225 K; *Hipp. Nat. Hom.* p. 70,13–15 Mewaldt; XV. 136 K; *Ord. Libr. Propr.* pp. 86,15–87,12 Mueller; XIX. 57–58 K; *Hipp. Praedict.* p. 20,11–12 Diels; XVI. 524 K. Satyrus, Galen tells us, was the "most accurate" (ἀκριβέστατος) expositor of Quintus' work (*Ord. Libr. Propr.* p. 87,12 Mueller; XIX. 58 K). The otherwise unknown figures of Aeschrion, Stratonicus and Aelianus Claudius were also some of Galen's teachers. Cf. *AA* II. 225 K; *De atra bile* p. 78,22–29 De Boer; V. 119 K; *Simp. Med. Temp. Fac.* XII. 356 K; *Musc. Diss.* XVIIIB. 926 K. Cf. Grmek and Gourevitch, 1988, 1994; Nutton, 1993a, 15–19; idem, 1987b, 235–9.

[13] *Lib. Propr.* p. 102,5 Mueller; XIX. 22 K. According to Galen, professional jealousy among Empiricist doctors led to Quintus being banished from Rome (*Opt. Med. Cogn.* p. 53,14–19 Iskandar; *Praecogn.* p. 70,2–25 Nutton; XIV. 602 K).

[14] Cf. *AA* II. 217–218, 225 K; *De antidotis*, XIV. 69 K; *De atra bile* p. 75,16–21 De Boer; V. 112 K; *De locis affectis* VIII. 194 K; *Lib. Propr.* p. 97,6–11 Mueller; XIX. 16 K. Smyrna was renowned as a teaching centre for philosophy and medicine, and possessed a Museum to rival that of Alexandria or Pergamum. There Galen composed for a fellow-student a work on the movement of the thorax and lungs which has not survived (*PHP* p. 124,21–6 De Lacy; V. 236 K; *Lib. Propr.* pp. 97,23–98,2 Mueller; XIX. 17 K. Cf. Nutton, 1993a, 28 n.68). Galen wrote three other works during this early period, *On the dissection of the uterus*, *Diagnosis of diseases of the eyes*, and *On medical experience* (*Lib. Propr.* p. 97,11–14 Mueller; XIX. 16 K). The first is extant (Nickel, 1971), and the third survives in Arabic translation (Walzer, 1944).

[15] *AA* II. 227, 261 K. Cf. *Hipp. Epid. VI.* pp. 286,26–288,3 Wenkebach-Pfaff; XVIIA. 274 K. Galen's abridgment of the anatomical works of Lycus has not survived (cf. *Lib. Propr.* p. 104,13–14 Mueller; XIX. 25 K), although his summary is preserved by Hunayn ibn Ishaq. Cf. Boudon, 2002.

[16] In the extant Greek text of *De anatomicis administrationibus* II. 217 K, it appears that Galen eventually tracked Numisianus to Alexandria and attended lectures by him. However, the Arabic textual tradition appears to deny this possibility. Cf. Garofalo, 1991, I, 83 n. 8; Nutton, 1987b, 237–8. In the Arabic part of *AA*, Galen states that Numisianus was "pre-eminent" in Alexandria, learned in anatomy, and wrote many books. None were widely distributed and Galen's trip to Alexandria also involved him in trying to persuade Numisianus' son Heraclianus into parting with this material, but Heraclianus allegedly burnt his father's texts (XIV.1, pp. 183–184 Duckworth; cf. Nutton, 1993a, 16, 18). Galen does not refer to Numisianus as one of his teachers, which he would have done had he attended his lectures in Corinth or anywhere else. Galen's account of Numisianus is to a man already dead. Pelops' books were also destroyed by fire; whether deliberately or by accident is not made clear (*AA* XIV.1, p. 184 Duckworth). Galen's frustration can be imagined. It recalls his anguish at the loss of his own works in Rome's great fire of 192, which engulfed the Temple of Peace. Cf. *Libr. Propr.* p. 99,23–25 Mueller; XIX.

journeys in such a way so as to create the impression of a line of succession in anatomical knowledge with himself at its head.[17] The best place for Galen to continue his medical studies, especially in anatomy, was Alexandria, where he spent four years (153–157).[18] Marinus (fl. 120 AD), who will be discussed in chapter 1, worked there and is singled out by Galen as the one who *recovered* (ἀνακτησάμενος) the study of anatomy.[19] In *De anatomicis administrationibus*, Galen refers to Alexandria as the place where it is easy for a student to study human osteology.[20] Given Alexandria's role in the instruction of some of Galen's teachers such as Satyrus and Pelops, it is reasonable to assume that by Galen's time the level of medical education remained thorough enough to warrant his attention.[21] Although Galen is reticent on his activities there, Alexandria undoubtedly formed a vital part of his medical education, and the knowledge that he studied there would do his developing career no harm.[22]

Galen returned to Pergamum in 157 with sufficient knowledge and experience to be appointed physician to the gladiatorial school by the *chief priest* (ἀρχιερεύς), a position he held for four years.[23] It pro-

---

19 K. Galen's attempts to recreate his writings in the wake of this catastrophic loss were incomplete (not surprising for a 63 year old man faced with the task of recreating his works from memory). Forgers would have benefited from such a loss, which may have acted as a further spur to Galen in his attempts to list his texts. Some also survived the conflagration without his knowledge (*De praenotione* is one example. Cf. Nutton, 1979, 50).

[17] Exactly when Galen formulated such a goal is uncertain. Roselli, 1999, 366, in discussing the doxographic material in some of Galen's Hippocratic commentaries, states: "In the list given in *De uteri dissectione*, a work of his youth, there is no mention of recent great anatomists. Evidently Galen had not yet developed that sense of belonging to a prodigious anatomical tradition . . ." The absence of figures such as Numisianus and Marinus could also mean that Galen is not yet familiar with their work. But their non-appearance does not allow the conclusion that Galen had not yet formulated his goal to identify himself as part of an anatomical tradition. For Galen to represent himself as master of such a tradition was a matter of experience and time.

[18] A *terminus ante quem* of 157 is accepted; the *terminus post quem* of 153 is open to interpretation. Cf. Nutton, 1993a, 12 n. 3.

[19] *PHP* p. 480,28–30 De Lacy; V. 650 K.

[20] *AA* II. 220–221 K.

[21] Cf. Nutton, 1972a, 173; idem, 1975, 1993a, 13–15.

[22] "Either he did not learn much, if anything, about anatomy that he did not already know—which is hardly consistent with his recommendation to the aspiring doctor to make certain of visiting Alexandria for its anatomical teaching—or he has carefully obscured his scholarly obligations to his Alexandrian teachers, perhaps in order to enhance the image of Marinus, Quintus, and their pupils who taught him in Asia Minor, or that of Galen the medical *Wunderkind*." Nutton, 1993a, 19.

[23] See Swain, 1996, 358; but cf. Iskandar. 1988, 164, and n. 2. For Galen's testimony of his observations during this period see *Comp. Med. Gen.* XIII. 564–565,

vided him with ample opportunity to develop further the practical skills necessary in order to produce his anatomical and physiological works and to reinforce his confidence in the value of publicly demonstrating his abilities.[24] In either his thirty third or thirty fourth year, Galen arrived in Rome, the obvious next stop for a person of ambition.[25] Rome, with a population of at least one million, was the largest city of the Empire.[26] There, three factors worked in Galen's favour. First, Greek medical practitioners were widely accepted.[27] By the advent of the Principate, in a form of colonialism in reverse, Roman medicine had been assimilated into a Greek paradigm, with the new model reflecting a wide range of diverse practices and practitioners.[28] However, in the Western half of the Roman Empire in the second century AD, only fifty percent of doctors enjoyed full citizenship rights.[29] Individual doctors could become quite rich in Imperial service, but these were exceptional.[30] Wealth and erudition did not necessarily

---

599–605 K. Cf. Nutton, 1972a, 170; Scarborough, 1971, 99–100, 103–5, 110–11. In *Opt. Med. Cogn.* pp. 103,10–105,19 Iskandar, Galen reveals how he was chosen for this position. His account invites comparison to the medical contests which took place at Ephesus. Cf. Nutton, 1995c, 7–8. See also Iskandar, 1988, 139 and 165 (but see also Nutton, 1990, 244 n. 41).

[24] Since *Opt. Med. Cogn.* was written around 178, that is after Galen's first stay in Rome, it is likely that his experiences there of public anatomical demonstrations coloured the account of his earlier anatomical triumphs in Pergamum.

[25] *Hipp. Artic.* XVIIIA. 347 K, Galen gives his age as 32. For the later figure, see *Lib. Propr.* p. 96,5–6 Mueller; XIX. 15 K. On the varying reasons put forth for Galen's departure from Pergamum, see Eichholz, 1951, 60–61; Nutton, 1993d, 66 n. 73; idem, 2000, 957; Swain, 1996, 358, 373.

[26] On the population of Rome, see Garnsey and Saller, 1987, 62, 83. For comments by Galen on the size of Rome and the types of medicine practiced there, see *De partibus artis medicativae* p. 29,13–24 Lyons; *Praecogn.* p. 92,29 Nutton; XIV. 624 K; *Opt. Med. Cogn.* pp. 47,6–9, 101,19–103,8 Iskandar. See also Swain, 1996, 363. For a general survey on the social, cultural and intellectual strands of the Roman Empire in this period, see Rawson, 1985; *CAH*^{2}, Va,b.

[27] On the introduction of Greek medicine to Rome and the Roman reaction, see the bibliography by Mudry, 1993, 787–799. Worthwhile studies include: André, 1987; Drabkin, 1944; Gourevitch, 1998; Jackson, 1988, 1993; Marasco, 1995; Nijhuis, 1995; Nutton, 1986, 1993d; Riddle, 1993; Scarborough, 1993, 22–29; Solin, 1995; Von Staden, 1996c, 1996d; Wallace-Hadrill, 1990. Allbutt, 1921, although dated, provides much information. On Galen as a practitioner and his interrelationships with his Roman colleagues see Ilberg, 1905, 276–312; Kollesch, 1965a; idem, 1965b; Nutton, 1977. Horstmanshoff, 1995, provides a reconstruction of a day in Galen's Roman practice.

[28] "[I]t is the medicine of the conquered that eventually becomes the dominant strain of learned medicine, not that of the Roman conquerors." Nutton, 1993d, 75. Cf. Langslow, 2000, 28–41.

[29] Nutton, 1992, 39.

[30] Pliny the Elder, *HN* 29.5.7–8; p. 22 Ernout, mentions incomes of 250,000 and 500,000 sesterces per annum which some Imperial physicians could command. Cf.

gain a physician entrance to the highest echelons of society.[31] Second, Galen benefited from the relative social and political stability of the period, although this should not be exaggerated. Galen does note however, that thanks to the *Pax Romana*, he was able to access pharmaceutical supplies from all over the world.[32] The third factor, and perhaps the most important, was the social and intellectual orientation of Rome's elite. The Rome of 162 AD was a Philhellenic capital.[33] Since Philostratus, the period has been termed the *Second Sophistic*.[34] Under this rubric varied forms of cultural and pedagogic activity coalesced and competed, and Galen's practice in Rome took place in this heady atmosphere of intellectual activity which involved sophists, rhetors, *littérateurs*, philosophers and physicians.[35] In this intensely competitive world, a "thorough philosophical training... was... above all a mechanism for acquiring social respectability in a society in which rhetors, sophists and rhetorically educated elite-members increasingly dominated urban politics."[36] Although Galen

Kudlien, 1976, 3ff.; Nutton, 1969, 37–48; idem, 1992, 45–7. Pleket, 1995, 31, has estimated that "not more than perhaps 5% of those same urban physicians belonged to the real elite".

[31] The highest office attained was that of equestrian procurator (cf. Nutton, 1992, 41). For all Galen's boastful talk of his relationship with Marcus Aurelius, he is not cited in the *Meditations*. That Galen is *not* mentioned should come as no surprise; the work, after all, is a deeply personal, albeit tendentious, philosophical discourse. Cf. Scarborough, 1981, 7. Marcus viewed doctors, together with architects, as "skilled craftsman" (βάναυσοι τεχνῖται). VI. 35, p. 110,25 Farquharson; p. 62 Trannoy). Such a comment, by no means disparaging, does not necessarily lessen the value of Galen's testimony concerning his perceived importance to the Emperor *qua* physician.

[32] *De antidotis* XIV. 7–9 K. Cf. Nutton, 1978, 211; Swain, 1996, 364 n. 20.

[33] In style certainty; its substance and pervasiveness are open to question. Cf. Woolf, 1994, esp. 130ff.

[34] *Vitae Sophistarum* pp. 2,26–3,7 Kayser. See Bowersock, 1969, esp. 59–75; idem, *CHCL* I.iv, 95–98. Brunt, 1994, in a persuasive article, queries the accuracy of closely linking Galen to the Second Sophistic (but cf. Von Staden, 1997a, 33 n. 1). See also Bowie, 1974; 1982. For Galen's relations with the Second Sophistic see Ieraci Bio, 1993; Kollesch, 1981; Pearcy, 1983, 1993; Von Staden, 1995a, 48–51, 1997a; Swain, 1996, 56–63, 357–379.

[35] "The intellectual stars of the Second Sophistic appear in the background of Galen's work or as case histories." Pearcy, 1993, 449. Among the constellation active in Galen's time may be mentioned the Academic philosopher Favorinus. The rhetors Polemo (*Hipp. Artic.* XVIIIA. 347 K) and Herodes Atticus, the latter consul for 143 and "the most able orator of our time" (*Opt. Med. Cogn.* p. 113,15–16 Iskandar). Aelius Aristides, whom Galen almost eulogises (*in Pl. Ti.* p. 33 Kahle). The rhetor Adrian of Tyre, (*Praecogn.* p. 96,16–17 Nutton; XIV. 627 K); Aelius Antipater, tutor to Caracalla and Geta, *ab epistulis Graecis*, consul and legate of Bithynia, whom Galen lauds for his rhetorical skills (*De theriaca ad Pisonem* XIV. 218 K).

[36] Pleket, 1995, 33.

disparages sophistic methodology, claiming it to be a "fraudulent device" (ὄργανον πανοῦργον),[37] his own endeavours, especially his public anatomical and physiological demonstrations, were framed as a carefully engineered response to the challenges raised by an intellectual milieu in which sophists, *qua* philosophers, formed a vital part.[38] Galen's great exercise in self-promotion, *De praenotione* (*On prognosis*), reflects this, being deliberately crafted using the techniques and "rhetorical devices that characterize second sophistic literature."[39] In this text Galen records his initial contacts with the Roman elite. Galen also ostensibly outlines his successful treatment of *quartan fever* (malaria) in Eudemus, the influential Peripatetic philosopher.[40] But Eudemus' treatment is secondary to a narrative in which Galen strives to make clear his medical qualifications and philosophical credentials. Eudemus is made to recognise Galen primarily as a philosopher and not as a physician, which would only augment Galen's emerging reputation.[41] Eudemus was part of a circle of friends and associates from Pergamum already resident in Rome at the time of Galen's arrival,[42] and it is no exaggeration to state that Galen skilfully exploited a second century old boy network.[43] Eudemus promulgated his cure by Galen among the social and intellectual leadership; in particular, Sergius Paulus (soon to be appointed *praefectus urbi*), and Flavius Boethus (ex-consul and future governor of Syria Palaestina). Boethus is introduced as having asked Galen to demonstrate on the nature of speech and of respiration, and, together with Paulus, is referred to as a student of Aristotelian

[37] *Preaecogn.* p. 74,4–5 Nutton; XIV. 605 K.

[38] "Galen's demonstrations, unlike the performance of the Sophists, were never *only* sophistic declamations. His public dissections and vivisections, however much rehearsed, demanded mastery of manual as well as verbal techniques and were a product of his private investigations." Siraisi, 1995, 6 (italics in original). Cf. Brunt, 1994, 43–5, 51–2.

[39] Swain, 1996, 359. This text was composed in 178 and was specifically written to maintain Galen's position in Imperial service (Nutton, 1972, 62; idem, 1979, 58–63).

[40] Eudemus may well have instructed Galen in philosophy in Pergamum and Galen does refer to him as his teacher (*Praecogn.* p. 82,12 Nutton; XIV. 613 K). On malaria in the ancient world, see Burke, 1996; cf. Arnott and Stuckey, forthcoming.

[41] Eudemus had thought Galen proficient "only in philosophical theory." *Praecogn.* p. 76,26–9 Nutton; XIV. 608 K. The Peripatetic Alexander of Aphrodisias took the trouble to write at least two refutations of Galen as a philosopher. Cf. Nutton, 1984b, 318ff.

[42] Only Ilberg, 1905, 286, had previously remarked upon the significance of this network. Cf. Nutton, 1979, 158.

[43] On patronage in the Roman world see for example Saller, 1982, chapter 3; Garnsey and Saller, 1987, 151–6.

philosophy.[44] In *De anatomicis administrationibus*, Boethus is described as possessing "an ardent love of anatomical observation".[45] When Boethus later invites Galen to give another demonstration, Galen states that Paulus remarks that he lacked the opportunity for "observing the phenomena shown by dissections."[46]

At this time, these demonstrations took place in public, in keeping with the dynamics of the "epideictic culture of Galen's century."[47] Many would turn up to hear a philosopher discourse in public or a group of physicians argue over disputed points in anatomy or physiology.[48] Galen's public exhibitions such as demonstrating the nerves of the voice were more than a painstaking exercise in identifying and isolating the recurrent laryngeal and intercostal nerves.[49] In these and other performances, Galen functioned in the knowledge that a

---

[44] *Praecogn.* pp. 80,15–82,7 Nutton; XIV. 612–613 K. In *Lib. Propr.* p. 94,18–21 Mueller; XIX. 12–13 K, Galen writes he dedicated *De causis respirationis* and his (lost) work on the voice to Boethus. Galen notes that his great exercise in teleological physiology, *De usu partium*, was greeted enthusiastically by the Aristotelians in Rome. *Lib. Propr.* p. 100,18–22 Mueller; XIX. 20 K. On the strength of Aristotelian philosophy in this period see Gottschalk, 1987.

[45] ἔρωτα τῆς ἀνατομικῆς ἐρασθεὶς θεωρίας. II. 216 K. When Boethus left to take up his post as governor, he took with him the first six books of Galen's *De placitis Hippocratis et Platonis*, and the first book of *De usu partium*. (*Lib. Propr.* p. 96,19–24 Mueller; XIX. 15–16 K).

[46] δεῖσθαι τῆς θέας τῶν κατὰ τὰς ἀνατομὰς φαινομένων. *Praecogn.* p. 82,1–2 Nutton; XIV. 613 K. Barbarus, uncle of the co-emperor Lucius Verus, requests similar instruction (p. 82,3–6 Nutton; XIV. 613 K).

[47] Von Staden, 1995a, 48 n. 3; see also 53–6; cf. Debru, 1995; Nutton, 1979, 187–8.

[48] In *De anatomicis administrationibus*, Galen provides several vignettes to illustrate part of an interactive process between performer and audience. One physician (a follower of Erasistratus), promises "to show" or "to demonstrate" (ἐπιδείξειν) that the aorta is empty of blood. Galen states that some "of the eager young men" in the audience place the doctor on the spot by bringing an animal for him to demonstrate his thesis, and that the physician prevaricates by demanding a fee. The youths oblige and the doctor proceeds, only to make a botch of the dissection. Another physician, described as "of the same class" (ἐκ τούτου χοροῦ), also makes a poor attempt at the initial dissection of the intercostal region, severing both artery and vein. The young men, who had also placed bets among the audience, then proceed to successfully carry out the entire procedure (II. 642–643 K). A little later, a similar claim from a physician who is presented as falsifying Galen's experiment in *An arteriis natura sanguis contineatur*, proving the arteries are not empty of blood, is similarly put to the test; again, the youths proceed to demonstrate, using a goat, the method by which the experiment should be conducted and the results interpreted (II. 645–646 K). Cf. *Art.Nat.Sang.Cont.* pp. 178–180 Furley-Wilkie; IV. 733 K. See also Harris, 1973, 379ff. Some of these youths are likely to have been Galen's students; they are surely more than ciphers introduced solely to ridicule an Erasistratean. They are employed to illustrate the vigorous format of thrust and counter thrust, typical of the competitive world in which Galen and his fellow doctors performed.

[49] On these experiments see *AA* II. 651–706 K; XI.4, pp. 81–87 Duckworth; XI.11, pp. 104–107 Duckworth. Walsh's 1926 account is not reliable.

part of his audience would be familiar with at least some of his experimental methodology, and could follow the steps in his reasoning.[50] Galen appealed to the intellectual aspirations of his target audience, using in part the devices and techniques which such a group, accustomed to public discourse from philosophers, rhetors and sophists, could relate to and appreciate.[51] In *De praenotione* for example, at the request of Flavius Boethus, a group of physicians and Stoic and Peripatetic philosophers gather for an anatomical demonstration by Galen. Among them are Alexander of Damascus, professor of Peripatetic philosophy at Athens, and the rhetors Adrian of Tyre and Demetrius of Alexandria.[52] However, the demonstration is abandoned by Galen, apparently incensed at Alexander's questioning of whether one should believe in the evidence of the senses.[53] Such criticism cuts to the heart of Galen's demonstrative methodology, but Galen's actions may

[50] Galen mentions one Martianus, septuagenarian physician and doyen of anatomists, in the context of one of his public lectures on the works of his predecessors during his first stay in Rome. Martianus accuses Galen of obtaining his prognostic skills not from medicine but "from divination" (ἐκ μαντικῆ). *Praecogn.* p. 84,5–8 Nutton; XIV. 615 K. At the time of writing this work, two of Martianus' anatomical works were in circulation, and his reputation was such that Galen wrote (the lost) *On the anatomy of Hippocrates* and *On the anatomy of Erasistratus* (*Lib. Propr.* pp. 94,21–95,2 Mueller; XIX. 13 K (where it is "Martialius", but there is no doubt that they are one and the same person). Galen's lecture is a set performance in which he is given part of Erasistratus' *On bringing up blood* to comment upon. The section, Erasistratus' rejection of venesection, is used by Galen to discomfort Martianus, an alleged Erasistratean. A friend suggests Galen's remarks be taken down. When Galen later returns to Rome after his self-imposed exile (166–169), he finds a version of this work now in circulation. This, according to Galen, led him to cease all public lectures and demonstrations (*Lib. Propr.* pp. 95,14–96,16 Mueller; XIX. 14–15 K). On Galen's handling of venesection see Brain, 1986.

[51] On the similarities between Galen's public performances and those of his sophistic peers, see Von Staden, 1995a, 59–60; 63–5; idem, 1997a, 47–51. Cf. Nutton, 1990, 257; Pearcy, 1983, 260.

[52] This has been taken as a reference to the better known Alexander of Aphrodisias (cf. Nutton, 1979, 189; Swain, 1996, 373 n. 54). On Adrian and Demetrius and their links to other intellectuals of the Second Sophistic, see Nutton, 1979, 190–1. On Demetrius, see Jones, 1967.

[53] *Praecogn.* pp. 96,5–98,8 Nutton; XIV. 627–628 K. Galen states, apparently without irony, that Alexander is known for his acrimonious disputation or "love of contentiousness" (φιλονεικία). Galen's anger with those who hold this sceptical position is well illustrated in his short but powerful treatise, *De optima doctrina*, composed as an attack on contemporary Academic philosophers like Favorinus of Arelate, whose sceptical viewpoint, according to Galen, makes the acquisition of knowledge by the sense organs invalid. Hence anatomical knowledge is not possible. But to Galen, Favorinus' position is inconsistent, for in not trusting the natural criteria (the evidence of the sense organs), he must simultaneously accept the very same criteria, otherwise he cannot hope to discriminate between opposing arguments. Cf. *Opt. Doct.* pp. 94,2–96,20 Barigazzi; I. 42–45 K. Cf. Ioppolo, 1993, 202–205. See also Holford-Strevens, 1997, 208ff.

also be interpreted as a stratagem to bolster his reputation, for the matter soon comes to the attention of the intellectuals in Rome, who demanded, according to Galen, that the whole process be repeated before an audience of physicians and philosophers. The subsequent demonstrations, which lasted several days, are a set-piece triumph for Galen. Boethus is so impressed that he sends scribes to take down Galen's anatomical observations.[54] These were not the only set of marathon exhibitions in which Galen takes part. In *De libris propriis*, he discusses another group which occurred after he returned to Rome in 169, when *De usu partium* was already in circulation. Galen's detractors maintained, in effect, that this text was fictitious, and Galen is persuaded by his friends to answer publicly this charge by a set of anatomical demonstrations, again held over several days.[55] With a receptive audience, their common intellectual interests, and the patronage which could be bestowed, Galen's success was assured. It was not surprising that he became one of the Imperial physicians. The advantages of a public anatomical display are encapsulated by Galen as follows:

> I know an intelligent and wise man who selected and honoured me when he saw a single act of mine: I dissected an animal by which I demonstrated the organs of the voice and the organs of locomotion. Two months earlier, that man had happened to fall from a considerable height, thus rupturing many organs in his body, and losing his voice altogether, so that his voice became like a whisper. His organs were treated, became sound, and recovered after several days: yet his voice did not return. When this man saw from me what he saw, he gained confidence in me and entrusted himself to me. I cured him in a few days because I knew where the affected place was and attended to it.[56]

A well executed demonstration concerning such an important function as the power of speech can help a potential client in choosing a physician. A prospective patient may also be impressed by a display of anatomical erudition, especially if such learning is directly

[54] *Praecogn.* pp. 98,9–100,6 Nutton; XIV. 629–630 K. Cf. Von Staden, 1997a, 48–9. These exhibitions did not always have the desired effect: "On many occasions I observed that when I did one thing, I won the admiration of some spectators, while others were unmoved and indifferent." *Opt. Med. Cogn.* p. 131,14–15 Iskandar.

[55] Pp. 100,23–102,10 Mueller; XIX. 21–22 K. Galen adds that some of the distinguished doctors present told him that to save time he should compare and contrast his findings with those of Lycus of Macedon, a disciple of Quintus. The result was the (lost) *On the ignorance of Lycus concerning anatomy.*

[56] *Opt. Med. Cogn.* p. 107,3–11 Iskandar.

applicable to his own illness.[57] Galen places this on a rather more elevated plane when he states that the first duty of a patient in choosing a doctor is to "find out how wide his knowledge is and how penetrative is his training in anatomy."[58] Such a request is hardly practical; but for Galen the anatomical science displayed on occasions such as these underscores the fundamental importance of such knowledge for both patient and physician. The best physicians are the best anatomists. Galen's statement that "physicians need anatomy to the highest degree"[59] is not simply a reiteration of what he regards as self-evident, but an indication that in Rome there are not only patients to be persuaded of this, but also medical colleagues.

In Galen's view, two medical *sects* (αἱρέσεις) contributed to a decline in anatomical investigation from the third century BC to the first century AD.[60] These were the Empiricists and Methodists.[61] In their denial of the epistemological validity of dissection, these groups govern a significant part of Galen's actions and responses. The Empiricists arose in Alexandria in the first half of the third century BC. Founded by Philinus of Cos, allegedly an apostate pupil of the groundbreaking anatomist and physiologist Herophilus of Alexandria, they were an affiliation sharing some, but by no means all Herophilean methodologies.[62] For Herophileans as well as for the medical sect later known

---

[57] It is obvious that building the patient's confidence in this way is itself a form of therapy.

[58] Ibid., p. 115,24–25 Iskandar.

[59] Ibid., p. 109,18–19 Iskandar.

[60] Glucker, 1978, 190, notes that Galen puts the philosophical sects on the same footing as the medical ones: "they are *schools of thought*, and people are criticized for adopting them without a sufficient examination of their *doctrines*." (italics in text).

[61] Although by the second century AD the strict demarcations between Empiricist and Methodist may in several instances have partly broken down and been replaced by more flexible methodologies. Here the boundaries between medical affiliations and their dependency or otherwise on various philosophical schools are perhaps not as clear cut as one might wish.

[62] Von Staden, 1989, 432. Herophilus' influence on Galen is discussed in chapter 1. For the Empiricists, see Deichgräber, 1965 130–2, 253–308; Frede, 1985, xx–xxxi; idem, 1987, 236–41; 243–260; Hankinson, 1987a; 1991c, xxvi–xxxiv; 1995; Lloyd, 1987b, 158–167; Matthen, 1988, 108–115; Stok, 1993. Frede, 1990, 234, 247–8, advances the case that there was considerable variance in what constituted an Empiricist. As Galen tells it, the famous Empiricist, Heraclides of Tarentum (*fl.* c. 75 BC), who had been trained under Mantias the Herophilean, undertook anatomical dissections and admitted the place of reason, albeit not in Rationalist terms (*Diff. Puls.* VIII. 720, 725 K; cf. Fraser, 1972, I, 361–2). Such anatomical work performed by Heraclides seems to have been done more as a means of improving surgical skills than for anatomical research.

as the *Rationalists* or *Dogmatics*,[63] the problem with Medical Empiricism lay in its denial of the epistemological validity of anatomical dissection and experimentation.[64] The Empiricists taught that experience of diseases and case histories provided all the information required for successful medical practice. They regarded as not so much unreliable as unattainable all efforts into elucidating hidden causes, instead basing their methodology on experience—that is, the evident symptomatology characterised by each individual patient.[65] Much of the information on the Empiricists comes from the first century AD Roman encyclopaedist Cornelius Celsus.[66] He records the Rationalist position that the dissection and vivisectional work of Herophilus and Erasistratus was necessary in order for knowledge to advance, and for lives to be saved.[67] However, according to Celsus, the Empiricist argues that such knowledge is unnecessary since what is obtained from vivisection and dissection is not really true knowledge of the body, since the very act of dissection produces significant changes in the appearance of the organs under investigation.[68] Celsus states that for an Empiricist, the best way to obtain useful information relevant to treatment is to take advantage of examinations on living

---

[63] The Pseudo-Galenic *Introductio sive medicus* (XIV. 683 K; T1 *Herophilus*), conveniently lists Herophilus as the head of the "Rationalist sect", with Philinus as head of the Empiricist Sect. Galen does state that there are two chief "schools", Rationalist and Empiricist. The former is highlighted to expose deficiencies in the latter. But Galen also knew of the limitations of such labels and their tendency to mislead (cf. *De sectis* pp. 1,16–2,11 Helmreich; I. 65 K). In any case, the notion of a monolithic (and exclusive) Rationalist Sect is weakened by doxographic accounts of its wide-ranging membership. Von Staden, 1989, 58, states that the term *Rationalist* is a "classificatory factotum" Cf. idem, 66, 115ff. This is correct as far as it goes, but it is more than simply a label of convenience. It is a working definition of a group of physicians opposed to the tenets of Empiricism. As Lloyd, 1987b, 158, points out, their label: "takes its origin from the objections of its opponents." There were doctors who were *Rationalists* in the narrow sense that they maintained anatomical knowledge required dissection.

[64] Exactly how such a sect may have influenced the direction of later Herophilean medical research is not known. But pressure from such a group may have persuaded a significant number of Herophileans to concentrate on ways to repel such rivals, one being to focus more on textual matters (including the codification and transmission of material) and practical medicine rather than devote time and effort to a sustained programme of anatomical research. Cf. Von Staden, 1989, 452.

[65] Cf. Hankinson, 1998, 37–43; Von Staden, 1989, 12.

[66] On Celsus as an historian of medicine see Von Staden, 1999.

[67] *De medicina. Prooem.* 23–26; p. 21,12–32 Marx; pp. 8–9 Serbat. Cf. Mudry, 1982, 106–111.

[68] *Prooem.* 40–3; pp. 23,28–24,14 Marx; pp. 13–14 Serbat. Cf. Mudry, 1982, 135–138.

persons as may come one's way.[69] The Empiricists therefore do not deny anatomy has a *use*, but it is to be learnt by fortuitous example; as Galen puts it, it is mere happenstance or "adventitious anatomy" (ἐπείσακτος ἀνατομή).[70]

The Methodists also saw no need for the researches of anatomy and physiology. They apparently originated in the first century BC under Themison, a pupil of Asclepiades of Bithynia, but became established as a distinct group under Thessalus of Tralles (fl. AD 60).[71] Their rise to prominence in the first century AD may represent a reaction to both Rationalist and Empiricist sects (as well perhaps as an endorsement of their moderate, common sense methods of treatment).[72] Methodist medical epistemology employed reason, not for use in the search for hidden causes, but in the acquisition of information about the body which is "obvious" to any thinking person.[73] They claimed to teach the method of medicine (hence their name) in six months. This standpoint is deeply inimical to Galen, whose own extraordinarily lengthy medical education is seen by him as the paradigm of a good doctor.[74] Galen considers the Methodist approach to medical education as "belief without demonstration".[75] However, like the Empiricists,

[69] *Prooem.* 44; p. 24,21–23 Marx; p. 14 Serbat. Cf. Mudry, 1982, 138; Von Staden, 1992, 235. See also *AA* II. 287–288 K.

[70] *Meth. Med.* X. 100 K. Cf. Cicero, *Acad.* II. 39.

[71] *Prooem.* 54–7; p. 26,9–29 Marx; pp. 17–18 Serbat; cf. Mudry, 1982, 153–166. See also Lloyd, 1987b, 158–9; Pigeaud, 1993b; Rawson, 1982; Temkin, 1956, xxv–xxx. A critical edition of the Methodist fragments is forthcoming in this series by M. Tecusan.

[72] Cf. Lloyd, 1987b, 161 n. 211.

[73] Cf. Frede, 1987, 262, 265–6.

[74] Thessalus is also guilty in Galen's eyes of *lèse majesté* in his denouncement of Hippocrates (*Meth. Med.* X. 7 K). Cf. ibid., 21–22 K for Galen's list of Thessalus' "vices". Galen's aim in attacking Thessalus is also to attack contemporary Methodists. In attacking these doctors, Galen is also prepared to deform his own source material. Thus Galen enlists an Aristotle in full agreement with Hippocrates as far as the study of Nature is concerned (X. 15 K); Plato and Hippocrates, needless to say, are, as ever in full accord (X. 14 K). Plato, Aristotle and the Stoics all hold Hippocrates to be "the father of their natural science" (X. 17 K). Galen also attacks the Methodists by abusing others whose relationship to Methodism is by no means easy to determine, and Galen's use of Asclepiades as a stick with which to beat all Methodists should be viewed partly in this light (cf. Vallance, 1990, 134f., 145). Yet Galen's anger should also be seen in perspective. As Asmis, 1993, 147, notes, "Galen was a highly polemical writer who often ignored differences among his opponents in order to attack what seemed to him a common position. This does not mean that he was unaware of the differences, or that he tried to hide them."

[75] *Meth. Med.* X. 76 K.

some Methodists did not altogether dispute the acquisition of anatomical knowledge obtained through dissection, even if one of its uses was apparently to disarm one's opponents.[76] In this respect, anatomy for a Methodist is more a rhetorical than epistemological tool. Dissection for the sake of acquisition of new knowledge does not seem to have been considered.

Galen's public anatomical demonstrations were important enough to attract an audience as well as influential enough to secure patronage.[77] It is not suggested that Galen owed his success to public demonstrations alone, for his therapeutic skills were much in demand.[78] But it cannot be denied that Galen's public exhibitions sought not only to establish his authority but to enhance the status of the anatomically proficient and philosophically trained physician.[79] It is this particular interplay of status, authority and knowledge that underlies Galen's approach to the study of the brain.

---

[76] Cf. Soranus, *Gyn.* I.5, p. 8,4–11 Ilberg. Cf. Temkin, 1956, xxxix.

[77] Galen never returned to public anatomical demonstrations after his return to Rome. Whilst plagiarism and jealousy undoubtedly played a part in his decision to cease public exhibitions, they were also no longer necessary: Galen had secured his reputation and his target audience. See n. 50 above.

[78] Galen's successful treatment of a severe eye ulceration is met by his being called "wonder-worker" and "wonder-teller" by his colleagues (*Opt. Med. Cogn.* pp. 59,17–61,16 Iskandar). See also Galen's *De nominibus medicis* p. 21,6–10 Meyerhof-Schacht. An important part of Galen's therapeutic success was his skill in the preparation of the polypharmaceutic *theriac* (opium is one of its constituents). Cf. *De theriaca ad Pisonem* XIV. 214–219 K; Swain, 1996, 368–70, 377. On Galen's preparation and use of theriac see Stein, 1997.

[79] Cf. Tarrant, 1998. Whether philosophy could also serve as an effective winding sheet for medical mistakes is another matter; there is evidence that it could (cf. Ierodiakonou, 1995).

PART ONE

THE FRAMEWORK OF GALEN'S ANATOMICAL METHODOLOGY

CHAPTER ONE

THE DEVELOPMENT OF THE HEGEMONIC CONCEPT: *THE MEDICAL AND PHILOSOPHICAL BACKGROUND*

One says the heart, another the meninges, and one that the brain contains the *hegemonikon* of the soul; therefore one will claim one use for the part in question here, one will claim another . . .

De usu partium.[1]

1.1 *Introduction*

The need to place the actions of the body under the command of a *ruling principle* (*principatus, principale*) or *regent part* (*hegemonikon*, ἡγεμονικόν) is a pervasive one in medicine and philosophy in Western Antiquity.[2] Regardless of whether this principle was identified completely with *soul* (ψυχή),[3] or, as later developed, was construed as the intermediary between soul and body, it is no exaggeration to say that the physical location of the *hegemonikon* was an important polemic in medicine and philosophy.[4] It was a debate that influenced, as much as it was informed, by the developing sophistication of medical and philosophical epistemology. The ruling principle may be thought to consist of (or be based in) air, and could be placed in the thoracic cavity, since it was recognised that air was taken into the lungs through the trachea.[5] The heart, through its connections with the lungs and blood

[1] ὁ μὲν γὰρ τὴν καρδίαν, ὁ δὲ τὰς μήνιγγας, ὁ δὲ τὸν ἐγκέφαλον ἐν ἑαυτῷ φησιν ἔχειν τὸ τῆς ψυχῆς ἡγεμονοῦν, ὥστε καὶ τῶν ἐν αὐτοῖς μορίων τὴν ὠφέλειαν ἄλλος ἄλλην ἐρεῖ . . . I, p. 15,2–5 Helmreich; III. 4 K.

[2] One who is ἡγεμονικός is a person suitable for leadership (LSJ 762–763). On *principatus* see Cicero, *Tusc.* I. 20, where, in commenting on Plato's tripartition of soul, reason (the sovereign part of the soul) is placed in the head as in a citadel: *Plato triplicem finxit animum, cuius principatum, id est rationem in capite sicut in arce posuit.* Cf. *N.D.* II. 29. On the use of *principale* see Seneca, *Ep.* 92. 1; 113. 23. See also Waszink, 1947, 221–222.

[3] Onians, 1951, 13–43, 93–122, provides a comprehensive survey of the historical and philosophical beginnings of the links between soul and body.

[4] Cf. Hankinson, 1991b; Mansfeld, 1990b; Runia, 1999a, b.

[5] According to Lucretius, *De rerum natura* III. 140, *animus* (soul or mind), is the governing council (*consilium*) of the body. It is located in the middle part of the

vessels, is another candidate. So is blood itself.[6] The brain may be chosen on analogous grounds, for the nasal passages were supposedly in direct contact with the brain itself. Some identified the hegemonic principle with the actions of the senses *tout court*, and the *hegemonikon* was therefore regarded as not being dependent on a single or fixed location.[7] The debate regarding the location of the *hegemonikon* developed into an often complex and interlocking set of theories which are partly embedded in the doxographic record, a record which must at all times be used with caution.[8] Their proponents can be divided

---

chest (*situm media regione in pectoris haeret*). However, in the doxographic accounts, only one source (Aëtius IV 4.6) has Epicurus (and Democritus) locating the rational part of the soul "in the chest" (ἐν τῷ θώρακι). Cf. Mansfeld, 1990b, 3088, 3114. On Democritus see below, n. 22.

[6] A view that goes back to Empedocles, who placed thought in the blood around the heart (αἷμα γὰρ ἀνθρώποις περικάρδιόν ἐστι νόημα). DK 31B105; cf. 31A84; 31A86. See also Harris, 1973, 18. Cf. n. 9 below.

[7] This is also an argument for the negation of the hegemonic concept, and was apparently the view of Asclepiades of Bithynia. According to Sextus Empiricus, Asclepiades was said to have *abolished* (ἀναιροῦντα) the *hegemonikon* (*M.* 7.202; cf. 7.380). Tertullian, *De anima* p. 15,2–3 Waszink, admittedly a hostile source, has Asclepiades and some others subvert the concept *hegemonikon* to the extent that they identify it with sensation, not with the soul (from Tertullian's Christian perspective, to deny the soul is to reject the *hegemonikon*.). Tertullian brackets Asclepiades in the company of two others, Dicaearchus (a pupil of Aristotle and contemporary of Theophrastus), and Andreas (a student of Herophilus and a personal physician to Ptolemy IV Philopator). Andreas was apparently one of the few Herophileans to have been interested in dissection (cf. Von Staden, 1989, 445 n. 1, 472–7; Waszink, 1947, 222). Tertullian states that Asclepiades used insects such as flies, wasps, and locusts, which were decapitated, and goats, tortoises, and eels, from which the heart was extracted. (*De anima* p. 15,2–19; T 139 *Herophilus*.) Asclepiades, Andreas and Dicaearchus may have attempted to elaborate a compromise position, concluding it impossible to reach consensus over the competing claims of cardiocentrists and encephalocentrists. Having the senses act not as agents of the *hegemonikon*—the governing tenet of cardiocentrists and encephalocentrists—but *as* the *hegemonikon*, could undercut both encephalocentrist and cardiocentrist claims. (Annas, 1992, 30, in her dismissal of the views of Dicaearchus on the soul as "frankly weird", is unhelpful).

[8] One of the main doxographic sources for the hegemonic debate is the collection of texts known as the *Placita*, attributed to the otherwise unknown Aëtius (c. first century AD), and which are a compilation of several later sources. Cf. Runia, 1999a, 40–1. See the comprehensive studies by Mansfeld, 1990a, 1990b, Mansfeld and Runia, 1997. For the medical aspects of the *Placita*, see Runia, 1999b. On the challenges of medical doxography for the historian see Van der Eijk, 1999b. The words of Mansfeld, 1990b, 3164, are apposite: "Doxography can be as amusing, and as confusing, as bureaucracy." Hegemonic theories also seemed to have been used for teaching purposes in antiquity (which might perhaps explain the manner of their codification). One example is the *Dictyaca* of Dionysius Aegaeus (Photii *Bibliothecae* cod. 211; pp. 336–340 Deichgräber) which, unfortunately, lists only the titles of the theses and antitheses concerning sites of the *hegemonikon* (employing the term ἀρχή). Cf. Caujolle-Zaslawsky, 1994, 863; Runia, 1999b, 242–3; Von Staden, 1999b, 177–185.

into two broad categories: those who maintained that the *hegemonikon* was found in the head (encephalocentrists) and those who argued that it was located in the heart or its immediate vasculature (cardiocentrists).[9] Apart from Galen, on the encephalocentric side can be placed, among others, Ptolemy, Herophilus and Erasistratus, Plato, the author of the Hippocratic text *The sacred disease*, and certain of the Presocratics. Among the cardiocentrists can be numbered Aristotle, the author of the late Hippocratic text *The heart*, and the Stoics. Stoicism devised the term *hegemonikon*, making it a Hellenistic innovation to the medical and philosophical lexicon.[10] For Galen the brain is the *hegemonikon* because that organ alone is responsible for *sensation* (αἴσθησις) and *voluntary motion* (προαιρετικὸς κίνησις), the chief distinguishing characteristics of the rational soul.[11] The rational soul functions, according to Galen, through the actions of its *first instrument* (πρῶτον ὄργανον), *psychic pneuma* (πνεῦμα ψυχικόν). Here Galen somatises not so much the soul but its agent. As chapter 2 will show, pneuma is an appropriate choice for Galen's physiological requirements. It allows him ample scope from which to construct an elaborate set of physiological arguments. In what precise manner psychic pneuma acts is however a matter of considerable dispute.[12] Galen also maintains that the site of the *hegemonikon* is the *source* or *origin*

---

[9] Cf. Manuli, 1977; Manuli and Vegetti, 1977, 179–82. The *diaphragm* was thought by some unnamed physicians or philosophers to be the site of the *hegemonikon* (Soranus B, ap. Caelius Aurelianus, *Acut.* I.VIII, p. 54,8–13 Bendz; p. 34 Drabkin). But this could represent a conflation of the views of Diocles of Carystus that *phrenitis* is an inflammation of the diaphragm (cf. Mansfeld, 1990b, 3106–8; Frs. 71–73 vdE). It is more likely to represent a recrudescence of older, Homeric notions concerning the importance of the diaphragm (cf. Langholf, 1990, 40ff.). In what can be said to represent a haematocentric position, the *aorta* (ἀορτή or ἀρτηρία παχεία), and the *vena cava* (φλέψ παχεία) are attested by unnamed physicians or philosophers as the site of the *hegemonikon* in one source (Soranus B, above), perhaps reflecting an Empedoclean influence. See n. 6 above.

[10] Long and Sedley, 1987, II, 313, point out that: "This term for the soul's 'commanding-faculty' may be Zeno's invention, though it may also have been coined or used independently by the Peripatetic Strato." (cf. 310–316). See also *SVF* 2. 836; L-S 53H, where the *hegemonikon* is also the part of the soul "which reasons" (λογισμός). Its usage by the Second Century AD was widespread and understood (cf. Tieleman, 1996a, xxiii, 26 and n. 74).

[11] Another term Galen uses for volition is ὁρμή (*conation* or *intention*); cf. *PHP* p. 110,1–2 De Lacy; V. 219 K): Ἔστι δὲ τὸ ἡγεμονικόν, . . . τὸ κατάρχον αἰσθήσεώς τε καὶ ὁρμῆς. Conation was used by the Stoics to distinguish animals from plants, and as a human being develops, it is placed under the influence of reason (cf. L-S 33I, 53P,Q).

[12] See chapter 2, n. 97.

(ἀρχή) of the nerves, for it is central to Galen's encephalocentrism that the nerves not only take their origin from the brain, but are composed of the same substance. To help validate such claims, Galen utilises the authority of the past as well as his experimental methodology. Hippocrates and Plato are made to speak with one voice in affirming that the source of the nerves is the brain and spinal cord, and that the origin of the spinal cord is the brain itself.[13] If Aristotle provided Galen with the methodology of dissection and experimentation, and Hippocratic authority legitimated his authority as a physician, then it was Plato who gave him the philosophic basis for the hegemonic status of the brain.[14]

Anatomy is Galen's chief weapon against his opponents. Those, like Aristotle and the Stoics, who maintain that the origin of the nerves is the heart are mistaken, according to Galen, because they do not seem able to learn the facts from dissection.[15] For Galen, the proof obtained from dissection enables the question of the location of the *hegemonikon* to be correctly resolved.[16] But the Stoics, as chapter 2

---

[13] ὅτι μὲν γὰρ ἀρχὴ νεύρων ἁπάντων ἐγκέφαλός τε καὶ νωτιαῖος καὶ ὡς αὐτοῦ τοῦ νωτιαίου πάλιν ἐγκέφαλος. *PHP* p. 68,20–21 De Lacy; Fr. V.

[14] For the general background to Galen's relationship to Plato, see De Lacy, 1972. Plato's argument for the primacy of the brain as the organ of the rational soul is teleological, not anatomical. As Lloyd, 1983a, 334, notes, "the significance of the *Timaeus* lies in its being the first clear and sustained statement of a teleological cosmology." Cf. Guthrie, 1978, *HGP* V, 313–314; Longrigg, 1993, 128–135. In analysing the importance of Plato to later anatomical studies, Cosans, 1995, 591 and 594, stresses the importance of teleology, but reads too much later anatomical and physiological accounts back into Plato, implicitly downplaying Aristotle's innovative empirical research methodology. Plato posits a tripartition of soul within the body, consisting of three parts, housed in brain, heart, and in the sub-diaphragmatic area. Each of these has different responsibilities, the brain housing the *rational* (λόγος) soul, the heart the *spirited* (θυμός), and the sub-diaphragmatic area the *appetitive* (ἐπιθυμητικόν). Cf. *Timaeus* 45a–b; 69d–70a; 70d–71b. Plato never explicitly states that the ἐπιθυμητικόν is in the liver (at *Tim.* 70d–e, he locates it between the diaphragm and the navel). Cf. De Lacy, 1988. The *locus classicus* of tripartition in Plato is *Republic* 441a–442c, 608c–611a, although there is no mention of the parts of the soul as construed in the *Timaeus*. The *Republic* is more concerned with the complex strands of an individual's behaviour, and how a tripartite model incorporating reason and emotions might help in understanding that behaviour (cf. Annas, 1981, 124f.). For Plato, the divine part of the soul is placed in the head, whose spherical shape reflects that of the universe, and is so constructed (*Tim.* 44d). No attempt is made to render the brain as an hegemonic organ, although the sense organs are placed in the head (*Tim.* 66a). The connections between them and the head are only mentioned in general terms.

[15] See for example, *UP* I, p. 414,15–17 Helmreich; III. 570 K.

[16] πόθεν οὖν τοῦτο δειχθήσεται; πόθεν δὴ ἄλλοθεν ἢ ἐκ τῶν ἀνατομῶν; *PHP* p. 110,5–6 De Lacy; V. 220 K.

will show, by interpreting the evidence from dissection another way, are among Galen's most formidable adversaries. In this respect, the empirical differences between encephalocentrist and cardiocentrist were perhaps not so far apart as Galen would wish.[17] Apart from anything else, the dispute between encephalocentrist and cardiocentrist may usefully be viewed as part of an ongoing debate concerning the relevance of anatomical investigation to medical and philosophical practice.

The following is no more than an attempt to provide a representative background to Galen's own contribution to the hegemonic debate.[18] It acknowledges the long period in which the amount of anatomical and experimental evidence for the brain to be regarded as an hegemonic structure became available and fully exploitable in Galen's hands.

1.2 *Presocratic knowledge of the brain*

Awareness of the vulnerability of the skull and its contents is as ancient as warfare; what functions were ascribed to those contents is a different matter.[19] Homer provides several graphic accounts of the destruction of "the brain", employing the term ὁ ἐγκέφαλος, which applies to the contents of the skull.[20] The capacity to view the brain

---

[17] Cf. *PHP* p. 170,23–27 De Lacy; V. 288–289 K. See also Tieleman, 1996a, 154ff.

[18] The studies by Edelstein, 1932, 1935, remain fundamental to any analysis of the history of anatomy in Western antiquity.

[19] The Edwin Smith Surgical Papyrus reveals knowledge of brain and spinal cord injuries. This Papyrus' conception has been estimated from the beginning of the Old Kingdom (Third Dynasty), that is, about 3000–2500 BC. But the Text (and Glosses) are much later, being composed in the Eighteenth Dynasty of the New Kingdom (about 1600 BC). Cf. Breasted, 1930, xiii; Nunn, 1996, 25–30. The cases represent wounds noted by surgeons in battle conditions (and possibly industrial accidents). Although this Papyrus contains the first extant use of the Egyptian word for *brain* (*ais*), it was not considered important, being removed and discarded during mummification. The word also means *viscera*. Cf. Dawson, 1932, 150; Iversen, 1947, 50; Grapow, Von Dienes and Westendorf, 1954, 28; Nunn, 1996, 50–51; Weeks, 1970, 28.

[20] Hippothous' brains "shoot up" (ἀνατρέχω) along the track of the spear point thrust through his helmet into his head (*Il.* XVII. 297–298). In the death of Erymas (*Il.* XVI. 345–350), the spear passes through the mouth and "under the brain" (νέρθεν ὑπ' ἐγκεφάλοιο), going on to shatter the bones of the skull. Cf. Fenik, 1968; Hanson, 1999, 213–214; Majno, 1975, 142; 166–175; Salazar, 2000, 13–15, 45–6; Saunders, 1999, 361–2; Thompson, 1952, 765. The very generality of terms employed by Homer in reference to any part of the body makes further elucidation of their function highly problematic. Cf. Ireland and Steel, 1975; Sullivan, 1981, 147.

as more than just another part of the body susceptible to injury had its origins among some of the natural scientists of the Presocratic period, where, for the first time, general theories were advanced concerning man and the environment.[21] These concepts are entangled in sometimes conflicting doxographic material, and are often given a vocabulary and interpretation they could not have possessed. Those Presocratics who represent a prefiguring of an encephalocentric hypothesis are Alcmaeon of Croton and Diogenes of Apollonia.[22]

Alcmaeon (fl. after 500 BC) bears the burden of a number of anatomical claims and insights, very few of which can be his own. His alleged status as a physician[23] may well have enhanced the nature of those claims later made of him concerning the brain and the nerves.[24] According to a fourth century AD account, that of Calcidius, Alcmaeon is said to have dissected the human eye.[25] But in making this claim, Calcidius (or his source) is likely to have used material from Herophilus.[26] There is no persuasive evidence for human dissection being practiced before Herophilus (the examination of animals is another matter). Regardless of what Alcmaeon's method of *autopsia* may have been, simple removal of the eye from an animal would have revealed the presence of the optic nerve at the back of the globe—or at least a structure later known to be the optic nerve,

---

[21] An analysis of the extant fragments and translation may be found in KRS. See also Long (ed.), 1999. Cf. Lloyd, 1975a; idem, 1979, 139–146; idem, 1992b, 118–122.

[22] At Aëtius IV 5.1, Democritus is said to have placed the *hegemonikon* in the head or brain. This contradicts Aëtius IV. 4.6, where Democritus (and Epicurus) have the regent part in the chest (cf. n. 5, above). Mansfeld, 1990b, 3088 n. 120, concludes: "we are not in a position to decide, on the basis of what is in Aëtius, whether Democritus spoke of a dominant part of the soul, let alone where he put it." (but cf. *HGP* II, 433–434, where it is stated that for Democritus, the reasoning part of the soul is located "in the head, as it was for Alcmaeon, Anaxagoras and Diogenes of Apollonia, and later for Plato."). According to Theophrastus, *De sensu* 58, Democritus regarded the *whole soul* as disseminated through the body, which perhaps better fits with what is known of Democritean atomism. Cf. Bicknell, 1968, 18–19. See also Laks, 1999, 253 n. 16. On the question of whether Democritus performed dissections, see Lloyd, 1975a, 132.

[23] Galen does not cite him as a physician (*Meth. Med.* X. 6 K). The evidence against his being a doctor is persuasively argued by Mansfeld, 1975, 26–38, 26–27.

[24] Among such claims, Alcmaeon is said to have regarded the brain as τὸ ἡγεμονικόν (Aëtius V 17.3).

[25] Cf. T 86 *Herophilus*. Longrigg 1998, 169, commenting on this passage, states that it is not likely Alcmaeon dissected or vivisected human beings, concluding that it is "more probable that the reference here is to a surgical operation upon a human subject."

[26] Cf. Lloyd, 1975a, 115ff.; Von Staden, 1989, 238.

for there is no persuasive evidence Alcmaeon regarded it as such.[27] Alcmaeon seems to have related the brain to vision, because of the link between the eye and the brain. This of itself is not enough to grant the brain hegemonic status. Although it is argued that Alcmaeon recognised the "pathophysiological and cognitive significance of the brain,"[28] it is safer to claim that "Alcmaeon's . . . preoccupation with these matters stimulated the interest of later philosophers in them to such an extent that after him psycho-physiological inquiries become almost a standard topic of investigation."[29]

A key theme in Greek physiology is the notion of πόροι (*passages* or *channels*), communicating within the body and with the outside air to provide sensation.[30] The idea of a system of passages in which air (or *pneuma*) may move within the body was perhaps first raised by Diogenes of Apollonia (fl. 440–430 BC).[31] The life-giving air is distributed about the body in a carrier system, the *blood vessels* (φλέβες).[32] Diogenes' own descriptions of vascular anatomy do not appear dependent on any detailed investigative methodology, and could have been made with little difficulty on sacrificial animals.[33] Diogenes' vascular system is also distributed to the head.[34] The brain is given a role, possibly for the first time, for according to the account of Diogenes in Theophrastus, air reaching the brain somehow enables cognition to take place.[35] Air resides in (or passes through) the brain and the

---

[27] Cf. Lloyd, 1975a, 123 nn. 39, 40. *Pace* Harris, 1973, 7, who states that "Alcmaeon is also credited, with some degree of probability, with the discovery of the optic nerve."

[28] Von Staden, 1989, 155. Nor is it true to say, as Guthrie does, that: "The explicit recognition of the brain as the central organ of feeling and thought was another striking contribution of Alcmaeon." *HGP* I, 349.

[29] Longrigg, 1993, 53.

[30] "In a sense all Greek medical theories are theories about πόροι; the human body is simply a system of πόροι. Πόρος is used in a general sense to cover all the channels of the body." Lonie, 1965, 128.

[31] On Diogenes, see *HGP* II, 362–381.

[32] Cf. Aristotle, *HA* 511b24–513a5. *HGP* II, 368–369. A comprehensive account of Diogenes' vascular anatomy and its influence is provided by Harris, 1973, 20–28. In this period, the designation φλέβες meant both veins and arteries, carrying both blood as well as air. Cf. Lloyd, 1975a, 126 n. 48; idem, 1983b, 152–153.

[33] Cf. Harris, 1973, 25.

[34] DK 64B6. Aristotle, in his account of Syennesis, Diogenes and Polybus, collapses the distinctions between them to the extent that he maintains all held that the blood vessels originated in the head (he adduces as evidence for this view other, unnamed, workers in this field. *HA* 513a8f).

[35] DK 64A19. Cf. 64B4–7. Cf. *HGP* II, 376–377. In the context of air conveying sensation to the brain, as well as a vascular system for its distribution, the similarities between Diogenes and the author of the Hippocratic text *Morb. sacr.* are striking (cf. Harris, 1973, 27; *HGP* II, 377 n. 2).

mediation of the brain in concert with air provides sensation and intelligence.[36]

Alcmaeon and Diogenes made several observations about the brain and its connections with other parts of the body and with the external environment. It is permissible to infer that dissections of a kind were performed to confirm some of the notions raised by Alcmaeon and Diogenes, although the most significant single step, opening the skull, was not made.[37] To these natural philosophers there is an awareness that the brain has connections with the body—however anatomically vague these connections are, and that a linking of the brain with the outside air allows (and mediates) intelligence. These concepts are important for they were later elaborated and formalised by those physicians who held an encephalocentric viewpoint.[38] In particular, the use of air (or *pneuma*) linked to an internal carrier system is an idea which will have increasingly important physiological considerations for encephalocentrist and cardiocentrist alike.

1.3 *The brain in the Hippocratic Corpus*

Evidence of anatomical investigation in the Hippocratic Corpus is sparse.[39] As Lloyd has noted, "references to dissection are rare, and their absence in some contexts where the authors in question are

[36] In Aëtius, IV. 5.7, Diogenes is cited as one who places the *hegemonikon* of the soul in the ventricles of the heart. However, this is most likely a reference to the Stoic Diogenes of Babylon, author of the lost *On the leading part of the soul,* whose cardiocentric arguments were known to Galen (*PHP* pp. 128,29–130,5 De Lacy; V. 241 K). Cf. Duminil, 1983, 57; Harris, 1973, 25 n. 2; Tieleman, 1991, 122–3.

[37] The story related by Plutarch, *Per.* VI, on the opening of the skull of a ram by Anaxagoras, if not apocryphal, may be interpreted as curiosity to explore an anomaly (the presence of a single horn growing from the middle of the forehead). The horn was interpreted as an excrescence of the brain. Cf. Onians, 1951, 237ff.; Lloyd, 1979, 24 n. 79, 161 n. 185.

[38] According to the doxography, the Presocratic Philolaus (c. 470–385 BC) in *On Nature* (Fr. 13 Huffman) apparently regarded the *head* (κεφαλή) as ἀρχή or "first principle" of the *intellect* (νόος). The brain is one of a four-fold system of ἀρχαί; the other locations being the *heart* (καρδία), responsible for "life and sensation" (ψυχὴ καὶ αἴσθηις); the *naval* (ὀμφαλός), responsible for "rooting and first growth" (ῥιζώσιος καὶ ἀναφύσιος τοῦ πρώτου) and the "reproductive organs" (αἰδοῖον) for "fertilisation as well as generation" (σπέρματος καταβολᾶς τε καὶ γεννήσιος). This fragment has more in keeping with Hellenistic medical theories, and should best be regarded as a summary of later medical and philosophical accounts of the body's ἀρχαί.

[39] The Hippocratic text *On anatomy* (*Anat.*) survives as a fragment (VIII. 538–540 L; pp. 208–209 Duminil). It is considered a late work, composed sometime after

discussing how to investigate certain anatomical and physiological problems suggest that these authors, at least, were not familiar with the method."[40] Several texts highlight the brain in cases of injury and therapeutic intervention.[41] But these reflect practical considerations and have nothing to do with an encephalocentric agenda. For example, in *Diseases II* (*Morb. II*) a condition which causes numbness to the head results, when the crisis is reached, in a partial loss of vision, the patient being able only to see half of a person's face.[42] In this text, although consciousness seems to be under the control of the heart (or is at least is located about the heart), there is also an attempt at establishing a causal connection between affections of the brain and some modalities of sensation.[43] The text *On wounds to the head* (*VC*) defines the brain topographically as lying more to the front of the head, under the *bregma* (βρέγμα) where the bone is at its thinnest and weakest.[44] Towards the back of the head there is said to be less brain, and thus wounds there are less fatal than wounds at or about the bregma.[45] Detailed instructions are given concerning when, where and how to trephine the skull.[46] Trepanation provides some evidence that several of the Hippocratic authors were

---

Praxagoras (cf. Harris, 1973, 83; Von Staden, 1989, 176 and n. 119). In this text, Craik, 1998b, 135, has noted that in the case of the lungs and liver there are "features peculiar to human organs." Galen's *On the anatomy of Hippocrates* is not extant. Cf. Ilberg, 1892, 496; see also Garofalo, 1992; Harig and Kollesch, 1975, 261.

[40] 1975a, 134. Cf. Von Staden, 1989, 156.

[41] For example, of the 75 chapters in *Morb. II*, 37 are concerned with descriptions of diseases of the head. The majority are not specifically named, and only a description of their effects is given. The author also points out that, technically, none of these diseases are due to vessels overfilling with blood; this is only an appearance, and the description which follows is a traditional one (bile or phlegm enters the blood vessels and accounts for their apparent overfilling with blood). Cf. *Morb. II*. VII. 10 L; pp. 133–134 Jouanna.

[42] VII. 20 L; pp. 142–143 Jouanna. At VII. 16 L; p. 139 Jouanna, pain experienced in the head is said to be greatest in the front, due to phlegm being attracted to areas where the blood vessels are said to be at their widest. As is the case in *VC*, here there is also said to be more brain at the front of the head than at the back.

[43] On sensation in the Hippocratic Corpus, see Ioannidi, 1992.

[44] III. 188 L; p. 64,9–12 Hanson. In the Loeb edition, Withington, 1928, 9, translates *bregma* as "dura mater". But it is the relative thinness and weakness of the bone over the bregma that is referred to, coupled with an observation that the flesh over the bregma is thinnest (cf. Adams, 1849, 446).

[45] *VC*. III. 190–192 L; p. 64,28–33 Hanson.

[46] *VC*. III. 258–260 L; pp. 90,18–92,9 Hanson. In *Morb. II* VII. 28 L; p. 150 Jouanna, the instructions are more general; one is only told to make the cut at the bregma. *Epidemics V* (*Epid. V*) lists several cases of trephining; most had fatal outcomes, attributed in large part by failure to trephine earlier or more thoroughly (V. 216, 226–228. L; pp. 10, 16–17, Jouanna).

aware of the outer meningeal covering of the brain; damage to it was often regarded as fatal.[47]

The Sacred Disease (*Morb. sacr.*) might be regarded as incontrovertible evidence of an encephalocentric hypothesis in the Hippocratic Corpus.[48] It states that the brain is the medium for thought, sensation and the emotions, and regards epilepsy as no more divine than any other disease.[49] Although not part of any empirical research protocol, there is textual evidence that some sort of dissection—or at least, an examination—of the brain (of a goat) took place, and that this was considered corroborative evidence for the effects of epilepsy on the brain by the author. The contents of the cranium were described as oedematous and foul smelling.[50] The "brain of man" (ἐγκέφαλος τοῦ ἀνθρώπου) is described as double and divided down

---

[47] It is dangerous to damage this membrane in any way (III. 256–258 L; p. 90,7–15 Hanson). Cf. Hanson, 1999, 117–119. There is a mention of two meningeal layers in *Places in Man* (*Loc.* VI. 280 L; p. 38,22–23 Craik), but although the description is accurate when interpreted with later knowledge, the passage is otherwise obscure. Phillips, 1973, 47, interprets it as follows, "Of the two membranes, or *meninges*, the outer is thicker (the *dura mater*), while the inner (the *pia mater*) is thin and in contact with the brain." There is no textual basis for this claim. *On Fleshes* (*Carn.*) notes the presence of the "thick meninx" (μῆνιγξ παχεῖα), the *dura mater* (VIII. 588 L). The spinal cord is described as similar to the brain, and also has a membrane. It has been claimed that in *Carn.* and *Morb. sacr.*: "the brain is likewise granted this traditional cognitive role." Von Staden, 1989, 249. Von Staden bases this on chapters 16 and 17 of *Carn.* But while those chapters discuss the brain in relation to smell and sight, they do not assign *control* of these functions to the brain.

[48] The date of composition has been estimated at 400 BC (cf. Temkin, 1971, 4–6). Jouanna, 1992, 549, puts it at the second half of the fifth century BC. Galen's questioning of the authenticity of *Morb. sacr.* has as its basis that text's claim that the spirited and appetitive elements of the soul, as well as the rational, are placed under the control of the brain. For Galen to cite this text as unambiguously authentic would undercut his claims for Platonic tripartition. Lloyd 1991a, 409, states that for Galen to ascribe Platonic soul partition to Hippocrates, he might make use of " a variety of texts . . . to suggest not, it is true, that the brain is the source of the nerves, but at least that it is the seat of intelligence . . ." Lloyd notes two examples, both from *Morb. sacr.*, as well as noting where Galen cites this treatise (*Hipp. Progn.* p. 206,13–15 Heeg). But what Lloyd does not mention is that in this text Galen places *Morb. sacr.* in a category of doubtful authenticity, a statement reinforced in one of the scholia attached to the MSS of *Hipp. Progn.* (cf. Tieleman, 1995c, 604). If Galen regarded *Morb. sacr.* as probably not by Hippocrates, it follows that for him the chief cardiocentric texts (*Cord.*, *Carn.*) should also be denied Hippocratic authorship, since they would not suit a Hippocrates in harmony with Plato. As Tieleman, 1995c, 605, notes, Galen solves this problem by simply not referring to them.

[49] The treatise never *denies* divine causation in epilepsy. It is equally as divine and as human. Cf. Prioreschi, 1992.

[50] VI. 382 L; p. 78 Grensemann.

the middle by a *thin membrane* (μῆνιγξ λεπτή).[51] This description could have been obtained from any sacrificial animal and does not indicate a human dissection was performed. All perceptions and sensations are said to come from the brain, but the brain is conceived, not as a controlling principle in its own right, but as a *mediator* or *interpreter* (ἑρμηνεύς) for the air itself.[52] Intelligence and sensation are in the brain precisely because these attributes have been passed to it from the outside air. It is the air received by the brain that is of crucial importance.[53] To put this physiology into effect, the author of *Morb. sacr.* makes use of two concepts. First, communications between the outside air and the brain through the mouth and nasal passages. Second, vascular connections between the brain and the body.[54] These blood vessels, whose source is the liver and spleen, also send branches to the heart and the lungs.[55] Air in the blood vessels reaches the diaphragm and heart, but these structures do not themselves directly mediate intelligence.[56] The result of these combined pneumatic and vascular connections is intelligence and movement.[57]

Although in the Hippocratic Corpus the notion of an organ as a discrete structure with its own powers is not a strongly held tenet,[58]

---

[51] VI. 366; p. 68 Grensemann. Cf. Von Staden, 1989, 156 n. 51. This should not be taken as a reference to the *pia mater*, the fragile inner lining of the brain which was discovered by Hellenistic anatomists. See chapter 3.7.

[52] VI. 390–392 L; p. 86 Grensemann. Cf. Solmsen, 1961, 155–7; Miller, 1948, 173–5.

[53] For example, the text informs us that climactic alterations in the outside air are directly linked to changes in the brain which can have pathological consequences (VI. 394 L; p. 88 Grensemann). Blockage of the vessels to the brain by phlegm is said to render the patient speechless and senseless (VI. 372 L; p. 72 Grensemann). Cf. Temkin, 1971, 53–54.

[54] VI. 372–374 L; p. 72 Grensemann. In this system, air inhaled by the mouth or the nose first goes to the brain, thence most passes to the *stomach* or *belly* (κοιλία) for the purpose of cooling. Some is also sent to the lungs and the blood vessels. Cf. Harris, 1973, 39–41.

[55] VI. 366 L; p. 68 Grensemann. No differentiation is made between artery and vein. This distinction was made on the basis of a detailed examination of the vascular anatomy of the heart, and appears to have been made by Praxagoras. Cf. Von Staden, 1989, 177, 266–267.

[56] VI. 392–394 L; pp. 86, 88 Grensemann. To disprove the notion that the diaphragm is important in these matters, the author claims that the word for *diaphragm* (φρένες), is due to *accident* (τύχη) and *custom* (νόμος), and not to any inherent *power* (δύναμις) which the diaphragm might possess in order for it to be regarded as the mediator of thought and intelligence.

[57] καὶ οὕτω τὴν φρόνησιν καὶ τὴν κίνησιν τοῖσι μέλεσι παρέχει. VI. 372 L; p. 72 Grensemann. Cf. Ioannidi, 1983, 327–30; Pigeaud, 1980, 418ff. See also Miller, 1948, 174–175.

[58] A persuasive case is offered by Gundert, 1992, 463. Harris, 1973, 33, sees it

the texts examined above illustrate, and to an extent augment Presocratic knowledge of connections between the brain and the rest of the body, characterised by a widespread system of vascular and pneumatic channels. There is no evidence of any detailed anatomical investigation of the brain, nor are the nerves recognised as discrete structures. There is however, increased speculation regarding the brain and its capacities as the *common sensorium*, although there is nothing to suggest that these conjectures either preceded or generated an anatomical research methodology of their own.[59]

1.4 *Aristotle on the brain and the heart*

Aristotle presents a clear and methodological programme of empirical research for the first time in Greek natural science.[60] According to Aristotle, the cause of his predecessors' mistakes in regard to the nature of the blood vessels lay in their inadequate observations; they relied on what they had observed externally; they were unfamiliar with the material (lack of experience in dissection); they did not have a proper methodology (failure to do dissections in the correct way).[61] Dissection may be unpleasant, but it is an opportunity to acquaint oneself with the purposefulness of nature.[62] Whilst Galen criticises Aristotle's views on the heart, he praises his ability as a dissector and his method of enquiring into nature.[63]

---

differently: "we may divide the physiological works into two classes, those which regard the head as the seat of intelligence, and those which assign this function to the heart." Singer, 1993, offers a more nuanced interpretation.

[59] *The Heart* (*Cord.*) is a Hellenistic work; its assigning of hegemonic status to the left ventricle of the heart *qua* organ is the singular exception (IX. 88 L). Cf. Harris, 1973, 83–91, Lonie, 1973; Von Staden, 1989, 174 n. 112. On a post-Erasistratean dating for this text, see Mansfeld, 1971, 34.

[60] See Lloyd, 1978b, 1987a, 1996a.

[61] *HA* 511b14f. Aristotle's text on anatomy (with diagrams) is not extant. Cf. Lloyd, 1978a, 216 n. 7; idem, 1979, 163 n. 194. See also Harris, 1973, 121–134; Lonie, 1964, 435; Shaw, 1972; idem, 1974, 533–534; Vegetti, 1979, 14–37.

[62] *PA* 645a25–35. On Aristotle's teleology, which is not as strong and pervasive as Galen's, see Balme, 1965; idem, 1987; Hankinson, 1988; idem, 1989; Moraux, 1976; idem, 1981.

[63] εἰ τῶν ἐξ ἀνατομῆς φαινομένων οὐκ ἀμελῶς ἔχων καὶ τῆς χρείας αὐτῶν οὐκ ἀμελέτητος ὢν αὐτός τε λέγων, ὡς τῶν προβλημάτων τὰ μὲν λύσεως δεῖται, τὰ δὲ κολάσεως, τὰ δ' αἰσθήσεως, *UP* I, p. 449,16–19 Helmreich; III. 620 K. Cf. Aristotle, *Top.* 105a3–9.

Restricting this examination to Aristotle's biological works,[64] it has been claimed that: "The opposition between brain and heart is one of the corner-stones of his physiology."[65] Aristotle's physiological priorities are centred on the heart, but although he regards it as the central organ in sanguineous animals, he does not otherwise ascribe to the heart the capacities and functions of an hegemonic organ.[66] Nevertheless, the arguments Aristotle advanced for a cardiocentric hypothesis are coherent and influential.[67] To function, animals *qua* animals must have a common sensorium.[68] Anatomically, this can be shown to be the heart, because the blood vessels arise from it;[69] it is situated in the centre of the body;[70] it is the first organ to develop.[71] The heart is also the centre of the vital heat, which plays a key role in the generation of the foetus and is the embodiment of the soul.[72] The action of vital heat on the blood in the heart produces *connate pneuma*.[73] This pneuma is also a carrier of the vital heat.[74] The carrier system is the blood vasculature. But for Aristotle, blood alone cannot carry sensation; the role of blood is nutritional.[75] Yet blood must be (and is) present in an organ for sensation to take place. Pneuma seems a way out of this impasse, for, as connate pneuma, it is carried in the blood vessels, and seems to be the

---

[64] In *Metaph.* Ζ 1035b25–27, Aristotle does not want to discuss the qualifications of either the brain or the heart for primacy. Cf. Pellegrin, 1986, 127 n. 17, at 201. See also Irwin, 1988, 292 and n. 33. Whether Aristotelian metaphysics drove his biological works or *vice versa* is controversial. Cf. Devereux and Pellegrin (eds), 1990.

[65] Lloyd, 1978b, 224. Cf. Moraux, 1984, 751; Verbeke, 1978, 198.

[66] Cf. Modrak, 1987, 72. Guthrie, *HGP* VI. 296–298, provides a good summary of why Aristotle chose the heart rather than the brain as the centre of sensation.

[67] Cf. Johansen, 1998, 78–81, 203–7, 211–2; Moraux, 1984, 751; Verbeke, 1978, 198.

[68] *Somn. Vig.* 456a1–5.

[69] This, says Aristotle, is "clear from the dissections" (φανερὸν ἐκ τῶν ἀνατομῶν). *Somn. Vig.* 456b1–2 (cf. *HA* 513a15ff.; 513b1ff.). At *PA* 670a25f., the heart is described as the acropolis of the body and the source of heat.

[70] *PA* 665b10ff.

[71] *HA* 561a10f.; *Juv.* 468b28ff.; *PA* 665a34; Lloyd, 1978b, 216. Cf. Lloyd, 1987a, 58.

[72] *GA* 735a14–29; 737a1f. For Aristotle, the heart, because it possesses innate heat, must be the more important organ. It is the hottest part of the body (*Sens.* 439a3–4).

[73] *GA* 742a14ff. On how this might occur, see Peck, 1965, lxiv and Appendix B, §§31, 32. Cf. Freudenthal, 1995, 121–126. At *GA* 736b37, connate pneuma in semen contains the generative heat.

[74] *Juv.* 469b9f.

[75] *PA* 647b2, 650a35 (blood is nutrition); *PA* 649b20f. (blood is not essentially hot). Cf. *HA* 520b12ff.; *PA* 656b19ff., 666a17f. Cf. Solmsen, 1961, 173.

medium for the transmission of sensation.[76] The use of blood and pneuma in this way helps in further understanding why Greek physiology concerns itself so much with channels, pores and blood vessels. By Aristotle's time, they had become the basic structural requirements for sensation.[77] On such grounds, for Aristotle, the brain cannot be the site of the common sensorium for in his schema the brain is bloodless.[78] Lacking a system of vascular channels, it can play no part in the process of sensation.

Aristotle does not appear to have dissected the brain in the same detail as he did the other organs of the body such as the heart.[79] In common with the author of *Morb. sacr.* he regarded the brain as double.[80] Aristotle also described a smaller brain behind the larger hemispheric mass, *the cerebellum* (ἡ παρεγκεφαλίς).[81] The back of the head was stated by him to be hollow and empty.[82] Aristotle provided an accurate description of the two meningeal layers of the brain.[83] He also regarded the brain as avascular, noted above. This might imply that Aristotle thought the brain to be essentially unimportant. But the brain has significance for him precisely because of its thermo-

---

[76] Although Aristotle only states that the passages allegedly connecting the eye to the blood vessels on the surface of the brain contain pneuma (*GA* 744a2ff.). Cf. Freudenthal, 1995, 127, 131–3; Solmsen, 1961, 183.

[77] Aristotle uses the term νεῦρα, but this was applied generically to sinews and tendons as well (*HA* 515a27ff.). The interior of the heart has νεῦρα (a reference to the *chordae tendineae*, small tendinous cords which connect the free edges of the atrio-ventricular valves to the papillary muscles, and the *trabeculae carneae*, muscular bands of tissue attached to the inner walls of the ventricles). Galen criticises Aristotle (and Praxagoras) for believing these to be nerves (*PHP* pp. 80,24–82,1 De Lacy; V. 187–188 K). Cf. Moraux, 1985a, 334–5. See also Lones, 1912, 108–9. On the extent of Galen's comprehension of Aristotle's zoological research programme, see Lennox, 1994, whose conclusion is pessimistic.

[78] *HA* 514a18. Aristotle also states that the eye develops from the brain, for, like the brain, it is moist and cold (*Sens.* 438b27–30).

[79] Cf. Clarke, 1963; Clarke and Stannard, 1963.

[80] *PA* 669b15.

[81] *HA* 494b30.

[82] *PA* 656b12f. (more sweepingly, it is τὸ δ' ἰνίον κενόν, *HA* 491a34). It is so for all animals, *HA* 494b33. Lloyd, 1978b, 224–225, states that Aristotle's interpretation "owes more to the fact that a similar view appears in some of the medical writers than to any observations he had carried out himself." Cf Preus, 1975, 26–8. Whether Aristotle may have inferred this hollowness from an examination of the endolymphatic sacs of reptiles has been raised as a possible explanation (cf. Clarke, 1963, 9). But it is problematic if this "hollowness" is used by Aristotle to facilitate his explanation of hearing (*PA* 656b15f.). Cf. Lones, 1912, 177; Johansen, 1997, 154–162.

[83] *HA* 494b25, 495a5. Cf. Peck, 1965, 59. There is also mention of three cranial nerves, although the nature of Aristotle's understanding of these is by no means on a par with later anatomists such as Marinus and Galen. Cf. Von Staden, 1989, 157.

regulatory capacity. The brain is the coolest part of the body.[84] If it were not for this, the organism would perish.[85] Galen also maintained that the brain and spinal cord were cooler than the heart, precisely because—like Aristotle—he considered the heart, containing the body's innate heat, as the hottest part of the body.[86] Yet Aristotle's statement that the brain is avascular is an easily understood observational error. The brain does appear bloodless to the naked-eye, especially when stripped of its coverings.[87] An inspection of the brain substance would meet with no such similar vessels as are found about the heart. Although Aristotle was well aware that the brain, in order to cool blood, had to receive blood vessels, he does not appear to have followed their course, which can only be elucidated by a systematic dissection of the neck and base of the brain. Aristotle's only reference to the internal anatomy of the brain is his note that the brains of some animals have a small central cavity.[88] Apart from this observation, there is no further information about the internal structure of the brain until Herophilus and Erasistratus.

1.5 *Hellenistic medicine and the* hegemonikon

The key anatomical and physiological events in the Hellenistic period were the continuing development of the cardiocentric hypothesis and the establishment of anatomical and experimental grounds for the rival encephalocentric one.[89] According to the doxographic account

---

[84] *Somn. Vig.* 457b27–458a10. Cf. Moraux, 1976, 142–3.

[85] Cf. Charlton, 1987, 414.

[86] In *De temperamentis* I. 570 K, Galen acknowledges that the heart is hottest, and the liver the second hottest of the body's organs. The subsequent thermal order of the remaining organs is not clear in this text, but the spinal cord has more moisture and is colder than the skin, and contains less moisture than the brain.

[87] A point well brought out by the nineteenth century comparative anatomist Richard Owen: "The Brain, says Aristotle, in all animals is without blood. This error, relative to a part to which more blood is transmitted than to any other in the whole body, is obviously owing to the extreme minuteness to which the blood vessels are subdivided before they penetrate the cerebral substance." 1837, §48 (quoted in Sloan, 1992, 101).

[88] ἔχει δ' ἐν τῷ μέσῳ ὁ τῶν πλείστων [πᾶς] κοῖλόν τι μικρόν. *HA* 495a6. That the cavity was reported as of a small size appears to vitiate any argument that Aristotle was familiar with the internal anatomy of the brains of higher animals. Whilst there are differences between ventricular configurations in higher animals, these cavities cannot be considered small in relation to the rest of the brain.

[89] Cf. Vegetti, 1993b, 76–84. For an analysis of the range and limitations of Hellenistic science, see Giannantoni and Vegetti (eds.), 1984.

of Aëtius, Strato of Lampsacus, successor to Aristotle after Theophrastus, is said to have placed the command centre between the eyebrows.[90] This does not necessarily allow Strato to be labelled an encephalocentrist, but what is more important is that he does seem to have pushed the Lyceum into further explorations of natural science.[91] Evidence of a cardiocentric hypothesis is to be found in Diocles of Carystus[92] and Praxagoras of Cos.[93] According to Galen, Praxagoras' anatomical investigations apparently led him to the conclusion that not only was the heart the source of the arteries, but when the arteries were traced to their distal ramifications they become nerves.[94] Galen may of course be reading his knowledge of the nerves into this account, but assuming he is representing Praxagoras' own observations, then Praxagoras was likely to have been following the paths of very small arteries to the point where they become indistinguishable to the naked-eye from nerves.[95]

It is also possible that Praxagoras' disquisition on the "nerves" of the heart was useful in buttressing Stoic argumentation for the *hegemonikon*, if his identification "as one of the main influences on Stoic physical psychology" is allowed to stand.[96] According to Galen, Chrysippus was aware of Praxagoras' views on the nerves.[97] Galen's answer to Praxagoras (and Chrysippus) is the one he gives to all cardiocentric

---

[90] Frs. 119–121 Wehrli; T 137 a–d *Herophilus*. Cf. Mansfeld, 1990b, 3093 n. 141.

[91] As Von Staden, 1989, 97–8, notes, "Aristotle's school was . . . the only philosophical school that played a significant role both in Athens and in Alexandria, and its possible role as a dual provider of models—for Stoics in Athens, for Herophilus in Alexandria—merits consideration." Cf. Annas, 1992, 11–12, 26–30; Long, 1986, 2, 9, 119.

[92] Diocles practised in the fourth century BC, although his precise chronology has been debated for more than a century. See the critical edition of the fragments and commentary by Van der Eijk, 2000, 2001. Diocles was probably the older contemporary of Praxagoras, who in turn was the teacher of Herophilus.

[93] For Praxagoras, see Steckerl, 1958, Frs. 7–9 (cf. Von Staden, 1989, 173 n. 106). See also Van der Eijk, 1999c, 315 n. 68; Lonie, 1973, 5–6; Von Staden, 1989, 58.

[94] *PHP* p. 82,1–10 De Lacy; V. 188–189 K. *Pace* Harris, 1973, 110, "we are still bound to admit that he (sc. Praxagoras) made a bad blunder in supposing that the heart was the centre of the nervous system."

[95] De Lacy, 1984, 619, states that Praxagoras, "was perhaps alluding to the relationship of arteries to muscles and tendons."

[96] Tieleman, 1996a, 189, who adds: "In Chrysippus' lifetime, Praxagoras' doctrine may already have represented a traditional and fairly authoritative paradigm of human physiology, with many centuries of influence still ahead, but also under pressure from the more recent discoveries made in Alexandria." Cf. 83ff.; idem, 1991, 122–123.

[97] Cf. *PHP* p. 82,11–14 De Lacy; V. 189 K; Fr. 11 Steckerl.

theories of the *hegemonikon*—the argument from dissection.[98] The Stoic counter to this is to argue that even the origin of the nerves does not in itself constitute proof of the ascription *hegemonikon* to that organ.[99] And if the arteries resemble nerves, as Praxagoras apparently thought, they can with little difficulty be made to carry the function of nerves, which is the very point on which the encephalocentrists set such store. There is also the possibility that some Stoics sought to bridge the gap between the research in Alexandria on the brain and the nerves with the physiological and psychological tenets of Stoicism.[100]

Diocles of Carystus, "second in age and in fame" to Hippocrates, according to Pliny the Elder,[101] is said to have referred to the heart as "the commander of the body."[102] He also seems to have used the expression "psychic pneuma" (ψυχικὸν πνεῦμα) for the first time.[103] According to a first century AD collection of medical texts known as the *Anonymus Parisinus*, Diocles and Praxagoras located physical and mental well-being in the heart and brain.[104] But it is the heart that is referred to in hegemonic terms, and is the source of psychic pneuma.[105] Diocles appears to have employed the blood vasculature as a carrier system for pneuma,[106] and is also credited by Galen with writing the first anatomical manual.[107]

If, like Aristotle, Dicoles' physiological and psychological interests were centred about the heart, the next generation of physicians

[98] Cf. *PHP* p. 110,1–14 De Lacy; V. 219–220 K, and the discussion of this passage by Tieleman, 1996a, 24ff.

[99] Cf. Tieleman, 1996a, 191–195.

[100] Cf. Mansfeld, 1990b, 3094, n. 145.

[101] *Secundus aetate famaque extitit.* Fr. 4 vdE.

[102] ὁ ἡγεμών τοῦ σώματος. Fr. 80 vdE; *Anonymus Parisinus* p. 30,17 Garofalo. Cf. Van der Eijk, 2001, 166.

[103] Fr. 78 vdE; *Anonymus Parisinus* p. 10,20–21 Garofalo. Cf. Frs. 80, 101, 107 vdE. Cf. Harris, 1973, 104–105.

[104] Fr. 72 vdE; *Anonymus Parisinus* p. 2,7–15 Garofalo. *Madness* (μανία) is caused by boiling of the blood in the heart (Fr. 74 vdE; *Anonymus Parisinus* p. 112,21–22,4 Garofalo). Cf. Van der Eijk, 1999c, 321–2.

[105] Lethargy is said to be caused by a chilling of the *psychic pneuma* about the heart and brain (Fr. 78 vdE; *Anonymus Parisinus* p. 10,20–22 Garofalo). In κεφαλαία (headache) which arises when the blood vessels to the brain are obstructed, Diocles states that this becomes dangerous if this obstruction affects the heart, since psychic pneuma is prevented from leaving the heart and moving about the body (Fr. 80 vdE; *Anonymus Parisinus* p. 30,15–19 Garofalo). Cf. the description of epilepsy (Fr. 98 vdE; *Anonymus Parisinus* p. 18,16–20).

[106] It is through the vessels that come from the heart that voluntary motion is sent to the body (Fr. 102 vdE; *Anonymus Parisinus* pp. 122,24–124,2 Garofalo).

[107] *AA* II. 281–282 K; Fr. 17 vdE. Cf. Van der Eijk, 2001, 30–31.

introduced a new perspective into the study of the brain. According to Galen, the pupils of Praxagoras (and Phylotimus) considered the brain to be "a kind of excessive growth and budding of the spinal cord."[108] Galen's citation is of further interest because although the accomplishments of Herophilus and Erasistratus are undoubtedly of great significance for the history of anatomy, they were not the only ones carrying out research on the brain and the nerves. Admittedly, this citation provides no information as to what *sort* of research the pupils of Praxagoras and Phylotimus (together with Phylotimus himself)[109] may have performed, but it is significant that Galen speaks of Phylotimus as a contemporary of Herophilus and Erasistratus and as a fellow pupil, with Herophilus, of Praxagoras.[110] It seems reasonable to conclude that Herophilus and Erasistratus were not working in isolation, although the evidence suggests that the most notable steps in understanding the brain were taken by them.

If Galen may be classified as an anatomist in the Alexandrian tradition, that practice in turn was based on Aristotelian animal dissection. Herophilus and Erasistratus probably performed more animal than human dissections and vivisections, although it is the latter that has given them their fame and notoriety.[111] The most striking fact about the work of Herophilus and Erasistratus is their use of humans for dissection and vivisection, these last being condemned criminals provided by the Ptolemaic State.[112] The sources for these practices are Celsus, Galen, and the Patristic writer Tertullian, who describes Herophilus as a "butcher" (*lanius*) who cut up bodies.[113] Celsus reports—via a Dogmatist argument he does not endorse, that this sort of practice, while admittedly cruel, might be considered useful in that future lives could be saved by the knowledge gained.[114] Galen,

---

[108] ὑπεραύξημα γάρ τι καὶ βλάστημα τοῦ νωτιαίου μυελοῦ νομίζουσιν εἶναι τὸν ἐγκέφαλον. Fr. 15 Steckerl. This does not allow one to decide whether they viewed the brain as an hegemonic organ or not.

[109] On Phylotimus, a pupil of Praxagoras, see Steckerl, 1958, 108–123.

[110] T 10 and 14 *Herophilus*.

[111] Nickel, 1971, 99–100, 101–103, notes that Herophilus' use of animals as well as humans often involved the same organ. Cf. idem, 1989, 27–8. On Herophilus' methodology of dissection see Simon, 1906, II, 30ff. (but note Longrigg, 1993, 199).

[112] T 63a, b, c *Herophilus*. Cf. Von Staden, 1989, 142–153.

[113] A "dissector of adults" (*maiorum prosector*) *De anima* 25.5, p. 36,8 Waszink; T 65 and 247 *Herophilus*. A butcher of people, *De anima* 10.4, p. 13,2 Waszink; T 66 *Herophilus*.

[114] *Prooem.* 23–26; p. 21,12–32 Marx; pp. 8–9 Serbat; T 63a *Herophilus*. See also Lloyd, 1985, 3–10. Scarborough, 1976, 32, argues that: "The vivisection tradition associated with Hellenistic Alexandria receives no support from a close examina-

whilst not mentioning human vivisection directly, states that Herophilus made most of his new advances in anatomy not by dissecting "unreasoning animals" (ἄλογα ζῷα) but from the dissection of human beings.[115] As far as can be determined on the available evidence, the practice of human dissection and vivisection in antiquity began and ended with Herophilus and Erasistratus.[116] As he himself did, Galen urged his own students to visit Alexandria if possible in order to augment their anatomical training in osteology, where specimens of the human skeleton were available for study.[117]

Von Staden has made a thorough analysis of the factors which enabled Herophilus and Erasistratus to undertake their ground breaking researches in anatomy and physiology, together with an understanding of the events which curtailed these activities.[118] These may be summarised as follows. Political patronage exercised by the Ptolemaic Court towards scientists was an important factor in attracting them to Alexandria and securing their services.[119] The work of Herophilus and Erasistratus was state sponsored. Such patronage, attests a perhaps wistful Galen, was mandatory for research into the human body.[120] On a more fundamental level, sponsorship of natural scientists and physicians was a means of increasing the prestige and status of the ruling dynasty. Galen is our chief source for the achievements of

---

tion of Celsus' account." (but see Von Staden, 1989, 144–153). The Roman attitude to (human) vivisection was overwhelmingly negative. Cf. Ferngren, 1982, idem, 1985. On Roman attitudes to the corpse, see Hope, 2000, especially 120–123.

[115] *De uteri dissectione* p. 42,26–29 Nickel; II. 895 K; T 114 *Herophilus.*

[116] Cf. Annoni and Barras, 1993. The attitude in the Byzantine era seems at times to have allowed human dissection and vivisection, at least of condemned criminals, according to this example from Theophanes (752–818 AD): "the physicians cut a person still living from the pubis up to the chest in order to observe the construction of human beings." (τοὺς ἰατρούς καὶ τοῦτον ἀνάτεμον ζῶντα ἀπὸ ἥβης ἕως τοῦ θώρακος πρὸς τὸ κατανοῆσαι τὴν τοῦ ἀνθρώπου κατασκευήν·) *Chronographia,* I, p. 436,18–20 De Boor). Cf. Miller, 1985, 187–8; 1999, 142. If such an account is not simply depicting a method of execution, then it may indicate an interest in anatomical investigation during this period; however, it is likely such work was done more to verify Galen's teachings than to further the cause of research. Cf. Bliquez and Kazhdan, 1984, 556. See also Browning, 1985; Miller, 1985, 212.

[117] Cf. *AA* II. 220 K. Cf. Vegetti, 1979, 40. On the possibility of Galen having performed human dissection, at least on human foetuses, see chapter 2 n. 99.

[118] Op. cit., 1989, 1–31, 138–53, 187–9; idem, 1992, 231–236; idem, 1993 and 1996.

[119] Among the scientists who worked in Alexandria during this period were Aristarchus, Euclid, Eratosthenes, and Ktesbios. Cf. Von Staden, 1992, 231–32; idem, 1996b, 85.

[120] *De antidotis* XIV. 2 K. Cf. Nutton, 1992, 30.

Herophilus and Erasistratus.[121] This creates obvious problems of interpretation, but because virtually nothing of the details of Herophilean brain anatomy are known outside Galen, this does not lead to the conclusion that Galen appropriates all of Herophilus' research as his own.[122]

It is unequivocally true that "the neuro-anatomical knowledge of the brain remained scanty until Herophilus performed his dissections."[123] Herophilus reaffirms the distinction, first made by Aristotle, between cerebrum and cerebellum.[124] He also gives a description of part of the venous sinuses of the brain, the so-called ληνός or *torcular Herophili*.[125] Another part of the brain vascular system that Herophilus seems to have described for the first time is the δικτυοειδῆς πλέγμα, the *retiform plexus*.[126] It is a relatively inaccessible, symmetrical arrangement of small arteries found in the base of the brain in certain animals. That the retiform plexus does not exist in man suggests that the nature of Herophilus' anatomical work relied far more on animals than on human beings. Where Herophilus departs from Aristotle is in his dissections of the brain, which were performed systematically for the first time to reveal, among the above-mentioned structures, the nature of the ventricles.[127] Within these cavities Herophilus noted a complex of arteries and veins which he thought resembled

---

[121] Cf. Von Staden, 1989, 157–8. See also Longrigg, 1981, 1988; Michler, 1968, 5–16; Potter, 1976, 58 n. 81.

[122] "While Galen was not uncritical of Herophilus, and while his vanity often draws a virtually opaque veil over his use of sources, Galen's willingness to concede the Alexandrian a place of honour helps render his accounts relatively straightforward and hence reasonably trustworthy." Von Staden, 1992, 221. Cf. idem, 1989, 257. Von Staden, 1993, 215, also assembles some evidence to suggest that Galen had "direct, unmediated access" to several of Herophilus' treatises, in particular *On Anatomy* and *On Pulses*. Whilst discussion of this cannot be entered into here, it is likely that Galen would have filtered such "direct" information to suit his own purposes. The question posed by Van der Eijk, 1995, 406, is pertinent: "To what extent were his (sc. Galen's) knowledge and appraisal of the achievements of the Hellenistic period the products of his own reading experience or determined by other factors or circumstances, such as historiographical or doxographic traditions he may have been dependent on, his teachers, or the general attitude towards the past as reflected in other writings of the second century?" Cf. idem, 1999b.

[123] Von Staden, 1989, 155–6.

[124] T 77a,b *Herophilus*.

[125] T 122 a,b *Herophilus*. The term ληνός is inaccurately rendered as *torcular*. See chapter 3.6.

[126] T 121 *Herophilus*. Simon, 1906, II, xxxixff., adduces this description as evidence that Herophilus did not perform dissections on the human brain. This is correct only as far as the base of the human brain is concerned. See chapter 6.2.

[127] For a general survey of the history of the ventricles see Sudhoff, 1913, 149–155; Leyacker, 1927; Manzoni, 1998, 104–110.

the foetal chorion, and so termed them *choroid-like knots* (χοροειδῆ συστρέμματα) or *choroid plexuses* (χοροειδῆ πλέγματα).[128] Herophilus seems to have regarded the posterior (or fourth) ventricle, the ventricle of the cerebellum, as the most important. He described in detail the floor of the fourth ventricle, likening it to a *reed-pen*, (κάλαμος), the *calamus scriptorius*.[129] There is no other information which specifies in what way Herophilus envisaged the remaining ventricles, and there is also no testimony concerning how Herophilus might have experimented on them in order to reach his conclusion concerning the importance of the fourth ventricle. Galen however, in a discussion which implies Herophilus knew of the other ventricles, states that Herophilus thought that it was not the middle (third) ventricle, but the fourth which was "exercising more control" (κυριωτέρα).[130] Given what was already known of the relationship between the brain and the spinal marrow (that it began in the region of the fourth ventricle), then Herophilus' choice of the fourth ventricle as the locus of the *hegemonikon* is perhaps not unexpected.[131] Observing how butchers kill animals by cutting through the base of the brain (in the region of the upper cervical vertebrae), together with the knowledge of the anatomical relationships of the fourth ventricle to the cerebellum and spinal cord, may well have provided Herophilus with the basis for his experimental work on the ventricles.

According to Rufus of Ephesus, Herophilus held that the motor nerves arise from the brain and spinal marrow.[132] Herophilus did not completely distinguish between nerves, tendons and ligaments, apparently placing all of them under the umbrella of structures which collectively exercise the same physiological function, and thereby creating,

---

[128] T 124 and 125 *Herophilus*.

[129] T 79 *Herophilus*.

[130] T 78 *Herophilus*.

[131] "The proximity of the hindbrain to the spinal cord and hence to the nerve tracts, and conceivably, the vivisectional discovery that the cerebellum is the centre responsible for the muscular activity and for maintaining the equilibrium of the body, may have led Herophilus to the conclusion that this part of the brain exercises the most significant control over the body. This conclusion became one of Herophilus' more influential physiological theories, because Galen accepted it as valid (on the basis of his experiments with animals)." Von Staden, 1989, 247–248. Solmsen, 1961, 193, however, is doubtful.

[132] *De partibus corporis humani* p. 185,5–6 Daremberg-Ruelle; T 81 *Herophilus*. See the discussion of this passage in Potter, 1976, 50f. Cf. Von Staden, 1989, 250–252. Such a distinction, at least of the motor nerves, can only be determined by vivisection. On Rufus see below n. 160.

whether deliberately or not, a partial physiological shift between this and the Praxagorean view in which pneuma carried by the arteries *alone* was the responsible agent for motion.[133] Herophilus seems to have used pneuma as a physiological medium for nervous action, but quite how it was used remains uncertain.[134] Herophilus did not depart from the Praxagorean concept that the arteries contained pneuma as well as blood,[135] and thus does not completely exclude the Herophilean vascular system from physiological consideration as a medium for sense transmission. The extant work of Herophilus shows first that the ventricles of the brain, or at least one of them, assumed anatomical and physiological prominence for the first time, and second that a causal relationship was established between the brain and the nerves on the basis of some form of empirical research.

Galen is also our main source for Erasistratus.[136] As far as can be determined, the work of Erasistratus was based on a similar, Herophilean programme of dissection and experimental vivisection.[137] However, it would appear that Erasistratus was the more wide-ranging and investigative physiologist of the two. Several sources speak of Erasistratus placing the *hegemonikon*, not in a ventricle or ventricles of the brain, but in the covering layers of the brain, or at least the *dura mater*.[138] The reason is linked to the question of the source of

---

[133] Cf. Fr. 27 Steckerl; T 141 *Herophilus*. See also Potter, 1976, 51–52.

[134] Whether Herophilus utilised "natural pneuma" or the specialised "sensory pneuma" (an analogue of psychic pneuma) is also problematic. "It therefore remains possible that Herophilus put the concept of pneuma to fairly limited use and failed to provide a satisfactory account of the mechanisms by which the command centre in the hindbrain transmits its decisions to the nerves and to other organs of voluntary motion." Von Staden, 1989, 258.

[135] Cf. T 145a *Herophilus*; Von Staden, 1989, 262–3.

[136] Galen wrote a three volume treatise on Erasistratus' anatomy, which has not survived (cf. *Lib. Propr.* p. 114,16–17 Mueller; XIX. 37 K). The fragments of Erasistratus are collected in an edition with commentary by Garofalo, 1988. Cf. Longrigg, 1988, 472–482; Von Staden, 1975, 180–184, 186–7. Galen is in some debt to Erasistratus as far as the anatomy of the brain is concerned; he is perhaps in more of a debt than he would wish to acknowledge in physiology. Nevertheless, Galen's general attitude to Erasistratus as a physiologist is coloured by the latter's alleged anti-teleological and corpuscular views. Cf. Von Staden, 1997b; Vegetti, 1999a.

[137] Fraser, 1972, I., 347ff., argues that Erasistratus did not perform human dissections in Alexandria. This has been effectively rebutted by Lloyd, 1975b; cf. Longrigg, 1988, 472–4.

[138] Cf. Pseudo-Galen, *De historia philosophica* XIX. 315 K; T 137d *Herophilus* (cf. T 137e, T 139). See also Mansfeld, 1990b, 3100 n. 172. In *Anonymus Parisinus* II. 1, p. 10,18–19 Garofalo, the psychic faculty is placed in the meninx (περὶ τὴν μήνιγγα ψυχικῶν δυνάμεων). Dionysius, *Dictyaca* p. 340,12 Deichgräber, asks whether the heart is the ἀρχή of the nerves or whether it is the meninx *surrounding* (περιέχουσα) the brain. No name is given, but Deichgräber suspects that this is a reference to Erasistratus.

the nerves. In common with Herophilus, Erasistratus held there were two kinds of nerves, motor and sensory. But, unlike Herophilus, Erasistratus appears to have been more forthright in his employment of *psychic pneuma* as the brain's physiological agent.[139] According to Galen, Erasistratus until late in life maintained that the nerves originated from the meningeal layers of the brain. In the following account, Erasistratus corrects himself. This passage is cited at length since it provides valuable information on Erasistratus' views of the structure of the brain:

> Erasistratus, who for a long time saw only the outer part of the nerve, (the part) that comes from the dura mater, thought that the whole nerve grows from that source, and most of his writings are full of statements that the nerves grow from the meninx that surrounds the brain. But when, late in life and at leisure to devote himself entirely to the study of the art, he performed his dissections with greater care he recognized also that the heart-wood so to speak, of the nerves grows from the brain. He writes as follows: "We viewed also the structure of the cerebrum, and it was bipartite as in all other animals, and there were ventricles lying there, elongated in form. The ventricles were united by a perforation at the point of contact of the parts. From this point a passage led to the so-called cerebellum,[140] where there was another small ventricle. Each of the parts had been partitioned off by the meninges.[141] For the cerebellum had been partitioned off by itself, and also the cerebrum, which is similar to the jejunum and has many folds; and the cerebellum, even more than the cerebrum, was provided with many varied convolutions. So the observer learns from these that as it is in the other animals—deer, hare and any other that far excels the rest in running being well provided with the muscles and sinews useful for this activity—, so in man, since he is far superior to the other animals in thinking, this (member) is large and has many folds. And the outgrowths of the nerves were all from the brain; and by and large the brain appears to be the source of the nerves in the body. For the sensation that comes from the nostrils passed to the member through the apertures, and also the sensations that come from the ears. And outgrowths from the brain went also to the tongue and the eyes." In these words Erasistratus admits that he then saw clearly a thing that he had not known earlier, that each nerve grows from the brain.

---

[139] ἀλλὰ μετὰ κατασκευῆς λόγων οὐκ ὀλίγων ἐκ μὲν τῆς κεφαλῆς φησι τὸ ψυχικόν, ἐκ δὲ τῆς καρδίας τὸ ζωτικὸν ὁρμᾶσθαι πνεῦμα, *PHP* p. 164,14–16 De Lacy; V. 281 K; Fr. 112B Garofalo. The context relates to the Stoic argument that psychic pneuma comes from the heart. Cf. T 145b *Herophilus*.

[140] ἐπεγκρανίς. Although another term for cerebellum is παρεγκεφαλίς, the former seems to have been Erasistratus' (cf. *UP* I, p. 488,14–16 Helmreich; III. 673 K).

[141] "Parts" refers here to the cerebrum and cerebellum, separated by the meningeal coverings. See chapter 3.4.

> And he wrote accurately about its four ventricles, which he had also failed to see the year before. And if he had also performed on living animals the experiment that we have performed not once or twice but very many times,[142] he would have known with certainty that the hard and thick membrane was formed in order to provide a protective covering for the brain, and that the soft and thin membrane also has this same function but serves even more to bind together all the vessels of the brain, the arteries and veins.[143]

This passage has been interpreted by Wellmann[144] and Verbeke[145] to mean that Erasistratus renounced his earlier contention that the nerves conveyed pneuma. Instead, they contain brain substance. This misinterpretation arises from Galen's expression that the *heart-wood* (ἐντεριώνη) of the nerves arises from the brain. As Solmsen has pointed out, all Erasistratus says is that the nerves are *outgrowths* or *processes* (ἀποφύσεις) from the brain.[146] It is Galen who compares the nerves to pith or heart-wood, with the implication that this is Erasistratus' opinion as well.[147] But Wellmann and Verbeke are correct to regard ἐντεριώνη as referring to brain substance. Galen, immediately preceding the above passage, states that nerve structure "is triple in nature" (τριπλοῦν τὴν οὐσίαν ἔστί). Each consists of a *core* (the ἐντεριώνη), surrounded by a meningeal envelope of pia mater and dura mater.[148] This need not take away from Erasistratus his contention that the nerves convey pneuma. Indeed, it is stated in *Anonymus Parisinus* that, according to Erasistratus, the effects of apoplexy are caused by phlegm blocking the nerves from the brain, impeding the passage of psychic pneuma.[149]

---

142 On the basis of this statement, Dobson, 1925, 26, clears Erasistratus of the charge of human vivisection.

143 *PHP* pp. 440,20–442,18 De Lacy; V. 602–604 K; Fr. 289 Garofalo (tr. De Lacy, 1980, pp. 441, 443).

144 1907, 343–344.

145 1945, 185. Phillips, 1973, 148, also misinterprets this passage.

146 1966, 188–190.

147 Cf. Longrigg, 1993, 213.

148 *PHP* p. 440.15–19 De Lacy; V. 602 K. Solmsen, 1961, 189, quotes this but ignores its implications.

149 P. 26,4–8 Garofalo. Paralysis has a somewhat different aetiology. The nerves, responsible for voluntary motion, are affected by fluid in the veins which moves into the nerve cavities, affecting their function (ibid., p. 122,19–23; Frs. 240–247 Garofalo). This occurs as a consequence of the Erasistratean *triplokia* of vein, artery and nerve, which Galen opposed on anatomical as well as physiological grounds (cf. *Anonymus Londinensis* XXI, pp. 37,25–38,28 Diels. See also *UP* I, p. 391,3–5 Helmreich; III. 538 K. On how Erasistratus might have handled pneuma, see Vegetti, 1993a, 72–3.

Erasistratus' knowledge of the ventricles builds upon that of his Alexandrian colleague, and is on a par with Galen's. From the above citation, Erasistratus observed that there is present a ventricle "elongated in form" (παραμήκης τῷ εἴδει) in each half of the cerebrum. These communicate with each other at their point of contact and from there a passage leads to a cavity in the cerebellum. This cavity is the fourth or posterior ventricle. According to Galen:

> Another passage, which is single and unpaired, empties into the first beginning of the spinal medulla; and here especially, when the dura mater is cut at this point, the entire passage is laid bare, along with the end of the posterior ventricle of the brain. This was not the least reason why Erasistratus mistakenly believed that the animal immediately becomes motionless when the meninx is cut; for he saw that oxen wounded at the first vertebra become motionless as soon as the meninx is severed. But this results not from the injury to the meninx but from the exposure of the posterior ventricle.[150]

Although the anatomy and physiology of this passage will be discussed in chapters 4 and 5, what can be noted here is that Galen's citation stresses the importance with which Erasistratus held the ventricular system, in particular the fourth ventricle, although Galen interprets Erasistratus' experimental observation differently. For Galen, it is the exposure of the fourth ventricle that results in loss of motion in the animal, and not, as Erasistratus is said to have maintained, the incision of its overlying meninx.

It is likely that the empirical methodology undertaken by Herophilus and Erasistratus took place with an awareness of the arguments put forward to advance the claims of the heart as the command centre of the body. This is not to suggest that Herophilus and Erasistratus set for themselves the goal of refuting cardiocentrism.[151] Rather, their

---

[150] *PHP* p. 446,20–27 De Lacy; V. 609–610 K; Fr. 42A Garofalo (tr. De Lacy, 1980, p. 447).

[151] According to Longrigg, 1993, 61: "Those who, on the contrary, favoured the heart as the seat of the intellect included Empedocles, Philistion, Aristotle, Diocles, Praxagoras and the Stoics. One might have expected that the discovery of the nerves by the Alexandrians and their demonstration of their origin in the brain would have settled this issue once and for all. But even after it had been demonstrated by dissection that the brain must be the seat of the intellect, the Stoic philosopher Chrysippus sought doggedly to defend his school's dogma by appealing to Praxagoras, a medical authority of about half a century earlier." First, dissection by itself cannot demonstrate that the brain is the *hegemonikon*. Second, it is unlikely Chrysippus was aware of the nature of the work of the Alexandrians, and if it were the case, is still no reason for Chrysippus to abandon his own theories which to

work, which emphasised the value of the practices of dissection and vivisection, provided enough empirical evidence—if not for themselves then for others—to place the *hegemonikon* in the brain. Whilst the extant material is not compelling, it is nevertheless sufficient to class Herophilus and Erasistratus as encephalocentrists.

1.6 *The second century AD: a recovery of anatomical knowledge*

Herophilus and Erasistratus set a benchmark for dissection and vivisection. Of those Herophileans who undertook anatomical and physiological research, Von Staden has stated that "the affinities are there, but they are occasional and do not reflect sustained continuity."[152] Gaius (fl. 100 AD?) apparently upheld Herophilus' work on the brain and nerves.[153] Eudemus is described as "Herophilus' fellow initiator of advanced anatomical investigations."[154] Galen states that Eudemus and Herophilus were the first—after Hippocrates, he is careful to insist—to record carefully their work on the anatomy of the nerves.[155] Hippocrates as an anatomist can of course be discounted, but Herophilus and Eudemus are also singled out by Galen as those who "increased anatomical theory most"[156] until the advent of Marinus and Numisianus. And it is Marinus of Alexandria (fl. 120 AD) whom Galen held in exceptional regard.

The period after Herophilus and Erasistratus—until the beginning of the second century AD—has been characterised, with very few

---

him were equally as empirically grounded. The Aristotelian and Stoic arguments for the heart as *hegemonikon* constitute an epistemology which Galen overcomes only by the continual reinforcement of his experimental evidence that the brain is the ἀρχή of the nerves. *PHP* p. 476,11–17 De Lacy; V. 645 K, provides a good example of Galen's labouring of this particular point. Cf. Tieleman, 1996a, 139, 194.

[152] 1989, 449. The Herophilean Hegetor (first century BC) stressed the importance of a knowledge of anatomy, especially in surgical cases. The clinical problems of the Alexandrian Empiricists could be greatly improved, according to Hegetor, if they had recourse to surgery (Apollonius Citiensis, *In Hippocratis De articulis commentarius* pp. 78,24–80,19; 94,4–8 Kollesch-Kudlien. Cf. Von Staden, 1989, 512–514. There is no evidence Hegetor continued research into the brain and nerves.

[153] Cf. Von Staden, 1989, 566–569.

[154] Von Staden, 1989, 63. Cf. T 14 and 220 *Herophilus*. Fraser, 1972, II, 1112–1113, speaks of Eudemus as Herophilus' pupil.

[155] *De locis affectis* VIII. 212 K; T 80 *Herophilus*.

[156] ἐπὶ πλεῖστον αὐξησάντων τὴν ἀνατομικὴν θεωρίαν. *Hipp. Nat. Hom.* p. 70,5–6 Mewalt; XV. 136 K. Cf. Nutton, 1993a, 16: "there can be no doubt that Galen knows nothing of anatomical discoveries or anatomical teaching between the third century BC and his own."

documented exceptions, as "a retreat from anatomy, the turning away from dissection."[157] Apart from a lack of opportunity for human dissection, this period of approximately four hundred years was one in which there seemed little need for anatomical study. The reasons for this are as varied as the ones adduced to explain why human dissection arose in Alexandria and nowhere else.[158] However, one factor that stands out is the increasing stress placed on philological and doxographic concerns. It was a time for taking stock of medical information already assembled. This is not to say that no medical advances occurred. On the contrary, there is evidence of an increasing preoccupation with the clinical and therapeutic aspects of medicine: pulse-theory, pharmacology, surgery and orthopaedics.[159] These took place in a milieu in which more and more emphasis was placed on the systematisation and codification of medical knowledge. Hippocratic and Herophilean exegetics for example, were promulgated at the expense of anatomical investigation.[160]

Galen personally knew of several doctors who were also anatomical investigators. He discusses the aged Martianus, who practiced in Rome. Two of his works on anatomy were extant in Galen's day.[161] Galen mentions the qualifications of his own teachers who were the first among the pupils of Quintus and Numisianus.[162] Numisianus instructed Pelops, Galen's teacher in Smyrna, and as noted in the General Introduction, Galen went to considerable trouble to learn more of Numisianus' anatomical contributions. Yet according to Galen, it is Marinus of Alexandria who alone was responsible for

---

[157] Von Staden, 1989, 445. Cf. Lloyd, 1975a, 142ff. As Nutton, 1984a, 5, remarks, "the tradition of anatomical research is a very fragile thing." To Edelstein, 1932, 299, "the necessity for dissections was by no means universally recognized."

[158] Cf. Edelstein, 1932, 296ff.; Kudlien, 1969, 90ff.; Von Staden, 1975, 185ff.; 1989, 445ff.; 1992, 234f.

[159] Cf. Von Staden, 1989, 445–453. Among those Herophileans were Aristoxenus (pulse-lore), Andreas (the construction of an instrument for reducing dislocations); Bacchius (pulse-lore and his classification of haemorrhage); Demetrius of Apamea (orthopaedics—in the original sense of the term). The material on pulse-lore is considerable and embodied at least several different theoretical standpoints and formulations.

[160] Whether Rufus of Ephesus (fl. 100 AD) can be classified as an anatomical researcher is doubtful. It is more likely that he utilised Herophilean and Erasistratean writings for his anatomical texts, of which nothing innovative in regard to the brain and nerves can be discerned. To Lloyd, 1983b, 167, Rufus illustrates "that by the second century AD concern was being expressed on the question of variations in anatomical vocabulary." Cf. Sideras, 1994, 1234–1235; Thomssen and Probst, 1994, 1255–1257; Ullmann, 1994, 1295–1296.

[161] Cf. General Introduction, n. 50. Although Martianus, as an Erasistratean, is not otherwise highly regarded by Galen.

[162] *AA* II. 660 K.

having *recovered* (ἀνακτησάμενος) the study of anatomy after it had fallen into neglect.[163] Coming from Galen, this is high praise. It might be said that Galen is attempting, and not for the first time, to present a concept of anatomical science as a specialised τέχνη that had fallen into disuse until its recovery by Marinus. Yet Galen's praise of Marinus goes beyond any mere appropriation or manipulation of his name and reputation. He states that Marinus, in common with several others, gave "their whole life" (τὸν ὅλον ἑαυτῶν βίον) to the task.[164] In that part of *De anatomicis administrationibus* which survives only in Arabic, Galen provides this encomium:

> For Marinus had accumulated no small experience in dissections, and it was he himself who had set his hand to (the work) and had observed everything that he explained in his writings.[165] In his great Work, he had established the record of the bones with the greatest thoroughness, and he had laboured to trace all the foramina which perforate the skull and the vertebrae, and to exhaust their study, although now and then we may discover him in error, as I have proved on repeated occasions in the city of Rome, in distinguished company in the presence of all the notable surgeons.[166]

Immediately prior to this, Galen compares Marinus to Lycus of Macedon, who also wrote on anatomy. However:

> In regard to these works of his which in this our own day I have seen in the possession of many, it is clear that they are constructed out of the writings of Marinus, but they are all full of errors, and are moreover less comprehensive even than the books of Marinus himself.[167]

In common with any anatomist after Herophilus and Erasistratus, the work of Marinus did not involve human dissection: "he seems to have systematically dissected apes and other animals and compared his results with those of his early Hellenistic predecessors."[168] This is precisely the combination of dissection and *Quellenforschung* which Galen also pursues. Thus Galen mentions that Marinus, in his anatom-

[163] *PHP* p. 480,28–30 De Lacy; V. 650 K; T 68 *Herophilus*. Cf. *Hipp. Epid. II* p. 312,20 Pfaff. Cf. Deichgräber, 1930; Grmek and Gourevitch, 1994, 1493–1503.
[164] *AA* II. 621 K. Marinus made mistakes, but these pale into insignificance besides those who performed no anatomical study at all. Cf. II. 283 K.
[165] In the extant Greek part of *AA* II. 280 K, Galen similarly praises him for writing on anatomical dissections.
[166] XIV. 1; p. 185 Duckworth. My gloss in round brackets.
[167] XIV. 1; pp. 184–185 Duckworth.
[168] Nutton, 1993a, 17.

ical textbook, discussed his predecessor's views on the subject of the meninges of the brain (presumably together with his own).[169]

Marinus must have served Galen as an important conduit for the transmission of past anatomical research. In the study of the cranial nerves for example, it is clear that Galen owes Marinus a great deal. In *De nervorum dissectione*, Galen states that Marinus was correct to have noted the fourth cranial nerve separating from the third.[170] According to Savage-Smith's detailed study: "Marinus seems to have laid the foundations for the division of the cranial nerves into separate pairs."[171] In *De libris propriis*, Galen writes that he composed a four volume work (not extant) summarising Marinus' twenty volume text on anatomy.[172] It is significant that in a text devoted to Galen's description of his own works, he devotes more space to Marinus than to any other physician or philosopher. Galen's description of his fourth volume is a synopsis of Marinus' last five anatomical books. Their contents are described in these words:

> [The sixteenth book] contains questions and observations regarding the brain, such as whether it has its own pulse-like motion, and if it breathes by itself; this is immediately followed by a discussion of the spinal cord and the meninges. The seventeenth discusses the *dominion* (κυριότητος) of the brain; that of the eighteenth, voluntary motion and the different parts of the nerves, and from whence some of them take their origin. The nineteenth discusses the nerves that come from the brain, the sense of smell and the source of its perception, and the nerves which lead to the eyes, which Herophilus and Erasistratus refer to as ducts . . .[173]

According to Galen then, Marinus thought the question of the *hegemonikon* sufficiently important to have devoted an entire volume to it. Nutton notes that: "the anatomical studies of Marinus influenced Galen in a variety of ways, by setting out a method of anatomy, as well as by posing questions that the anatomist ought to be able to

---

[169] *AA* II. 716 K.

[170] II. 837 K. The fourth nerve for Marinus and Galen seems to have been the palatine, or, if the account in *AA* XIV. 3; p. 192 Duckworth is used, the sympathetic. Cf. Savage-Smith, 1971, 177–180, 191.

[171] Op. cit., 177 n. 70. Cf. *AA* IX. 9; pp. 9–10 Duckworth; T 82 *Herophilus*. Cf. XIV. 4; pp. 195–6 Duckworth; *De nervorum dissectione* II. 837–9; T 83 *Herophilus*. See also Savage-Smith, 1971, 181–3.

[172] Galen's summary of all the books on Marinus' anatomy is preserved in Arabic by Hunayn ibn Ishaq. See Boudon, 2002.

[173] P. 108,2–14 Mueller; XIX. 29–30 K. A textual hiatus follows, and no more information on Marinus is available.

answer."[174] Admittedly, it was an influence orchestrated by Galen himself, but the importance of Marinus also lies in his use as a reference source for Galen's anatomical methodology by implicitly (and perhaps explicitly) endorsing the validity of the work of Herophilus and Erasistratus on the brain. The full extent of Galen's debt to Marinus can never fully be recovered, but by the beginning of the second century AD several anatomists were retracing and expanding the work of Herophilus and Erasistratus on the brain and the nerves. For Galen, the most significant of their number was the "most excellent Marinus."[175] For this reason, perhaps above all others, Galen *praises* (ἔπαινος) him.[176]

1.7 *Conclusion*

Galen's achievements were made possible because Herophilus and Erasistratus had established the basic epistemological framework crucial to any investigation of the anatomy, and to a lesser extent, the physiology of the brain. Together with Quintus, Martianus, Numisianus, Pelops, Lycus, and especially Marinus, they represent a watershed in the study of the brain as an hegemonic organ.[177] Yet it would be misleading to think of Galen as holding undisputed sway over the hegemonic debate. In closing, one further upholder of the cardiocentric claim should be mentioned. Alexander of Aphrodisias has been called a witness "to the appeal and scientific respectability of the cardiocentric theory well into the second century CE."[178] It has been suggested that the final part of Alexander's *De anima*, which details his arguments that the heart is "the *hegemonikon* of the soul"

---

[174] 1993a, 17.

[175] κράτιστος Μαρῖνος, as he is referred to in *De semine* p. 200,23 De Lacy; IV. 646 K.

[176] *AA* II. 280 K. Here, Marinus is discussed in connection with a putative earlier oral tradition of the transference of anatomical knowledge. Cf. Vegetti, 1999b, 351f.

[177] Ptolemy of Alexandria (c. 83–161 AD) also thought fit to discuss the question of the *hegemonikon*, doubtless regarding the subject to be within his brief as a scientist. In *On the Kriterion and Hegemonikon*, he places the *hegemonikon* in the brain (p. 210, 15.2,1–3 *L-MS*). While Ptolemy does not use Galen's expression "*hegemonikon* of the rational soul", he states that "the faculty of thought" (τὸ διανοητικόν) is housed in the brain (p. 208, 14.3,1–3; *L-MS*). That part of the soul concerned with life and with "living well" (εὖ ζῆν), is also to be found in the brain (p. 210, 16.2,2–3 *L-MS*). Cf. Long, 1988, 205–206.

[178] Tieleman, 1996a, xxxvii.

(τὸ ἡγεμονικὸν τῆς ψυχῆς), were deliberately framed with Galen in mind.[179] In several places Alexander appears deliberately to employ Galen's arguments for encephalocentricity, albeit for his own purposes. Alexander reiterates the main arguments from anatomy, embryology, and position in support of his cardiocentric claims.[180] However, in what might reflect a weakness on the part of cardiocentrists for this form of speculation or investigation, Alexander does not spend any time discussing the special properties of the nerves, nor does he differentiate them to any appreciable extent from muscles and tendons. Like Galen, Alexander relies on the Stoic notion of *sympathy* (συμπάθεια), in which the actions of one part of the body may have effects in another.[181] According to Alexander, sympathy accounts for sensation *appearing* to be felt in the head, but its transmission point and reception centre resides in the heart.[182] In this way, by denying Galen the central tenet of his concept of encephalocentrism—that the nerves are derived from the brain and transmit its power—Alexander seeks to undermine the importance of Galen's methodology of dissection and experimentation. Alexander's *De anima* is a reminder that the cardiocentric position had adherents in Galen's era as well as a valid epistemological grounding of its own. Regardless of whether they are considered separately or together, the arguments Galen advances against his medical and philosophical opponents illustrate that the dispute on the physical location of the *hegemonikon* was far from settled, even if Galen thought that this polemic, thanks to his own empirical endeavours and those of certain of his colleagues, had been resolved unquestionably in his favour.

---

[179] Pp. 94,7–100,17 Bruns. Cf. Tieleman, 1996b, 266–7, 281; see also Todd, 1995. Alexander's anatomical and physiological writings have long been neglected and demand a critical examination.

[180] Anatomy (pp. 94,26–95,6; 96,25–97,8 Bruns); embryology (p. 95,6–16 Bruns); position (pp. 95,16–96,8 Bruns).

[181] See for example Hahm, 1977, 163; Pohlenz, 1949, I, 102; II, 58, 108. Cf. *De locis affectis*, VIII. 30 K (one organ changed by sympathy in another); 221–222 K (the eye affected by disturbances in other organs). On the links between Alexander and Stoicism see Todd, 1976, 21–29.

[182] Pp. 99,30–100,13 Bruns. Galen also uses the notion of sympathy in an attempt to negate the Erasistratean idea of the hegemonic importance of the meninx. According to Galen, if the meninx is removed, no initial harm comes to the animal. But ill effects may occur later "by sympathy." Cf. *PHP* pp. 446,34–448,3 De Lacy; V. 610–611 K. Galen's argument is a disingenuous attempt to gainsay the Erasistratean thesis of the *importance* of the meninges, a hypothesis with which Galen is in broad agreement.

CHAPTER TWO

GALEN'S METHODS AND MATERIALS FOR THE INVESTIGATION OF THE BRAIN

Thus after you have acquired operative practice in the way which I have mentioned, on the bodies of dead animals, then proceed to those of living animals, and apply to these the method which I will here explain to you.

De anatomicis administrationibus.[1]

2.1 *Introduction*

Galen's hegemonic claims for the brain derive from anatomical and physiological considerations that are for him superior to Peripatetic and Stoic notions of a cardiac control centre. These epistemological criteria also serve to separate Galen from other medical sects, especially the Empiricists and Methodists, for whom anatomical demonstration was deemed unnecessary for medical practice.[2] An examination of how Galen elaborates his anatomical methodology and his attempts to graft it to an Aristotelian system of first principles comprises the first part of this chapter. Galen's physiology of the brain is dependent upon pneuma. Although the details of this physiology will be explored in chapters 5 and 6, it is appropriate at this time to discuss the strengths and weaknesses of pneuma as a physiological agent, and this forms the second part of this chapter. The third will examine the ways in which Galen employed animals for his studies of the brain, and will seek to explain why some of his descriptions of the brain have remained enigmatic. The cause of this lies in Galen's selective use of certain animals. Whilst it is not always possible to be completely certain which animal Galen employs in every instance, there is a discriminatory method in his use of several types of animal to construct a coherent picture of the anatomy of the brain.

---

[1] XIV.7; p. 208 Duckworth.

[2] Nor was anatomical knowledge used to enhance any claims to medical authority by the Empiricists or Methodists. See pp. 11–14.

2.2 *Galen's claims for anatomical demonstration*

Galen wishes to show that due to the knowledge gained from anatomical dissection, his first *premise* (λῆμμα)[3]—that the brain is the *hegemonikon* of the rational soul—can be regarded as established.[4] In his discussion of the parts of the brain in the extant Greek part of *De anatomicis administrationibus*, Galen states that "those who perform a dissection badly not only make mistakes, but introduce those mistakes into Nature's discourse. Indeed from this it necessarily follows that, wonderful as is the utility of those parts observed in dissections, if they are incorrectly observed, it is difficult to give a correct discourse."[5] According to Galen, the argument concerning the *hegemonikon* is formulated from "the method of apodeictic proof".[6] However, according to him, the function of some anatomical organs can only be regarded as *plausible* (πιθανός) and not *demonstrative* (ἀποδεικτικός).[7]

---

[3] Restricting the examination to the *PHP*, Galen's references to apodeictic or scientific λήμματα, for example, imply they are to be taken as propositions or even axiomata. Galen also uses πρότασις (at times used by Aristotle in the sense of a minor premise, *EN* 1143b3) interchangeably for λῆμμα. Cf. p. 480,19–21 De Lacy; V. 649 K, where Galen's encephalocentric statement, "where the origin of the nerves is, is also the *hegemonikon* of the soul", is referred to as his "principal premise" (κυριωτάτη πρότασις); At *PHP* p. 112,3–8 De Lacy; V. 222 K, Galen says it makes no difference to the argument whether one employs "premise" (πρότασις), "axiom" (ἀξίωμα), or "statement" (λόγος). See also *PHP* p. 118,2–8 De Lacy; V. 228 K, where Galen examines the number of προτάσεις concerning the position of the heart. In *Institutio Logica*, I. 5, p. 4,13–22 Kalbfleisch, Galen distinguishes πρότασις from ἀξίωμα; the former is employed in proof based on sense perception, the latter is a self-evident truth. Cf. Hankinson, 1994a, 62–66.

[4] Galen uses animal models to seek to establish just such a premise. However, Galen never addresses the broader question of in what way dissection of irrational animals can demonstrate that the human brain is the *hegemonikon* of the rational soul.

[5] τοιαῦτα μὲν οὖν οὐ μόνον εἰς αὐτὴν τὴν ἀνατομὴν πλημμελοῦσιν οἱ μὴ καλῶς ἐγχειροῦντες ἀνατέμνειν, ἀλλὰ καὶ τοὺς φυσικοὺς λόγους αὐτούς. ἀνάγκη γάρ, ὥσπερ τῶν ἀληθῶς ὁρωμένων ἐν ταῖς ἀνατομαῖς αἱ χρεῖαι θαυμαστῶς ἔχουσιν, οὕτω καὶ τῶν παροφθέντων ἄπορον γίγνεσθαι τὸν <τῆς> χρείας λόγον. *AA* II. 727–728 K.

[6] *PHP* p. 110,8–9 De Lacy; V. 220 K; cf. p. 480,7–8; V. 648 K.

[7] In *De semine*, Galen acknowledges that his description of the function of certain glandular structures (either the seminal vesicles or the prostate) is only a plausible one. Cf. De Lacy, 1991, 304. Yet Galen also acknowledges that the account he has just given is also *true* (ἀληθής), but that perhaps a *truer* (ἀληθεστέρα) account will one day be provided (p. 204,15–16 De Lacy; IV. 649 K). In a related example at the beginning of *De semine*, Galen notes that, in order to work out the *use* (χρεία) and *power* (δύναμις) of semen, one must not make recourse to "plausible arguments" (πιθανοῖς λόγοις), but employ those of demonstration (p. 64,4–11 De Lacy; IV. 512–3 K). Cf. Donini, 1992, 3488. Elsewhere, however, Galen makes deliberate recourse to just such a strategy of plausibility; this particular form of argumentation features prominently, as it must, in his physiological claims. For

Significantly, the brain and the nerves are exceptions to this. Galen's demonstrative terminology is instructive; ἀπόδειξις means *show, demonstrate* or *prove.*[8] For him, experimentation "gives proof" (ἐνδείκνυσθαι) of the function of a nerve, in the same way that dissection *demonstrates* the nature of a nerve and its relationship to other structures. Galen uses this aspect of dissection as part of his refutation of Aristotle's premise that there are nerves about the heart, claiming that these are really "nerve-like bodies" (νευρώδη σώματα).[9] However, Galen also wishes these apodeictic demonstrations to be seen not only as empirically based accounts (which they are), but also in the sense of geometric statements which proceed from indemonstrable premises, via valid arguments, to incontrovertible conclusions. Galen's demonstrations move between these two different methodologies, and an awareness of this is important in any discussion concerning the nature of his anatomical or physiological proofs of the brain.

There is no doubt that Galen's encephalocentric claims are dependent on a level of anatomical knowledge of a high order.[10] His demonstrations of the anatomy (and to a lesser extent the physiology) of the brain, as this and subsequent chapters will show, are all examples of first-rate empirical research. Galen's project for the anatomy of the brain is both a serious and painstaking undertaking.[11] Lloyd has stated that: "The results of such empirical anatomical investigations were as clearly demonstrated as anyone could wish."[12] Galen's exact anatomical knowledge is the foundation of many of his diagnostic skills.[13] However, such anatomical claims, although deeply impressive

example, in *De usu respirationis* p. 108 Furley-Wilkie; IV. 492 K, Galen states that we breathe for the sake of innate heat. This does not admit of scientific demonstration, nor is it a necessary proof; it is a *plausible* one. Cf. Debru, 1993, 37–38; Frede, 1985, xxxii.

[8] Von Staden, 1997a, 37–39, argues in part that Galen frequently uses the term ἐπίδειξις in the sense of a public display or exhibition of his prowess as a dissector, and the term certainly underscores the notion of something being shown or revealed for public view.

[9] *PHP* p. 162,13–16 De Lacy; V. 278 K.

[10] On Galen's anatomy see Rocca, forthcoming.

[11] As Fleck, 1935, 35, observed, "the path from dissection to formulated theory is extremely complicated, indirect, and culturally conditioned."

[12] 1996b, 277.

[13] To take one example. Galen's account of what is today termed a median nerve palsy in Pausanias the Sophist is a clinical masterpiece based on a thorough knowledge of the distribution of the nerves of the arm and hand. Cf. *AA* II. 343–345 K; *Loc. Aff.* VIII. 56–59, 138, 213–214 K; *Opt. Med. Cogn.* p. 109,17–19 Iskandar. For Pausanias see Diller, 1955, 272–273.

and empirically testable, cannot be supported by (or viewed as) demonstrations *more geometrico*. Consider the second of Galen's encephalocentric propositions, that the nerves originate in the brain. The cardiocentric thesis holds that the nerves arise in the heart. This latter premise, Galen holds, is false on empirical grounds.[14] Nevertheless, even if there is broad agreement with Galen's encephalocentrism, the argument that follows from such a false premise may be valid, although the conclusion reached, that the *hegemonikon* is located in the heart, would also be false. This, after all, is the logical outcome derived from the initial proposition. On these grounds the cardiocentric standpoint is an internally consistent deductive one, but to Galen it does not provide a valid account of the anatomical relationship of the nerves. In this way Galen asserts that Chrysippus, his chief cardiocentric target, departs from apodeictic and scientific premises.[15] Yet his stance, that the initial cardiocentric premise is falsifiable, need not concern the cardiocentrist with the logic of his position. Further, it is apparent that Galen's own encephalocentric standpoint may be cast in a similar logical form, and may be presented syllogistically as follows:

(I) Where the nerves have their origin is also where the *hegemonikon* of the soul resides.[16]
(II) The nerves originate from the brain.[17]
(III) Therefore, the *hegemonikon* of the soul is in the brain.[18]

Whether "brain" or "heart" is inserted as the object of the second premise, either syllogism is valid. Their relevance lies in what Galen *claims* from his encephalocentric axiomata, and what restrictions he places on those of cardiocentrism. Galen wants his axiomata to be considered not only as self-evident truths (since to deny them would

---

[14] This is not to say that a cardiocentric (or for that matter an encephalocentric) proposition is without meaning; it has meaning only if shown to be verifiable or falsifiable.

[15] Cf. Tieleman, 1996a, 47–52.

[16] *PHP* p. 66,19–20 De Lacy; Fr. III. This is Galen's major premise. It may be refuted by a cardiocentrist, not least by denying the nerves the complete set of functions Galen allots them. A cardiocentrist may also deny this premise entirely, as Chrysippus does, by pointing out that even if the head is the source of the nerves, this does not necessarily imply the *hegemonikon* is also to be found there. Such a premise does not have universal agreement and therefore is not a self-evident premise or first principle.

[17] *PHP* p. 66,22 De Lacy; Fr. III.

[18] Cf. Barnes, 1991, 84–5; Tieleman, 1996a, 27f.

involve a self-contradiction), but also as empirically testable, that is, verifiable through experience. Furthermore, Galen rejects the logical argumentative apparatus of cardiocentrism on empirical grounds: their premise that the nerves originate in the heart, is, says Galen, "manifestly refuted by dissection".[19] Yet that there are nerves about the heart may be as incontrovertible a premise as Galen's that there exist nerves about the brain. One could argue on empirical grounds alone that since dissection has shown there to be nerves around the heart, this implies that nerves originate from the heart, and that other nerves are relayed to and from the brain. Galen's anatomical axiomata can never lead to incontrovertible conclusions, not because they are based on anatomy *per se*, but because they cannot be subject to axiomatic-deductive proof. In the case of his encephalocentric axiom concerning the origin of the nerves, Galen proceeds to demonstrate it by dissection and by experiment; his arguments for encephalocentrism are well formulated and effectively argued, and may even be regarded as logical (in the weak sense of rational or sensible). Yet Galen also wants his empirical demonstrations to be regarded as logically irrefutable as those evoked by Aristotle's *Posterior Analytics*, and equivalent to a geometric proof. Since Galen makes such a claim, it is necessary to try to understand the reasons for it. These have as their basis Galen's overwhelming need to demonstrate comprehensively the superiority of his encephalocentric hypothesis over that of his rivals.[20] Galen holds his encephalocentric propositions to be derived from the only legitimate hegemonic arguments and therefore encapsulate an exact science. To comprehend Galen's motivations, it is necessary to examine how he constructs his empirical and logical methodology in relation to his encephalocentrism.

In *De Placitis Hippocratis et Platonis*, Galen singles out the Stoic Diogenes of Babylon for his assertion that the source of nourishment

[19] ἐναργῶς ὑπο τῆς ἀνατομῆς ἐλεγχθέντος, *PHP* p. 428,22 De Lacy; V. 587 K.

[20] The arguments Galen advances in support of this hypothesis are played out in several texts, but perhaps nowhere as clear (or with as much condescension) as in *De placitis Hippocratis et Platonis*. There, as Vegetti, 1999b, 336, puts it, "on the plane of tradition Galen faces the great rivals of the Hippocratic-Platonic line of thought, namely the Aristotelians and the Stoics. On the theoretical level the permanent adversary is the cardiocentrism proper to these traditions. In commenting on the reception of the first part of the work, Galen acknowledges a certain success on both fronts: after the best Stoics and Peripatetics had studied it at length, it had, so he says, made them less "arrogant" than before and more inclined to accept the localisation of the rational *hêgemonikon* in the brain."

and the cause of voluntary motion are the same. The argument is complex, but what is noteworthy is Galen's statement that Diogenes prefers straight *affirmations* (δόγματα) rather than "the appearances from anatomy."[21] At the core of Galen's dispute with Diogenes of Babylon is not merely that nourishment and voluntary motion have different sources, but that these sources can only be understood by the proper use of the apodeictic method. Galen's method is not dependent on opinions that, in any case, do not always correlate with the information obtained from dissection. The starting points of Galen's encephalocentric hypothesis do not rely on such commonly held beliefs, but on epistemic premises which form the distinctive feature of the apodeictic method.[22] The notion of what is apparent in a dissection is elevated by Galen to evidence-based status.[23] Thus, the characteristics of the nature of the brain are established by what is apparent through dissection.[24] Moreover, such information is not obtained simply through examination: one must know how to perform a dissection and what to look for; however, repeated *observation* (αὐτοψία) of dissected material is useful for gaining some *experience* (ἐμπειρία) with the material.[25] It is, after all, possible "to be mistaken" (καταψεύδεσθαι) about the mere appearance of a dissection.[26] Galen's students were trained to avoid such mis-

---

[21] τὸ φαινόμενον ἐκ τῆς ἀνατομῆς, *PHP* p. 166,22–23 De Lacy; V. 284 K. Cf. Tieleman, 1991, 106–25. In Aristotle, the term φαινόμενον carries several meanings, but in the *Prior Analytics* he uses it in the sense of empirical observations necessary to secure proof (*An. Pr.* 46a17–22). See the fundamental study by Owen, 1961.

[22] ἐδείχθη δὲ καὶ ὡς αὐτὸς ὁ λόγος ἀρχαῖς μέν ἐστιν ἀρκούμενος οὐκ ἐνδόξοις, οὐ μὴν ἀλλ' ἐπιστημονικοῖς τοῖς λήμμασι κεχρημένος, ὅπερ ἴδιόν ἐστιν ἀποδεικτικῆς μεθόδου. *PHP* p. 430,5–7 De Lacy; V. 588 K. For Aristotle, ἔνδοξα may be useful at the onset of an empirical inquiry. Arguing from commonly-held beliefs is a characteristic of the method of dialectic (*Top.* 100a18–21, *Metaph.* 995b22–24. Cf. Irwin, 1988, 29–31, 36–39, 296).

[23] *PHP* p. 98,7 De Lacy; V. 207–208 K; cf. p. 82,14–15; V. 189 K; p. 386,10–11; V. 536 K; p. 388,8–9 V. 638 K; p. 394,39; V. 548 K; p. 404,9–10, 34–35; V. 558–560 K. The importance of believing (as well as understanding) what is apparent though the senses lies at the heart of Galen's dispute with the Peripatetic Alexander of Damascus, when the latter, having witnessed Galen's public demonstration of the function of the recurrent laryngeal nerve, asked if one could believe the evidence of one's senses (*De praenotione* pp. 96,27–98,8 Nutton; XIV. 628 K). Cf. Todd, 1995, 125f.

[24] Cf. *PHP* p. 446,29–33 De Lacy; V. 610 K.

[25] This furnishes part of the criteria of truth for the Empiricist—although the anatomical application of the criterion does not for them play as prominent a role as it does for Galen (cf. *Subf. Emp.* III, pp. 47–48 Deichgräber).

[26] *PHP* p. 150,34 De Lacy; V. 266 K, where those who have misrepresented what is evident from anatomy (i.e. that the nerves come from the brain), are respon-

takes.[27] For Galen, if the correct empirical methodology is employed, then such errors will be infrequent, and what appears to the trained senses can therefore be relied upon as a clear apprehension or *impression* (φαντασία).[28] Galen states that we possess natural criteria for elucidating information. Their presence allows him to establish what he calls a proficient or "technical criterion" (τεχνικὸν κριτήριον).[29] Nature provided us with a double gift: the criteria themselves, and our "untaught trust" (ἀδίδακτος πίστις) in those criteria.[30] The criteria consist of the organs of sense perception and their *faculties* (δυνάμεις). How well they are employed is a matter of training and judgement,

---

sible for the length of Galen's exposition (cf. p. 480,26–7; V. 650 K). It is a self-serving statement, yet Galen is determined to resolve such διαφορά and to prevent his pupils from making similar errors. See n. 27 below.

[27] Galen's student could always check his preconceived notions against that of a dissection itself. Today, whilst there are abundant visual aids, the problem of interpretation remains essentially the same as it was in Galen's time: "In assessing the 'truth' of such work, the anatomist should, in principle, take little account of theories; he simply carries out a dissection, and compares the drawings with what he can see with his own eyes. In practice it is difficult for him not to see what he has learnt to see, under the accepted paradigm of his subject. But within that limitation of vision, the messages that anatomists communicate to one another, and store as 'objective knowledge' in the scientific archives, are drawings and photographs, to which the accompanying text is merely a commentary." Ziman, 1978, 49–50. More generally, see Hanson, 1958, chapter 1.

[28] Cf. *De dignoscendibus pulsibus*. VIII. 770–771 K, where the concept of training the dissector's perceptive faculties is stressed. There is a wider strategy at work. At *PHP* p. 586,16–30 De Lacy; V. 778–779 K, Galen seeks to collapse the distinctions between how the Fourth Academy defines φαντασία, and the Stoic conception (cf. p. 606.20–1; V. 802 K). For Galen, apprehension is similar to the Stoic notion of καταληπτικὴ φαντασία, that is, conveying the direct (and accurate) apprehension of an object (cf. *SVF* 2, 26f.). Galen's separation of what is apparent to the senses and to the intellect seems pretty much in keeping with Fourth Academy teaching (cf. Tarrant, 1985b, 109 and 168 n. 35). See also Garcia-Ballester, 1994, 1651–9; Von Staden, 1978.

[29] *PHP* p. 542,7 De Lacy; V. 723 K. De Lacy's translation of τεχνικόν as "scientific" is not strictly accurate; τέχνη refers to a skill or expertise, which may differ in degree (ἐπιστήμη refers to a body of knowledge from which a skill may be derived). Galen's use of τεχνικόν seeks to emphasise the practical nature of the criterion used for obtaining knowledge, in deliberate contrast to the *theoretical* (θεωρητικός) concerns of philosophers. Galen's distinction of the two criteria seems from our perspective a terminological nicety: both ultimately perform the same thing. But for Galen, the polemical value of such a distinction is important. For the Hellenistic epistemological background to the criterion, with which Galen was thoroughly imbued, see the fundamental study by Striker, 1974.

[30] *PHP* p. 544,10–12 De Lacy; V. 725 K. Galen here employs πίστις in the Aristotelian sense of "confident belief." (cf. *De anima* 428a20–22; see also Todd, 1995, 126). At *PHP* p. 482,21–26; De Lacy; V. 652 K, Galen describes the "firm belief" of those who, properly trained in the logical method, acknowledge as correct the arguments of Hippocrates and Plato concerning the *hegemonikon* of the rational soul.

and Galen's own programme of education was a rigorous and demanding one.[31] The information obtained from one's senses is reliable, but must be also be subject to discrimination and verification, which is why Galen also refers to the criteria as comprising sense perception and *understanding* (γνῶσις).[32] Galen's methodology relies on accepting the evidence available through sense perception, thanks to these natural criteria, which enables discrimination and interpretation of the evidence.[33] In so doing, Galen "holds evident perception to be the fundamental, non-negotiable criterion of truth."[34] The natural criteria enable *reason* (λόγος) to be applied to an experiential set of observations. Such an application is mandatory since even repeated observation of a particular anatomical structure may not give an *indication* (ἔνδειξις) of the function of that part, although it may help infer it.[35] For Galen, the cumulative evidence gained from repeated dissections, in combination with a trained and discriminating mind, leads to an empirically based and testable method of *scientific* (ἐπιστημονική) demonstration.[36] Galen's combined use of natural and technical criteria secure for him the scientific criterion of proof. Yet Galen is also determined that his anatomical demonstrations should be seen in an Aristotelian context as a concatenation of geometric or math-

[31] In *De constitutione artis medicae* I. 244 K, the would-be physician must possess certain moral, social, and intellectual qualities in order to qualify for instruction at Galen's hands. Cf. Mansfeld, 1994, 119ff., 167–173.

[32] *PHP* p. 542,27–30 De Lacy; V. 724 K.

[33] Galen's conceptualisation and handling of epistemology is a large topic which has yet to be comprehensively addressed. See Frede, 1981; Hankinson, 1994b.

[34] Hankinson, 1998, 15–16; cf. idem, 1991a, 271ff.; idem, 1992, 3510; idem, 1997, 161–166.

[35] The term ἔνδειξις derives from Hellenistic theories of sign-inference (cf. Sedley 1982). In *Institutio Logica* XI.1, p. 24,14–17 Kalbfleisch), ἔνδειξις follows from observation of the thing under consideration; *per contra*, "logical demonstration" (ἀπόδειξις) obtains its conclusion through true or incontrovertible premises (cf. *Meth. Med.* X. 126–7 K). See also Hankinson, 1991d, 202–7; Kieffer, 1964, 100–101; Kudlien, 1991; De Lacy, 1991, 293; Von Staden, 1997, 39 and nn. 25–27. According to Galen, the Empiricist defines ἔνδειξις as knowledge based on rational consequence (*Subf. Emp.* II, p. 44 Deichgräber). It is used by the Methodists in the sense that each disease is indicative of its own treatment, but this is *independent* of reason or experience. Galen upholds the Rationalist claim that through the indication of what is observed one may make legitimate inferences on what is hidden. Cf. Frede, 1987, 263–266, 289, 293.

[36] Cf. *PHP* pp. 576,27–578,2 De Lacy; V. 766 K, where Galen points out that one of the problems of philosophical disagreements is that they cannot be subject to discrimination by empirical testing. Cf. Barnes, 1991, 61–62, 70–71. On the similarities between Galen's training of the senses and the Stoic view, see Hankinson, 1991a, 296–299.

ematical propositions.[37] Why does Galen seek this? Examining this question provides additional information on the importance of Galen's method of anatomical dissection and the wider claims he makes for the function of the brain.

Galen wishes to present his case for the function of the brain as that which can also be established *more geometrico*, through premises or first principles agreed by all physicians and philosophers.[38] For Galen, the study of logic is an absolute necessity for a successful physician. A doctor trained in logic is a good practitioner, and Galen repeatedly stresses the importance of the Greek mathematical or geometrical model in the teaching and practice of medicine.[39] To Galen, his colleagues lack logical training.[40] Even some philosophers do not employ the method of geometrical proofs in their arguments.[41] The critical rider is that such physicians are also indifferent to dissection, and thus their entire training in empirical methodology is equally suspect.[42] A significant part of Galen's drive to establish anatomy on logical grounds is underscored by such polemics. Yet Galen must show that the anatomical axiomata he has chosen are axiomata in the manner of geometrical proofs.[43] But as Lloyd, in a important study of this question, observes: "Everything turns, therefore, on the question of the extent to which and the manner in which geometrical-style proof *is* appropriate in the fields in which Galen principally worked."[44] In a logical demonstration, an axiom must be primary, non-demonstrable, and universally agreed.[45] But as far as the structure and function of the human body is concerned, any proposition claimed

[37] Cf. Barnes, 1991, 72 n. 73; Irwin, 1988, 118; Lloyd, 1996b.

[38] *PHP* p. 480,20–21 De Lacy; V. 649 K.

[39] Galen's text on logic, *De Demonstratione,* in fifteen books, is not extant. On Galen's wide-ranging claims for logic, see Barnes, 1991; idem, 1993; Edlow, 1977; Hülser, 1992; Kieffer, 1964, 16–17. Galen points out that the mathematical method of proof saved him from Pyrrhonian ἀπορία (*Lib. Propr.* p. 116,20–26 Mueller; XIX. 40 K). Cf. *Meth. Med.* X. 469 K. See also Barnes, 1991, 56–60; Eichholz, 1951, 67; Hankinson, 1992, 3507–3508; Pearcy, 1983, 264.

[40] In *De curandi ratione per venae sectionem* XI. 256 K, Galen remarks that the trouble with the other medical sects is they never study geometry, instead building on unproven theories. Cf. Barnes, 1991, 53 n. 13, 93–4; Dillon, 1982, 60–75.

[41] For example, at *PHP* p. 486,11–13 De Lacy; V. 656 K, where from the context it is clear that these philosophers are cardiocentrists.

[42] Cf. *PHP* p. 98,5–8 De Lacy; V. 208 K.

[43] "Galen saw that the axiomatisation of a practical, empirical science required him to admit a class of practical, empirical first principles." Barnes, 1991, 72.

[44] 1996b, 261.

[45] Cf. Barnes, 1991, 70–2; Lloyd, 1996b, 272–3.

as *axiomatic* becomes problematic in the sense that none can be universally agreed upon. And, as noted earlier, Galen's major premise, that the site of the origin of the nerves is also the locus of the *hegemonikon*, may be disputed by the cardiocentrists simply on the grounds that it does not necessarily follow that from the ἀρχή of the nerves, the *hegemonikon* must also be in the same place. Galen does not have the advantage of Euclid, who "can and does build on a substantial consensus among mathematicians for which there is no real parallel in medical theory."[46] Indeed, according to Galen, there is consensus on only one aspect of the hegemonic debate amongst physicians and philosophers: that the *hegemonikon* of the rational soul is responsible for sensation and voluntary motion.[47] Nevertheless, as also noted above, Galen holds all his encephalocentric propositions to be axiomatic, while simultaneously maintaining that such axiomata are not self-evident, but accessible only to one trained in the proper comprehension of the evidence of the senses.[48] For Galen therefore, indemonstrable first principles are not only incontrovertible to the trained medical scientist, but may also be demonstrated.[49] This is of course contrary to the Aristotelian definition of a first principle as one that is self-evidently true in order to avoid circularity and infinite regress.[50] Galen thus attempts to link the empirical method of epistemic demonstration with that of axiomatic-deductive proof, but his encephalocentric propositions can only be dependent for their validation on an empirically based body of knowledge.[51]

[46] Lloyd, 1996b, 264.

[47] *PHP* p. 480,7–9 De Lacy; V. 648 K; a little further, Galen is also bold enough to state that his major premise (the origin of the nerves is also the site of the *hegemonikon* of the soul) has similar universal agreement (p. 480,19–21; V. 649 K).

[48] As Barnes, 1991, 78, remarks: "The axioms must be accepted at first impression—yet they are also the product of trained reflexion." Cf. 79, 90–2; Lloyd, 1996b, 261.

[49] An apodeictic demonstration may reveal non-evident truths but this is based on first principles that are themselves indemonstrable. Cf. Sextus Empiricus, *P.* 2. 140.

[50] Cf. Lloyd, 1996b, 263–4, 270–1. Although in *An.Post.* Aristotle has very little to say on how the premises come into being.

[51] Tieleman, 1996a, 36, maintains that: "The addition of empirical testing to the logical method (i.e. analysis/synthesis) reflects a familiar Galenic scheme, viz. the duo reason/experience. Galen's claims to originality in this respect may not be wholly unjustified." This may be so, and one can demonstrate a number of practical applications which derive from geometric axiomata, but an empirical methodology cannot be grafted on to the logical method, and even if some statements are empirically secure (that the nerves originate in the brain), they can never be *certain* in an axiomatic-deductive sense.

2.3 *Galen's pneumatic physiology*

Temkin observed that Galen's use of *pneuma* (πνεῦμα) is the focal point of his "biological and medical doctrine."[52] Galen conceptualised a pneuma-based physiology of the brain and sought experimental proof for it. His ventricular experiments, as will be discussed in chapter 5, are dependent on a pneuma-based physiology. The following examines the concept of pneuma, the lack of any truly alternative physiological models, the debt Galen owed his predecessors, and his acknowledgment of the limitations of pneuma as a complete physiological agent.

Galen's employment of pneuma is dependent upon anatomical structures whose functions are close-fitted to a methodology of empirical investigation.[53] Pneuma is a theoretical construct that may be used in a physiological system without undue difficulty.[54] It is arguably one of the best examples of continuum theory in late antiquity, and pneuma exerted a significant influence on physical and physiological theories.[55] Pneuma is a broad church. Galen's often violent duelling with his peers—past and present—takes place against a background in which the deterministic, qualitative, and teleological aspects of pneuma theories are set against the indeterminism of other matter theories—principally atomistic or particulate ones.[56] Given Galen's strong teleological stance, his holding of any indeterminate matter theory would have been out of the question.

---

[52] 1951, 160. Cf. Souques, 1936, 212.

[53] "L'anatomie (la cavité des ventricules cérébraux, le plexus réticulé) et l'expérimentation donnent au *pneuma* invisible un statut épistémologique beaucoup plus sûr que celui de "substance" de l'âme." Debru, 1996b, 147.

[54] A comparison can usefully be made between pneuma and the employment of later terms such as ether, "life or motive force", electricity, and the comparisons of circuit theory models to contemporary neuroscience. These are as good a set of analogical models for physiological systems as much as Greek pneuma served Galen and his predecessors.

[55] At *PHP* p. 170,9–10 De Lacy; V. 287 K, Galen states that Chrysippus holds that the "soul is connate pneuma" (ψυχὴ πνεῦμά ἐστι σύμφυτον), which exists as a continuum throughout the body. For an overview see Sambursky, 1956, chapter 6; idem, 1959, chapter 1; idem, 1962, 3, 37–9.

[56] For example, those espoused by medical writers such as Asclepiades of Bithynia. Behind Galen's vehement and unrelenting attacks on Asclepiades and his kind lies a critical issue: the denial of qualitative change by the Asclepiadeans. In order to fight this battle, Galen is quite prepared to blur the distinction between Asclepiadeans and Epicureans in respect of atomism. Cf. Casadei, 1997; Pendrick, 1994. For a reconstruction of Asclepiades' corpuscular theory see Vallance, 1990.

Examined either as a general or a specialised concept, pneuma is a good choice with which to underwrite a physiological system. It is indeterminate, invisible, and malleable.[57] It can therefore be defined in several, different ways.[58] It is not possible to state with any certainty when and by whom pneuma began to be used as an explanatory device to account for physiological processes. Anaximenes apparently equated πνεῦμα with the outside air and identified it with the life-principle.[59] Diogenes of Apollonia seems to have made the first explicit equating of *air* (ἀήρ) with the soul and intelligence. As discussed in chapter 1, this life-giving air is distributed about the body in a carrier system, the *blood vessels* (φλέβες). The air so distributed with the blood throughout the entire body somehow enables cognition to take place.[60] None of these notions are doctrinal.[61] Pneuma in the Hippocratic Corpus reflects its assumed role in physiological processes, whether as ἀήρ or as πνεῦμα. As also noted in the last chapter, the author of *Morb. sacr.* maintained that certain blood vessels are responsible for the intake of most of the outside air, distributing it through the body. This breath cools the body and some also goes straight to the brain, where it is responsible for intelligence and motion. These texts present certain general themes which will easily lend themselves to further refinement: inspired air as the source of pneuma; blood vessels as its carrier system; the use of air or pneuma to account for sensation and voluntary motion.

On physiological grounds alone, the Stoic contribution to pneuma theory is arguably the most significant. Balme has noted that: "At Aristotle's time 'pneuma' was the ordinary word for wind; when used of animals it meant breath and also internal wind. But in medical theory it was fast developing special connotations, which led to the Stoic conception (soon after Aristotle) of an indwelling divine spirit."[62]

---

[57] Cf. Sambursky, 1956, 133.

[58] For a general survey of the association of breath with life and the soul see Onians, 1953, 51, 54–56, 76–77, 168–172, 250–252. Cf. *RE* Supp. XIV, cols. 387–412, Saake; Cuncliffe, 1924, 333, 424. See also L-S § 47 and commentary in Vol. I, 287–298. Wellmann, 1895 65–84, remains indispensable. See also Tanner, 1985; Verbeke, 1945, 206–220.

[59] Cf. Aëtius I 3.4. "That the air which we breathe should be the life itself which animates us is a common idea, and the breath-soul a world-wide conception." *HGP* I, 128. See also Verbeke, 1945, 6.

[60] Cf. DK 64B4–7.

[61] Longrigg, 1993, 98, in his discussion of Empedocles and Diogenes of Apollonia, refers to pneuma as an example of: "Certain of the physiological doctrines (which) exercised a lasting influence."

[62] 1972, 161; cf. Furley, 1989, 199.

Pneuma was regarded either as a special compound of fire and air[63] or as vital heat itself.[64] Pneuma in Stoic hands shows its adaptability as something capable of underpinning a variety of explanations, from material causation to the psychic attributes (and qualitative functions) of the soul itself.[65] On inspiration it provides the vital force and psychic agency of the body, and operates as the fundamental principle of coherence on a microcosmic and macrocosmic scale.[66] To be sure, early Stoic terminology did not provide the user with a set of types of pneuma, nor—and perhaps more importantly—to an exposition of a specialised physiological theory. But Stoic pneuma was nevertheless capable of *alteration* (ἀλλοίωσις) or *change* (μεταβάλλειν), and under later Stoic embryology there was developed a more formalised differentiation of *pneumata*.[67] According to Plutarch, Chrysippus held that pneuma in the foetus was changed at birth by the outside air to become "vital pneuma" (ζωτικὸν πνεῦμα), the equivalent of soul.[68] As shall be noted below this broad notion of pneumatic differentiation is one which Galen exploits in his schemata of elaborative change from "vital" to "psychic pneuma" (ψυχικὸν πνεῦμα) but, unlike Galen, the Stoics did not use the concept of *qualitative* change in pneuma.[69] The pneuma of the growing Stoic embryo is

---

[63] *SVF* 2, 439. The explicit identification of pneuma as a compound of fire and air appears Stoic in origin (*SVF* 2, 310, 2, 389, 2, 444, 2, 471, 2, 774, 2, 786; cf. *PHP* p. 306,23–27 De Lacy; V. 447 K). Pneuma, however, was sometimes said to be just composed of fire. Cf. Nemesius, *De natura hominis*, p. 16,15–16 Morani. Sorabji, 1988, 86–9, argues that the Stoics did not invariably *define* pneuma as a compound of fire and air (but cf. Sedley, 1993, 326 n. 44).

[64] Zeno, according to Diogenes Laertius, describes the soul as hot pneuma, (*SVF* 1, 135). In the early Stoa, "the soul in the form of pneuma (breath) permeates the whole body; body and soul are material and their combination is described as total blending, that is to say both occupy the same place. So intimate is the relationship that what affects the body affects the soul and vice versa." Erskine, 1990, 71.

[65] Galen, in *De plenitudine* VII. 525 K, in reference to the Stoic view of pneuma as a sustaining cause, speaks of the pneumatic substance as that which sustains. Cf. L-S II, 47F, 280 for further discussion of this passage. See also Hahm, 1977, 163–166. On causation in Galen see Hankinson, 1998.

[66] "Stoic pneuma was not so much a structured force as a structuring one: it structured all things in the cosmos, lifeless, vegetative, animate, and rational." Tarrant, 1985a, 57.

[67] As noted in n. 55, above, Chrysippus defined the soul as "innate pneuma" (πνεῦμα σύμφυτον). The equating of σύμφυτον with φυσικόν was not a difficult step to take. But as Stoic pneuma is both soul and connate throughout the body, it is arguably too general a concept for Galen to accommodate without modification and qualification.

[68] Plutarch, *St. rep.* 1052F; *SVF* 2, 806. Cf. Long, 1982, 34–57.

[69] This is not to say that the Stoics did not maintain a belief in the different qualities of matter. Cf. Sambursky, 1962, 37–38. The respective qualities of various

natural (φύσις). At gestation it changes its status (becoming finer) to vital.[70] This change perhaps reflects the notion that the foetus does not breathe, and thus lacks the capacity to acquire pneuma. What is now referred to as *psychic pneuma* is held to be responsible for sentience.[71] The mechanism behind these changes is ascribed to an alteration in the *tension* (τόνος) of pneuma, which in some way accounts for all the activities of the living being.[72] Yet although the term psychic pneuma was used by the Stoics, and changes in its tensional state were said to account for individual action, pneuma was not linked to any specific *part* of the body.[73] Pneuma then, does not have to change from vital to psychic to underwrite the actions of the Stoic *hegemonikon*, as it must in Galen's account. Whatever the wider intentions of Stoic philosophy and its relationship, dependence, or affinities to Hellenistic medical theories,[74] Stoicism underwrites the transformation of pneuma from a purely general principle of animation to one adaptable to more specialised physiological needs, even if the physiological criteria were not—or perhaps could not—be met entirely by Stoicism.[75] Together with the work of Herophilus and Erasistratus,

substances are due to their possessing an innate *tenor* (ἕξις), pneumatic in origin. Cf. Plutarch, *St. rep.* 1053F–1054B; *SVF* 2, 449.

[70] τὸ πνεῦμα μεταβάλλειν καὶ γίνεσθαι ζῷον (Chrysippus ap. Plutarch, *St. rep.* 1052F; cf Galen, *Foet. Form.* IV. 665 K). Cf. Long, 1982, 43–44.

[71] *SVF* 2, 716.

[72] Cf. *SVF* 2, 393, 458. The Stoics make use of a "third" pneuma, that which sustains (*SVF* 2, 716). This sustaining cause is linked to tenor and tonicity. Cf. Von Staden, 2000, 100–101.

[73] Cf. L-S II, 47K, and the discussion in L-S I, 286–289 (but cf. Von Staden, 2000, 101 n. 61).

[74] How much Stoicism "owed" to the physiological theories of Herophilus and Erasistratus is moot. Von Staden, 2000, 104–105, puts it this way: "Affinity is, of course, no guarantor of influence, and the tendency to account for all affinities by means of a 'genealogy of influence' can be methodologically insidious. Furthermore, when dealing with a given Stoic theory that bears resemblance to a medical theory, it should also not be overlooked that 'Stoic' in the ancient sources can cover a considerable chronological spectrum, that the question whether a given theory is attributable to Zeno, Cleanthes, Chrysippus, or a later Stoic is often intractable, and that Zeno (331–262 BC), for example, was an almost exact contemporary of Herophilus, whereas Chrysippus (281 or 277–208 or 204 BC) was a good forty to fifty years younger than Erasistratus and Herophilus."

[75] The Stoics may also have used pneuma as a way of bridging the gap between their own and the Platonic theory of the soul. As Sedley, 1993, 326, notes, "we are at liberty to read the Stoic adoption of the *pneuma* theory not as a rejection of Platonic psychology, but as an attempt to update it in the light of the latest science." According to Harger, 1982, 97, "the later Stoics had little interest in physics." Later than whom is not stated. But even if some Stoics after Chrysippus were disinterested in physics, that does not exclude them from appropriating physical or medical concepts into their philosophy.

Stoicism helped popularise and make acceptable the notion of pneuma as a general physiological theory.[76]

According to the author of the Pseudo-Galenic *Introductio sive Medicus*, it would seem that by the Hellenistic period a distinction was being drawn between *natural pneuma* (responsible for all life functions) and *psychic pneuma* (responsible for all nervous activities).[77] Herophilus and Erasistratus regarded pneuma as acquired through respiration.[78] How they used it is another matter. In an admittedly problematic passage discussed in chapter 1, Herophilus apparently refers to the optic nerves containing *natural pneuma*.[79] He may have only used "natural pneuma" to reflect that agency's link with the *soul* (ψυχή).[80] Apart from this, Herophilus does not seem to have used the term *psychic pneuma* in any way.[81] Exactly when psychic pneuma entered the medical lexicon is not known, although, as noted in chapter 1, Diocles of Carystus and Praxagoras of Cos are possible medical sources. Galen states that the distinction between a vital pneuma (located in the heart), and a psychic pneuma (located in the brain) was known to Erasistratus.[82] Whilst Erasistratus' handling of pneumatic elaboration appears broadly similar to that of Galen, it is nevertheless sufficiently different to warrant examination. One difference seems to be that, for Erasistratus, psychic pneuma is elaborated in the meninges of the brain. To be sure, this is not precisely spelt out by Galen. In *De usu respirationis*, Erasistratus and his school are said to have maintained that the pneuma in the heart was the source of the brain's psychic pneuma. They are placed in deliberate polemic opposition to the "school of Hippocrates" who maintain psychic pneuma is sent *straight* (εὐθύς) to the brain via the nostrils.[83] The

---

[76] On Galen's reactions to Stoicism see Manuli, 1993, although the nature and extent of Galen's anti-Stoic views is perhaps not as wide-ranging as Manuli infers.

[77] XIV. 697 K. Cf. Von Staden, 1989, 253–254; idem, 2000, 102–103. The Pneumatist medical sect, founded by Athenaeus of Attalia in the first century BC., utilised pneuma as an explanatory agency in health and disease. Cf. Kudlien, 1968; Verbeke, 1945; Wellmann, 1895. A new study of the Pneumatists is a desideratum.

[78] Cf. *Art. Nat. Sang. Cont.* pp. 146–148 Furley-Wilkie; IV. 706–707 K. Cf. Hahm, 1977, 161ff.; De Martini, 1964, 43; Von Staden, 1989, 254 n. 53; 527 n. 23.

[79] *Ad oculorum cavernas meent naturalem spiritum continentes.* T 86 *Herophilus*. Cf. Von Staden, 1989, 253–254.

[80] Cf. Von Staden, 2000, 88.

[81] Cf. Von Staden, 1989, 254–255; idem, 2000, 89.

[82] *Diff. Puls.* VIII. 714 K. Cf. *PHP* p. 164,13–16 De Lacy; V. 281 K.

[83] ἀλλ' οὐδ' ἐκ τῆς εἰσπνοῆς ὁμοίως οἱ περὶ τὸν Ἐρασίστρατον τοῖς <περὶ τὸν> Ἱπποκράτην τρέφεσθαί φασι τὸ ψυχικὸν πνεῦμα. τοῖς μὲν γὰρ ἐκ τῆς καρδίας διὰ τῶν ἀρτηριῶν ἐπὶ τὰς μήνιγγας, τοῖς δὲ εὐθὺς διὰ τῶν ῥινῶν εἰς τὰς κατὰ τὸν ἐγκέφαλον

consequences for Galen's physiology of such an external source of pneuma will be discussed in chapter 6. Galen's polemical stance aside, it would appear that he may have taken up the general Erasistratean thesis when he discusses the transformation of vital pneuma (via the heart) to psychic pneuma in the brain, albeit via the elaborating agencies of two other vascular structures there—the retiform and choroid plexuses—rather than the meninges.[84] The major point of contention between Galen and Erasistratus lies in the latter's regarding pneumatic differentiation as a *quantitative* process, with pneuma becoming increasingly finer.[85] By contrast, Galen's physiology is based on a pneuma that is subject to qualitative change.

Galen never rules out an external source of pneuma, and as will be considered in chapter 6, it plays a role in his generation of psychic pneuma. Nevertheless, the mainstay of Galen's pneumatic physiology lies in a progressive elaboration of pneuma by several parts of the body. According to Galen, the creation of psychic pneuma begins when "inspired air" (ἔξωθεν ἀήρ) enters the lungs, which alter it into a "pneuma-like" (πνευματῶδες) substance.[86] In this use of the air-like quality of pneuma, Galen implicitly acknowledges the Stoic contribution to pneuma theory.[87] From the lungs, this new entity enters the left ventricle of the heart where it is fully elaborated into vital pneuma. This change is made possible by *innate heat* (ἔμφυτον θερμόν) within the left ventricle, acting in concert with altered venous blood from the right ventricle.[88] Although Galen's treatise on innate

---

κοιλίας ἔρχεσθαι τὸ πνεῦμα δοκεῖ. p. 122 Furley-Wilkie; IV. 502 K; Fr. 112 Garofalo. Presumably, for the Erasistrateans, pneuma reaches the meninges via the arteries that are found there.

[84] In any case, for Erasistratus, as for Galen (and Herophilus), the meninges are vascular structures. Thus, Erasistratus' position concerning the elaboration of psychic pneuma is not all that different from Galen's: both site its production within the head. Their point of departure seems to be what part of the brain elaborates it.

[85] Cf. Harris, 1973, 225. On Erasistratus' physiology examined in the context of a teleological perspective, see Von Staden, 1997b, 203–208; cf. idem, 2000, 92–96.

[86] Cf *PHP* p. 528,28–32 De Lacy; V. 707–709 K. Pneuma in the blood is sometimes referred to by Galen as "the air-like substance of the blood" (ἡ ἀερώδης οὐσία τῷ αἵματι, *De differentiis febrium* VII. 277 K). In *De usu pulsuum* p. 206 Furley-Wilkie; V. 161 K, Galen notes that the expansion of the arteries draws the airy substance in through the pores of the skin. Cf. *De causis pulsuum* IX. 192 K: τὰς διαστολὰς τῶν ἀρτηριῶν ἕλκεσθαί τις ἀερώδης οὐσία.

[87] But this origin is deliberately obscured by Galen; for him it is a source of pneuma established by Hippocrates himself. Cf. *Usu. Puls.* pp. 202, 212 Furley-Wilkie; V. 157, 166 K. *PHP* p. 164,19–20 De Lacy; V. 281 K.

[88] Blood and pneuma are equally innate: οὐδὲν δὲ κωλύει καὶ τὴν αἱματικὴν οὐσίαν καὶ ἀερώδη θερμὸν ἔμφυτον ἀκούειν ἅμα τῷ πνεύματι. *Simp. Med. Temp. Fac.* XI. 731 K.

heat is not extant,[89] enough is known from other references to regard it as an indispensable part of any Galenic elaborative process, whether of blood, humour or pneuma.[90] To emphasise the link between blood and pneuma, the left ventricle is described by Galen as *pneumatic* (πνευματική); it is where the yellow, warm, and fine pneuma-like blood is generated and sent out to the rest of the body via the arteries.[91] The left ventricle is able to do this as it is the chief repository for innate heat (which also accounts for the thickness of its walls relative to that of the right).[92] For Galen, innate heat is not pneuma but that which is necessary for the (now modified) inspired air to be elaborated into pneuma. The ingredients to elaborate vital pneuma are blood, πνευματῶδες, substance, and innate heat, but how they are combined is not made explicit.[93] The entire process, however, is analogous to coction, and Galen invokes established authority when he states that Hippocrates, together with Plato concur that innate heat is necessary for coction.[94] Galen's vital pneuma now has access to the arterial system, affording it entry to the brain where it infuses two vascular structures, the *retiform plexus* (δικτυοειδὲς πλέγμα), a network of fine arteries at the base of the brain, and the *choroid plexuses* (χοριοειδῆ πλέγματα) of veins and arteries in the ventricular system, which complete the transformation of vital to psychic pneuma.[95]

---

[89] Cf. Durling, 1988, 210–212; Nutton, 1987a, 41.

[90] Innate heat is responsible for growth (for example, of the child, *Hipp. Nat. Hom.* pp. 79,16–80,4 Mewaldt; XV. 155–156 K). It is linked to the soul. (*Trem. Palp. Conv. Rig.* VII. 616 K). Innate heat in the liver transforms nutrient fluid sent from the stomach into blood and the liver substance (*PHP* pp. 412,30–414,1; V. 570 K). For innate heat to function properly, it must be well-balanced (*PHP* p. 524,19–22 De Lacy; V. 703 K; *Us. Resp.* p. 128 Furley-Wilkie; IV. 507–508 K; *Caus. Puls.* IX. 129 K). Innate heat is also present in the kidneys (*Hipp. Epid. VI* p. 148 Wenkebach), and in the brain (*Us. Resp.* p. 130 Furley-Wilkie; IV. 509 K).

[91] αἵματος δὲ ξανθοῦ καὶ θερμοῦ καὶ λεπτομεροῦς καὶ πνευματώδους ἡ μὲν πρώτη γένεσις ἐν τῇ τῆς καρδίας ἀριστερᾷ κοιλίᾳ, διανέμουσι δὲ καὶ παράγουσιν εἰς ὅλον τὸ ζῷον αἱ ἀρτηρίαι τὸ τοιοῦτον αἷμα. *PHP* p. 414,30–33 De Lacy; V. 572–573 K.

[92] The left ventricle is described as ἱκανῶς παχύ τε καὶ σκληρόν. *UP* I, p. 355,1 Helmreich; III. 487 K. Its thickness is a reflection of the arteries that arise from it; they too must be thick-walled to retain their pneumatic contents, the famous exception being the pulmonary vein. Cf. Debru, 1996b, 109–111.

[93] One might add vital pneuma to this recipe. There seems no reason why, once it is elaborated, it is not permitted to assist in further elaborations of its own kind.

[94] *PHP* p. 528,11–14 De Lacy; V. 707 K.

[95] Cf. Manzoni, 2001, 39–44. There are no grounds for postulating a "tripartite pneumatology" in Galen's physiology. Siegel, 1968, 186, has stated that a tripartite division of pneuma, "does not fit well into Galen's physiological system", and that his "basic ideas appear fully represented by the two, *pneuma zotikon* and *psychikon*."

However, the ventricles of the brain do not derive all their psychic pneuma in this way. Galen also allows the ventricles to elaborate outside air through the nasal passages. The ventricles are the final repository of psychic pneuma which then continues—in a way not fully determined by Galen—through the nerves and thence to the rest of the body, providing sensation and voluntary motion. With the exception of the cranial nerves, which for Galen are direct prolongations of the cerebrum, all other nerves take their origin from either the spinal cord or the cerebellum. The fourth ventricle is associated by Galen with each of these two structures, since, for nerves to function, they must be provided with a large supply of psychic pneuma.[96] Although psychic pneuma in the ventricles of the brain is made to account for sensation and voluntary motion, the chief attributes of the *hegemonikon* of the rational soul, how pneuma functions is something Galen never resolves.[97]

Vital and psychic pneumata are able to carry out all of Galen's physiological requirements. Galen makes but a single mention of natural pneuma, apparently housed in the liver and veins; but he never seems to utilise it, describing its existence in deliberately vague terms: τοῦ μὲν δὴ ψυχικοῦ πνεύματος ἐναργῶς ἐδείξαμεν οἷον πηγήν τινα οὖσαν τὸν ἐγκέφαλον, ἀρδομένου καὶ τρεφομένου διά τε τῆς εἰσπνοῆς καὶ τῆς ἐκ τοῦ δικτυοειδοῦς πλέγματος χορηγίας. τοῦ δὲ ζωτικοῦ πνεύματος οὐχ ὁμοίως μὲν ἐναργῶς ἡ ἀπόδειξις ἦν, οὐ μὴν ἀπιθανόν γε κατά τε τὴν καρδίαν αὐτὸ καὶ τὰς ἀρτηρίας δοκεῖν περιέχεσθαι, τρεφόμενον καὶ τοῦτο μάλιστα μὲν ἐκ τῆς ἀναπνοῆς, ἤδη δὲ καὶ τοῦ αἵματος. εἰ δέ ἐστί τι καὶ φυσικὸν πνεῦμα, περιέχοιτ' ἂν καὶ τοῦτο κατά τε τὸ ἧπαρ καὶ τὰς φλέβας. *Meth. Med.* X. 839–40 K. To emphasise its importance in his system, only the source of psychic pneuma is "clearly demonstrated". Cf. Manuli and Vegetti, 1977, 241 n.266; May, 1968, 49, 111 n.72; Singer, 1993, 47.

[96] 'Αλλ' ἐπεὶ πάντα τὰ κατὰ τὸ σῶμα νεῦρα τὰ κάτω τῆς κεφαλῆς ἢ ἐκ τῆς παρεγκεφαλίδος ἢ ἐκ τοῦ νωτιαίου πέφυκεν, ἐχρῆν καὶ τὴν ταύτης κοιλίαν ἀξιόλογόν τ' εἶναι τὸ μέγεθος καὶ τὸ προκατειργασμένον ἐν ταῖς προσθίοις ψυχικὸν πνεῦμα μεταλαμβάνειν. *UP* I, p. 482,14–18 Helmreich; III. 665 K.

[97] Cf. Lloyd, 1987, 213. n. 143; Von Staden, 2000, 111–112, 115–116. Galen's handling of the anatomy and physiology of the nerves, together with the problems he faced in explaining sensation and voluntary motion, merit a separate study. However, it is appropriate to say something here. Galen's conceptual dilemma is well brought out in his examination of the physiology of vision (*PHP* Book VII, chs. 4–7), where several explanations are provided based on pneuma as *substance* (the explanation for the changes in the size of the pupil is given strictly in the amount of pneuma present in the aqueous humour) and pneuma as *quality* (the nature of the visual image itself). Galen's text on vision is not extant (cf. *UP* I, p. 465,3–4 Helmreich; III. 641 K. See also Cherniss, 1933, 154–161; Katz, 1890, 1–9). Although Galen's handling of optics is impressive, it does not help in understanding how in pneumatic terms the image is sent to the eye and how that process operates continuously as long as the image is viewed. This uncertainty is mirrored in Galen's very brief discussions of the remaining sense organs. These, by a process of qualitative alteration similar to that of the eye, somehow assimilate their respective sensations. The organ of hearing is ἀεροειδής, that of smell, ἀτμοειδής; of taste,

2.4 *Galen's empirical substrate: animal material*

The nineteenth century comparative anatomist Richard Owen believed that Galen's comparative anatomical knowledge was "arranged with all the correctness of a Critic, with all the abilities of the first and most accomplished scholar of his time."[98] This encomium is a little misleading, for the picture that emerges from examining Galen's handling of animals for dissection and vivisection is of Galen as a comparative anatomist, albeit in a weak sense. It is weak since, for Galen, comparative anatomical studies only had value insofar as they provide information about, and could serve as models for human anatomy. All of Galen's anatomical findings in animals were extrapolated to that of man, and his work represents a synthesis of higher vertebrate anatomy grafted on to a human framework. This is not to exclude the possibility that Galen had at one time or another access to human bodies. He informs us that some physicians dissected (or attempted to dissect) humans, either of enemies slain in battle or of

ὑγρός; of touch, γεώδης (*PHP* p. 462,1–3 De Lacy; V. 627 K). These cannot help but be weak postulates. In what way pneuma acts in each of these organs is left unsaid. But since these organs (with the exception of the sense of smell) have nerves which come from the brain, it follows that pneuma must be responsible for some stage of the respective sensory process. Even though Galen processes his predecessors' knowledge through a vastly enhanced amount of sophisticated anatomical demonstration, the problem of visual perception, and of sense perception in general, was not capable of complete resolution in late antiquity. Aristotle does not entirely resolve the question of visual perception, and Galen cannot advance much further. As far as voluntary motion is concerned, Galen is equally as indeterminate. In *De motu musculorum* IV. 371 K, Galen cites a *power* (δύναμις) flowing from brain to muscle, with the nerves acting as passive carriers for this power (cf. 402–403). In this text, pneumatic transmission along a nerve is not mentioned. Galen postulates instead a rather general ψυχικὴ δύναμις as being responsible for voluntary action. This power seems to alter the innate tension in a muscle, allowing activity (On Galen's handling of muscular action see Bastholm, 1950, 87–95). These references to powers are of course, perfectly compatible with pneuma. Yet the maintenance of this essential *tension* (τόνος) for voluntary action is never fully explained. Vegetti, 1993a, puts forward a thesis that Galen utilised Hellenistic mechanical theories to explain how pneuma might act in terms of effecting voluntary muscular movement. It is helpful for an understanding of the concept of muscular pneumatic tension that it may be compared to the essential tension a spring possesses in order to function. But Galen did not seem to adhere to a doctrine of pneuma as a species of hydraulic fluid moving through nerves and acting on muscles like water in a piston pump (as Vegetti implies). And Galen did not take up another analogy offered by Hellenistic mechanical theory: the notion that the heart has valves (cf. French, 1994, 356). On Hellenistic mechanical theories see Drachmann, 1973; Schürmann, 1991, 36–48; Von Staden, 1996a, 147–172.

[98] 1837 (quoted in Sloan, 1992, 118).

the fortuitously exposed dead after burial. But these are occasions of haste and fleeting observation, and are reminiscent of the Empiricist approach to anatomical knowledge. Galen's opportunities to study human material remained limited, and there is no firm evidence that he engaged in human dissection.[99] For all practical purposes, human dissection and vivisection in Antiquity began and ended with Herophilus and Erasistratus. Animals constituted Galen's dissection and vivisection subjects. As Galen puts it, in reference to his work on the origin of nerves, it is a subject whose validity he has demonstrated on *innumerable* (μυριάκις) occasions with animals.[100] Yet Galen's extrapolation of animal anatomy to that of humans has led to understandable confusion.[101] His anatomical descriptions can result in difficulties in interpretation, unless it is made clear which animal Galen is dissecting and how he dissects it. But it is not possible to know with complete certainty which animal Galen refers to in every case. For example, in a text with such a wide physiological brief as *De usu partium*, it is understandable that, given the complexity of the topic, "Galen seldom mentions what animal he is describing or notes differences between different animals or between animal and human structures."[102] That

[99] May, 1958, adduces a passage in *De compositione medicamentorum per genera* XIII. 604 K, where Galen notes that dissection was carried out on the bodies of the Germans in the Macromannic wars. Yet this was limited due, says Galen, to the dissectors' own lack of adequate anatomical knowledge. Taken together with *AA* II. 385–386 K, where Galen notes that one should take advantage of the body of an exposed brigand, it suggests that Galen enjoined his pupils to utilise these and similar opportunities, provided of course that they were properly trained. It is possible Galen might also have had access to human foetal and infant material. He was certainly aware of the practice of dissecting the bodies of exposed children. Galen specifically cites a practice of dissection of the bodies of exposed children, which enabled those anatomists to conclude that man had the same structure as an ape: καὶ παιδία δὲ τῶν ἐκτιθεμένων νεκρὰ πολλὰ πολλάκις ἀνατέμνοντες ἐπείσθησαν, ὡσαύτως ἔχειν κατασκευῆς ἄνθρωπον πιθήκῳ. *AA*. II. 386 K. This is cited in the third person plural, perhaps to draw attention away from Galen's own use of this material. His knowledge could of course have been second hand, although there were subjects available since, as Balsdon, 1974, 195, points out, "miscarriages were distressingly frequent", and "infant mortality was extremely heavy." Cf. Scheidel, 1999, 266–272. Infanticide (as well as exposure, although the two practices are different) was practiced in second century AD Rome, although the extent and degree of tolerance are disputed. As King, 1998, 155, notes, "infanticide—which could be seen as a very late, and sex-specific, abortion—was practiced, although it is difficult to know how widely it occurred." Cf. Fox, 1986, 343. See also Boswell, 1988, 41–45, 60 n. 16, 89, n. 132; Hopkins, 1983, 225–6. On the philosophical background to abortion and infanticide see Carrick, 1985, 101–125; Eyben, 1980/1981, 12–19, 43–48.

[100] *PHP* p. 82,17–19 De Lacy; V. 189 K.

[101] Cf. Woollam, 1958, 15.

[102] Savage-Smith, 1971, 78.

said, it should be added that Galen's anatomical texts do provide sufficient evidence of which animals are dissected, and Galen's students knew to peruse the anatomical and physiological works together.[103] Galen states that his studies of the brain were largely based on the ox, but he also used apes, sheep, pigs and goats, and it would be remarkable if their brains were not examined by Galen in some way.[104] However, there is no evidence that Galen utilised the brains of these animals with anything like the frequency with which he made use of the bovine brain.[105] In what follows, two reasons for Galen's choice of the bovine brain will be given, availability and size. It cannot be emphasised too strongly that Galen's anatomical work on the brain was carried out using the bovine as paradigm for the human brain.[106] However, it is helpful to examine how Galen used other animals, especially the ape, in his investigations of the brain.

Galen's *On vivisection* (*De anatomia vivorum*) is not extant. There is, however, sufficient evidence from other works to determine which animals he employed, and for what purpose.[107] Galen's vivisectional descriptions make unpleasant reading. Yet he did not engage in cruelty for its own sake, and his remarks at the beginning of his account of brain vivisection, that one not use an ape for vivisection of the brain to "avoid seeing the unpleasing expression of the ape", shows

---

[103] Cf. *Ord. Lib. Propr.* pp. 83,23–84,25 Mueller; XIX. 54–55 K.

[104] *AA* II. 708 K. Cf. Woollam, 1958, 14.

[105] Savage-Smith, 1971, 78–9, believes that the reason Galen refers almost always to ungulate anatomy in *De usu partium* is that he was not so familiar with ape anatomy as he was later when it came to composing *De anatomicis administrationibus.* There are drawbacks to such a chronological analysis; because Galen relied on ungulates in one text, this does not imply that he lacked knowledge or experience with other animals, such as apes. Galen's anatomical experience with various animals surely developed over time (although he refrains from expressing this development on paper), but his anatomical and physiological texts were complete in themselves, geared to specific audiences and addressed clearly defined issues.

[106] Cf. Rocca, 2002b. Although in *De anatomicis administrationibus* for example, Galen clearly states he was dissecting ox brains, the loss of his relevant anatomical and related physiological texts to the West until the sixteenth century was one reason this distinction could not be promulgated. In any case, Galenic animal anatomy was regarded as human anatomy until Vesalius (although there were exceptions, as Scarborough, 1985, 131 n. 44, notes).

[107] Cf. *Ord. Lib. Propr.* p. 84,17–20 Mueller; XIX. 55 K, referring to this text and another, *On dissection* (*De anatomia mortuorum*), both now lost in Greek. The Giuntine *De anatomia vivorum* may, according to Ivan Garofalo, comprise genuine elements (G. Strohmaier, personal communication). Whether these represent part of the lost works, or have been compiled from other extant texts and then inserted is another matter. See also Ormos, 1993.

at least an anthropomorphic sensitivity.[108] Many of Galen's dissections were carried out on primates, specifically, according to him, on five types of "ape".[109] Apart from πίθηκος (the Barbary ape of North Africa, *Simia sylvanus* or *Macacus inuus*), the list includes λύγξ (an unknown tailed ape);[110] σάτυρος (which is not the gibbon but is perhaps the Rhesus monkey, *Macaca mulatta*); κυνοκέφαλος (dog-headed baboon, *Papio hamadryas*);[111] and κῆβος (sometimes employed as a synonym for σάτυρος, but possibly the North-East African *Cercopithecus pyrrhonotus*).[112] The one most commonly used by Galen (and known to Aristotle) was the Barbary ape.[113] Galen made use of the ape in a way not dissimilar to how anatomy is usually taught today—that is, by prosected specimens, which can also be examined over a few days (allowing for varying rates of decomposition depending on the season and the tissues involved). The five types of apes formed for Galen part of a group of six classes of animals which he held were "not far removed from the nature of man".[114] According to Galen,

---

[108] *AA*. IX.11; p. 15 Duckworth. Galen recommends instead an animal such as a pig or goat, since they have louder voces which the experiment demands. On the one hand, Galen could be hiding his anthropomorphic sensibilities in this need for sustained vocalisation; on the other, he might be reflecting the lack of regular availability of apes for vivisection. Galen seems to have leant towards the Stoic view that animals were non-rational beings. On this reading, Galen seems to have also maintained that animals suffer less than humans (cf. *AA* II. 631–632 K). On the treatment of animals in antiquity see Passmore, 1975, 198, 206–7. See also Maehle and Tröhler, 1987, 16, although their notion that Galen also avoided apes for vivisection in order to escape a charge of human vivisection is weak.

[109] I employ Galen's generic term for primates, πίθηκος, or "ape", since he did not differentiate between what is today distinguished as ape and monkey. Cf. Hill, 1953, 3–4; idem, 1966, 2–10, 211–212; idem, 1970, 7–9; idem, 1974, 194–6. See also Jennison, 1937, 21; McDermott, 1938, 77–78, 95–100; Simon, 1906, II, xxxii. McDermott's account does not mention the Rhesus monkey by name in connection with Galen. In contrast, Singer, 1956, 240, n. 22, holds that although Galen "preferred the Barbary ape . . . it is probable that he relied chiefly on the Rhesus monkey." Singer assembles some evidence in support of this claim, but no anatomical description of Galen's can be exclusively applied to *Macaca mulatta* (cf. Hill, 1966, 9–10). Savage-Smith, 1971, 79, states that Galen "did not use the Rhesus monkey but rather the then plentiful Barbary ape." See Hartman and Straus (eds), 1933, who discuss the types and distribution of the *Macaca* species.

[110] Cf. Hill, 1974, 195.

[111] Cf. Hill, 1970, 7–9.

[112] For this last see *UP* II, p. 114,16 Helmreich; III. 844 K, and the mention by Aristotle in *HA* 502a17.

[113] *HA* 502a16–b26. Aristotle also cites—but without a complete description—the κῆβος and the κυνοκέφαλος. Cf. Jennison, 1937, 20–21, 127–129; Hill, 1966, 9, 213; idem, 1974, 194–196.

[114] ὡς οὐ πόρρω τἀνθρώπου φύσεως ὄντα, *AA* II. 423 K. Garofalo, 1993, 86, con-

this classification was known to the older anatomists.[115] The six classes comprise (i) apes (a *parody*, μίμημα γελοῖον, of humans),[116] and ape-like animals; (ii) bears; (iii) pigs; (iv) saw-toothed animals; (v) horned two-hoofed ruminants;[117] (vi) hornless, smooth-hoofed animals.[118] Such a classificatory system clearly gave Galen enormous leeway, not only in what he could dissect but in enabling him to claim that the anatomical findings made from such animals could validly be applied to humans. Garofalo adduces three possible reasons for Galen's use of such a variety of animals, citing availability, the relative differences in size, and the importance such a wide range of dissections had for the purpose of teleological argumentation. Teleological grounds might explain a significant part of Galen's motivation, but the ready availability of some animals is surely the crucial factor.[119] For the detailed dissections of the brain Galen made extensive use of the ox (*Bos taurus*), an ungulate, a member of his fifth class. One reason for this choice is given by Galen at the beginning of the discussion of the brain in the Greek part of *De anatomicis administrationibus* Book IX, when he specifically mentions that the brains he is dissecting are ox brains, which, in the large cities at least, were usually for sale.[120] Availability of dissection material is, obviously, crucial for an anatomist. In the case of goats, for example, Toynbee notes that they were "relatively cheap and easily obtainable sacrificial victims."[121] That certain animals such as cattle, goats and pigs were kept in sufficient numbers for such a purpose is an important factor in their ready procurement for other uses.[122] When Galen mentions that a dissector should be prepared to dissect other animals if there is a shortage of apes, it is an acknowledgment that apes were at times unavailable.[123] Apes, especially the Barbary ape, although known in

---

cludes that the six-class classification was devised by Galen: "The Ancients merely spoke of, and alluded to, the six classes, but did not institute the group." See also Vegetti, 1980.

[115] *AA*. XI.2; p. 72 Duckworth.

[116] *UP* II, p. 273,9–10 Helmreich; IV. 126 K. The ape is γελοῖος with respect to the hand (*UP* I, p. 59,6–8 Helmreich; III. 80 K).

[117] τὸ κερασθόρον καὶ δίχηλον καὶ μηρυκάζον. *AA* II. 430 K.

[118] *AA* II. 431 K.

[119] 1993, 85–86.

[120] ἕτοιμοι δὲ τοὐπίπαν ἐν ταῖς μεγάλαις πόλεσιν ἐγκέφαλοι βόειοι πιπράσκονται τῶν πλείστων τοῦ κρανίου μερῶν γυμνοί. *AA* II. 708 K.

[121] 1973, 166. Cf. 55–60 (apes); cattle (148–62).

[122] See Keller, 1909, 336–7; 402–3 and fig. 140. Cf. Ryberg, 1955, Chapter viii; *OCD* 1456, *suovetaurilia*.

[123] *AA* II. 227 K. Cf. Garofalo, 1993, 85.

the Roman Empire, may have been hard to come by.[124] Galen's own instructions to drown and not strangle the ape in order to avoid injury to the neck might also imply that the ape is a specimen that one does not readily encounter on a daily basis and thus all due care should be taken to see that no part of it is unnecessarily damaged.[125] In such a friable and quickly decomposing material as the brain, the ready procurement of dissection specimens is above all else crucial, and once more indicates why ox brains were Galen's preferred subject matter.

The other critical factor is that of size. Of all the animals Galen dissected on a regular basis, the brain of the ox was by far the largest. The question of *comparative* size is a critical factor in Galen's approach to dissections in general: some points of anatomy *are* better seen in larger animals than in smaller ones.[126] Galen dissected many types of animals, ranging from elephants (albeit an incomplete dissection) to insects.[127] Conversely, useful generalisations on certain anatomical points drawn from larger animals could be employed to verify what cannot readily be observed in smaller, related ones.[128] As Galen remarks: "A profound knowledge of what lies within their body cavity is impossible to attain since one cannot dissect them because of their small size."[129] The larger size of the ox brain relative to that of the other animals Galen regularly dissected, combined with its ease of availability, made it his uncontested choice. Of the apes Galen used, it is significant that their brains, on average, weighed half

---

[124] Pliny gives evidence for a *cynocephalus* and a *sphingion* form of ape, as well as a reference to varieties of ape in general (cf. 8.80, 215–6; p. 99 Ernout). But there is nothing here to indicate a breeding of domesticated monkeys, as Jennison, 1937, 128, claims.

[125] *AA* II. 423 K. For the neck dissection of an ape, see *Nerv. Diss.* II. 845 K; *AA* XI.1; pp. 67–72 Duckworth.

[126] In studying the branch patterns of the abdominal aorta, Galen reflects that it is in larger animals that some parts of the body are better seen. *AA* XIII.9; p. 171 Duckworth. Cf. Scarborough, 1985, 124.

[127] Galen states that he dissected not just those animals belonging to the six classes, "but also animals of the kind which crawl, those which move forwards by bringing the abdomen to their aid, water animals, and those which fly. And if I complete this work that I have started, as is my intention here, I want to dissect those animals also and to describe what there is to see in them." *AA* XI.12; p. 108 Duckworth. This project would have secured Galen's reputation as the foremost anatomist of Antiquity. Whether this would have been accompanied by a revision of Aristotle's classificatory system is unknowable. Cf. Garofalo, 1993, 85 n. 68.

[128] *AA* XV.2, pp. 227–8 Duckworth. Cf. Cole, 1944, 47; Scarborough, 1985, 124 n.6.

[129] *AA* XV.2, p. 228 Duckworth.

as much as that of the bovine.[130] This is not to say that Galen entirely ignored the brain and adnexa of the ape. How then, did the ape serve Galen's investigations of the brain? The way in which Galen utilised this resource is instructive. Galen employed the brain and skull of the ape to demonstrate a number of discrete structures and to impart a specific set of practical instructions. In other words, Galen's use of the ape brain is deliberately restrictive.[131] It is his intention to delineate:

> . . . the method of dissecting the parts of the brain while it remains in its place in the animal body. The dissection is best made in apes, and among the apes in such a one as has a face rounded to the greatest extent possible amongst apes. For the apes with rounded faces are most like human beings.[132]

Behind this statement lies the fact that Galen bases his understanding of the anatomy of the human brain on the ox, not the ape. Yet the *intention* behind Galen's claim is that the ape brain serves as a learning template for the student, enabling him to gain experience before embarking on the more detailed anatomical investigations of the (bovine) brain. This approach may also have the effect of partially lessening the tension between the use of the "human-like" ape brain (the subject of superficial dissections) and that of the ox (the focus of the more comprehensive anatomical research programme). But before proceeding in this manner with the brain of an ape, Galen enjoins the student to gain experience by performing dissections on the relevant areas of *other* animals:

---

[130] The average weight of the brain of *Homo sapiens sapiens* is 1350g. The following may be added: Lemur (*Lemur mongoz*), 20.0g; Spider monkey (*Ateles geoffroyi*, 61.5g; Gibbon (*Hylobates* sp.), 93.0g; Mandrill (*Mandrillus sphinx*), 178.0g; Chimpanzee (*Pan troglodytes*), 364.0g; Domestic cattle (*Bos taurus*), 470.0g; Domestic sheep (*Ovis aries*), 117.0g; Domestic hog (*Sus scrofa*), 98.0g. See Igarashi and Kamiya, 1972. Although not all the species of primates Galen dissected are cited in this atlas, the lemur and mandrill have approximately similar brain weights to that of the Barbary ape. The brain of the ox is larger than that of any ape Galen utilised (only the brain of the young adult male of the Lowland Gorilla (*G. gorilla gorilla*), at 440.0g, is comparable to the ox brain. There is no evidence that this species of ape was ever brought to Rome (no primate is truly indigenous to the Mediterranean).

[131] In Galen's use of the elephant, for example, it is not the *entire* animal Galen dissects; rather he examines its heart for two particular points of information: its similarity to the heart of other animals, and to solve the question of the presence or absence of the so-called heart bone. *AA* II. 616–622 K. Cf. Scarborough, 1985, 125.

[132] IX.10; p. 10 Duckworth. The last sentence is an extraordinary claim: a superficial external resemblance is thought by Galen to indicate a supposed internal similarity. The above description fits πίθηκος, the Barbary ape, and σάτυρος (perhaps the Rhesus monkey). Cf. *UP* II, pp. 114,26–115,3 Helmreich; III. 844 K.

> Should you have become practised previously in this mode of procedure on the carcase of a dead animal, then it will not prove difficult for you to carry it out well and correctly on a living animal, all around the greater part of the bones of the skull, without tearing away the dura mater along with the bone of the skull.[133]

The reference to using other animal specimens (here not named by Galen), underscores the anatomical value of the ape, and implies that such specimens are not as readily available as one might suppose. That Galen stresses the need to find as close an approximation as possible to man is also obvious: Galen reminds his audience that the information obtained from such a source is directly applicable to that of man. This is crucial for an understanding of human osteology, where Galen shows some familiarity with the human skull. In discussing skull foramina, Galen remarks that:

> All these foramina you will see with your own eyes in a cadaver in which all that overlies the bones is decayed and the bones alone remain, in their connections with one another, without separating from each other. These can be seen in such human cadavers as you happen to look at . . . and also in the bodies of apes when we have buried them for four months and more in earth that is not dry.[134]

The phrase, "in such human cadavers as you happen to look at" should not be taken as evidence that Galen had recourse to a human skull for daily study when composing his anatomical works (it does not rule it out either). As noted in chapter 1, an important part of Galen's own anatomical study in Alexandria involved the human skeleton, the only place where complete specimens were available for study. The situation in Rome was different. It is not unlikely that Galen, at the time of setting down his observations of the brain, made use of notes from his Alexandrian studies on the human skull. Of more immediate interest in the above citation is the mention of the ape skull. It is presented as an important source of osteological information and the careful preparation of an ape body by interment for four months underscores the value of such material.

In the Arabic part of Book IX of *De anatomicis administrationibus*, Galen proceeds with an exposition of the cranial bones in the ape, stressing their names, position, and sutures. The purpose of this exercise is to emphasise the importance of the sutures as a landmark in

---

133 *AA* IX.10; p. 13 Duckworth.
134 *AA* XIV.1, p. 182 Duckworth.

the procedure of *trepanation* (ἀνάτρησις). Galen highlights this in the following way. After the entire skull of a living ape has been exposed,[135] Galen instructs the dissector to:

> Go on until you come to the sagittal suture and to the two limbs of the suture which resembles the letter Λ of the Greek script. For in this place the enveloping cranial membrane and the dura mater enter into close partnership. Their combination and partnership are clearly recognisable at the meeting place of any two bones, and between them a suture is interposed. In these places only should you leave the skull unstripped and unbared. Next cut away the whole of the bone between the two sutures. That is the bone called parietal, one on each side. Four lines limit it and mark it out, two of these travelling in the longitudinal diameter of the head, and two in the transverse diameter. As for the two lines running in the longitudinal direction, these are the so-called median suture [*sagittal*] and the scale-like [*squamous*] suture. The transverse ones are the lambdoid suture and the coronal. If you cut away these bones, as you are accustomed to do when you pierce through the skull, either with the perforator or with the instrument called the lens, then you see, when you take stock of and apply your intelligence to what is visible, how in the whole of that region round about, the dura mater is pressed down upon the brain, and you see how in the region of the suture running straight in the longitudinal direction it [*the dura mater*] attaches itself to the skull. And if, in addition, the animal is already very aged, then you see also that the part of the dura mater which I said may be pressed down upon the brain has also fallen away markedly from that part of it which is fused with the median suture. Similarly when you cut away the whole of that part of the skull behind the suture which resembles the letter Λ of the Greek script, and after you have preserved in this region the attachment of the dura mater to the skull, you see that all the remaining subdivisions of the dura mater have fallen away from the part at [attached to] the suture.[136]

Galen's account of the meningeal membranes will be examined in the following chapter. Here, the relationship between the cranial bones and the outer meningeal layer (*dura mater*) of the brain is linked to the arrangement of the cranial sutures. These serve as landmarks in trepanation in order to avoid contact with the dura. It was known that if the integrity of this membrane was compromised by skull fracture or during trepanation, the results were often fatal.[137] It is

---

[135] That it is a living animal follows from Galen's statement in IX.10; p. 13 Duckworth, n. 133, above.

[136] IX.10; p. 11 Duckworth. Square brackets Duckworth.

[137] See the discussion in chapter 5.3.

clear to Galen that procedures such as trepanation must be rehearsed many times until proficiency is gained.[138] Moreover, to perform this procedure on a living animal enables the student to learn the importance of blood loss and to minimise trauma if possible. For Galen, the best way to gain such expertise is by utilising the skull bones and the dural relationships of an animal whose skull most resembles man. Thus he chooses "apes with rounded faces". They provide the best means for the most accurate application of a practical but potentially fatal technique. The skull and exposed brain of living apes therefore serve a different set of epistemological requirements compared to that of the bovine brain. They provide exact knowledge of vital anatomical landmarks for trepanation; secondly, they give the aspiring practitioner valuable preparation for later understanding the more detailed anatomy of the brain, since the key to such knowledge rests on experience, on learning the relationships between the skull, the membranous coverings of the brain and its underlying surface.[139] The skull of the ape is for Galen a teaching tool. Its preparation must be meticulous, otherwise errors of interpretation may arise.[140] Where there was neither a tradition of (nor, perhaps, a need for) anatomical illustration, the brain and especially the skull of an ape served both documentary and pedagogic purposes.[141] They comprise a set of source materials to which the student may repeatedly refer during the more detailed investigations of the brain and its nerves.

[138] Galen also instructs his students in a thought experiment: "construct for yourself in imagination this osteoclastic instrument which is commonly used and which everyone knows." In this way, one may with the mind's eye employ a bone chisel to, "open up a place for inspection, through which you can make an entry and so inform yourself accurately on the gap between the dura mater and the brain." *AA* IX.10; pp. 12–13 Duckworth.

[139] According to Galen, to understand the topographic relationship of the fourth ventricle of the brain to the upper vertebrae in the neck, an examination of the relevant bones of an ape is ideal. *AA* IX.10; p. 14 Duckworth. In tracing the course of the optic nerve through the skull, it is best to use the skull of an adult ape. *AA* X.1; p. 28 Duckworth. Cf. X.6; p. 54 (mandible); XIV.1; p. 182 (cranial nerve foramina and preparation of an ape skeleton); XIV.4, pp. 196–197 (cranial nerve foramina and the importance of animal skulls in general as learning guides).

[140] *AA* XIV.1; pp. 182–3 Duckworth. Cf. Garofalo, 1991, III, 1038–1039.

[141] There is no evidence that Galen used pictorial representations of the brain in his demonstrations either to students or colleagues. He does, it is true, make use of a diagram in his description of the insertion and origin of the deltoid muscle (*AA* II. 273–4 K) and the cervical part of the trapezius muscle (*AA* II. 445–6 K). These are geometric idealisations which are meant to aid, not substitute for dissection. Similarly, the depiction of visual rays is given in strict geometric terms (*UP* II, pp. 94,22–103,19 Helmreich; III. 815–828 K). Cf. Herrlinger, 1970, 9–24.

2.5 *Conclusion*

The highly competitive medical and philosophical world of the second century AD involved Galen in public demonstrations of his anatomical repertoire. The success of these exhibitions partly depended on a thorough knowledge of the debating skills and philosophical knowledge of his rivals. Galen finds it necessary to refute these opponents not only with the empirically based method of anatomical dissection, where he has arguably few rivals, but in asserting that his anatomical axiomata are incontrovertible (in the geometric sense) and therefore unquestionable. Galen seeks the authority of Aristotelian logic to reinforce his case for encephalocentrism, claiming for it the mandate of an exact science. In this way he may well have hoped to put an end to the disagreements over the role of medical science in the location of the *hegemonikon* of the rational soul.[142] As the following chapters will show, Galen provides coherent and convincing arguments for an encephalocentric hypothesis based on empirical research. Yet far from silencing the debate between encephalocentrist and cardiocentrist, Galen's inappropriate application of the concept of proof *more geometrico* simply maintains this polemic, deflects attention from the powerful epistemology of anatomical investigation, and thereby renders a disservice (albeit unintentional on Galen's part) to that Alexandrian anatomical tradition represented by Herophilus and Erasistratus. Galen's handling of the physiology of the brain illustrates both the adaptability of pneuma as a continuum theory as well as the limitations of that theory to supply a complete account of physiological action. As far as a single explanation for the actions of psychic pneuma is concerned, Galen concludes that he has no ready answer to hand.[143]

Galen makes more effective use of his anatomical materials. His extensive use of apes for dissections of the rest of the body and the selective use of ape brains for dissections *in situ*, does not extend to a detailed examination of the deeper parts of the ape brain. The head of the ape that is *most like man* is unrivalled as a learning template. The internal anatomy of the ape brain, is, however, another matter.[144] In any case, Galen does not need to rely on the brain of

[142] Cf. Lloyd, 1996b, 274 (but cf. Hankinson, 1991c, 86).

[143] ἐγὼ μὲν οὐκ ἔχω προχείρως ἀποφήνασθαι. *PHP* p. 448,11 De Lacy; V. 611 K. Cf. Rocca, 1998b, 1998c.

[144] The lack of attention to the base of the ape skull may further be inferred

an ape for detailed dissections of the structure and substance of the brain. He has a larger, more readily available subject for his studies in the ungulate, and it is this class of animal which provides the model for his construction of the anatomy of the human brain.

---

because Galen never provides an explicit citing of the lack of the retiform plexus in any dissection involving the brain of an ape. See chapter 6.2.

PART TWO

THE ARCHITECTURE OF THE BRAIN

CHAPTER THREE

THE EXPOSURE AND DISSECTION OF THE BRAIN

. . . for here is the place where I must describe to you the method of dissecting the parts of the brain while it remains in its place in the animal body.

De anatomicis administrationibus.[1]

3.1 *Introduction*

Galen's regards the brain as the sole organ of the *hegemonikon* of the rational soul and therefore responsible for perception and voluntary motion. The brain is granted this status by Galen because it has unique properties and abilities which are assessed by dissection and experimentation. Yet these investigations are also capable of interpretation from a cardiocentric perspective. To a cardiocentrist, the brain is not necessarily unimportant; it may fulfil a peripheral role as supporting organ for the nerves which ultimately obtain the source of their power the heart.[2] The impressively executed and anatomically sophisticated experiments Galen performs on the spinal cord and the peripheral nerves are open to just such an interpretation. If Galen is to make his case for what might be called *complete encephalocentrism*, he must not simply stress the importance of the brain *qua* organ: he must also show that the brain possesses distinct characteristics that uniquely qualify it, and not the heart, for hegemonic status. This is only possible by performing experiments upon those parts of the brain where, according to Galen, the hegemonic qualities reside. Those parts are the ventricles. But to study them in isolation gives only an incomplete picture. A proper understanding of the ventricles begins with a detailed examination of the anatomy of the surrounding brain. The purpose of this chapter is to provide a

[1] IX.10; p. 10 Duckworth.

[2] This is the essence of Chrysippus' argument in his *De anima*, from which Galen quotes verbatim. *PHP* p. 140,24–32 De Lacy; V. 254–255 K. Cf. Tieleman, 1996a, 51–52.

general orientation of the brain as understood by Galen, noting in particular the terminology he employed and developed to identify some of its key components. This will be followed by an examination of Galen's handling of the meningeal coverings of the brain, an analysis of the venous sinuses, and will conclude with an investigation of the deeper structures which support the ventricles.

3.2 *The parts of the brain: orientation and nomenclature*

As chapter 2 has noted, the bovine brain is the basis of Galenic human brain anatomy. The external characteristics, which it has in common with other higher vertebrates, are as follows. The brain consists of two cerebral hemispheres (*cerebrum*) and the hind-brain or *cerebellum*, which is below the cerebrum. The cerebellum overhangs the part of the brain stem known as the medulla oblongata, whence the *spinal cord* (νωτιαῖος μυελός) begins. To Galen, the term *encephalos* (ἐγκέφαλος) denotes the cerebral hemispheres, although Galen also uses the expression for the entire contents of the cranial cavity as well as to distinguish the cerebrum from the *cerebellum* (παρεγκεφαλίς).[3] For the sake of convention, ἐγκέφαλος will be transliterated throughout in its neuter form, *encephalon*. Galen states that the followers of Herophilus referred to the cerebral hemispheres as ἐγκέφαλος, and the cerebellum as παρεγκεφαλίς (Galen's usual term for it).[4] For Galen none of these terms are an immutable set of definitions.[5] Galen's relative fluidity in encephalic terminology is understandable, since Greek anatomical terminology was not codified under anything approximating an agreed set of terminological rules and definitions.[6] For example, Galen sometimes refers to the cerebellum as *epenkranis* (ἐπεγκρανίς), noting that it is inconsequential whether this part of the

---

[3] The word ἐγκέφαλος dates from Homer (cf. Cunliffe, 1924, 104; Irigoin, 1980, 249–50; Scarborough, 1992, 146). "Brain" is from the Old English *brægen* (via Middle Dutch, *brein*), which probably derived from βρέγμα, the front of the head (cf. Onians, 1951, 193 n. 3; 23–38).

[4] *UP* I, pp. 482,22–483,4 Helmreich; III. 665 K. On the use of this term, see Garofalo, 1991, II, 801 n. 11.

[5] This notwithstanding Galen's strictures on the ambiguity of language in logical relations (cf. Edlow, 1977). The crucial factor, so far as an anatomical term is concerned, is that its use is consistent (cf. Hankinson, 1994d, 171ff.).

[6] Cf. Simon, 1906, II, vii; Ulrich, 1919, 20–22; Lloyd, 1983b, 149–167.

brain is called *enkranion* or *parenkephalis*.[7] Galen remarks that Erasistratus also used *epenkranis* for cerebellum, and while Galen criticises Erasistratus' physiology and the application of his anatomical knowledge, he nowhere faults Erasistratus for his lack of understanding of the anatomy of the brain.[8] To emphasise that the *nature* (οὐσία) of an organ does not derive from its position in the body,[9] Galen also states that the brain may be referred to as *cerebrum* (κέρεβρον), which is what the Romans call it.[10] Equally, one could also give it a meaningless appellation, say σκινδαψός.[11] Galen states that "no organ has a particular nature because of its position, even though it is named from it."[12] Nevertheless, a description of an anatomical structure sometimes assists in indicating its possible function.[13] Although Galen's arguments for the brain as *hegemonikon* rest on dissection, it is obvious that he also requires a set of generally agreed terms for the parts of the brain. As this chapter will illustrate, on several occasions Galen attempts to create a more precise anatomical terminology. In other instances, Galen clarifies the meaning of some terms already in use.

In *De usu partium*, Galen summarises what for him are the chief features of the brain:

> The substance (of the brain) is similar to the nerves, of which it was meant to be the source, except that it is softer, and this was fitting for

---

[7] ὀπίσθιον δὲ λέγειν ἐγκέφαλον, ἢ ἐγκράνιον, ἢ παρεγκεφαλίδα, διαφέρει οὐδέν. *AA* II. 714 K. In the Arabic continuation of this text, at IX.12; p. 18 Duckworth, the cerebellum is described as the "hind-brain or occipital brain." Cf. *UP* I, p. 483,5–9 Helmreich; III. 665–666 K.

[8] *UP* I, p. 488,14–24 Helmreich; III. 673 K; cf. *PHP* pp. 440,20–442,18 De Lacy; V. 602–604 K, cited in chapter 1, pp. 39–40.

[9] οὐσία is a term with several meanings. In the case above it is employed by Galen in a more restricted sense to mean the essential characteristic of an organ. Of course, it also means "substance" in general; for example, the material "air-like substance" (ἀεροειδής οὐσία), found in the gut, veins and flesh (*PHP* p. 494,14 De Lacy; V. 666 K).

[10] *UP* I, pp. 455,21–456,2 Helmreich; III. 629 K. Cf. Cicero *Tusc.* I.19. On the development of Latin medical terminology, see Langslow, 1991; idem, 2000.

[11] *UP* I, p. 456,2–8 Helmreich; III. 629 K. May, 1968, 395 n. 35, translates this as "thing-a-ma-jig."

[12] οὐ γὰρ ἐκ τῆς θέσεως ἕκαστον τῶν ὀργάνων τόδε τι τὴν οὐσίαν ἐστι, κἂν ἀπὸ τῆς θέσεως ὀνομάζηται. *UP* I, p. 455,12–14 Helmreich; III. 628 K.

[13] The *corpus callosum* and *fornix* are two examples. See 3.8, below. The meticulous description of the *retiform plexus* is a more problematic example of Galen virtually arguing function from a structural point of view. As will be discussed in chapter 6.3, the descriptive imagery employed for the retiform plexus becomes in Galen's hands a powerful argument for the validation of its physiological status.

a part that was to receive all sensations, produce images[14] and understand thoughts. For a substance readily altered is most suitable for such actions and affections,[15] and a soft substance is always more easily changed than one which is harder. This is the reason why the encephalon is softer than the nerves, but since there must be two sets of nerves, . . . the brain itself was also given a double nature, that is, the anterior part is softer than the rest, the latter being known by anatomists as *enkranion* as well as *parenkephalis*.[16] These two parts are separated by an enfolding of the hard meninx [*tentorium cerebelli*] and are connected only by the passage and its surrounding bodies which lie beneath the crown of the head.[17] And the anterior part of the encephalon had to be softer because it was intended to be the source of the soft nerves which lead to the instruments of sensation, and the posterior part had to be harder, since it is the source of the hard nerves which are sent to the entire body. Since the close association of a soft body with a hard one would be not harmonious, Nature sets each apart, placing between them the hard membrane that will also surround the whole encephalon which is made up of these two parts [*cerebrum* and *cerebellum*]. Moreover, the parts of the anterior encephalon which is next to the covering known as the hard and thick meninx were with good reason made harder, and the parts in the middle of the encephalon, enclosed by those parts above, were made softer; for the outer parts had to be constructed in order to resist injury and give off the harder nerves, whereas the middle part of the encephalon, protected from injury by position, became a proper source for the soft nerves.[18]

---

[14] Galen is using this term in a specialised sense to mean a mental image or impression, which picks up the use of νοήσεις (as part of διανοητικαὶ φαντασίαι).

[15] This is very likely an allusion to the discussion in *Theaetetus* (191c–196c), where Plato likens memory to a form of imprinting. It may be Galen's intention to compare the soft nature of the brain to a wax tablet on which an impression may easily be left (cf. *Tht.* 191d; cf. 194e–195a). Galen does not discuss cognition to any appreciable extent, except to state, as here, that it takes place in the brain. See chapter 5.6, and Appendix One.

[16] Galen now has the three terms for the cerebellum: ἐγκράνιον, ἐπεγκρανίς, and παρεγκεφαλίς.

[17] Galen's description of this passage as lying beneath the crown of the head is topographically accurate. The reference to a passage is that of the communication between the middle ventricle and the posterior ventricle, which is today known by anatomists as the *Aqueduct of Sylvius*. See chapter 4.6.

[18] οὗτος οὖν τὴν μὲν οὐσίαν ὁμοιότατός ἐστι τοῖς νεύροις, ὧν ἔμελλεν ἀρχὴ γενήσεσθαι, πλὴν ὅσῳ μαλακώτερος αὐτῶν ὑπάρχει. καὶ γὰρ καὶ τοῦτ᾽ ἔπρεπε τῷ πάσας μὲν εἰς αὐτὸν τὰς αἰσθήσεις ἐκδεξομένῳ, πάσας δὲ φαντασίας φαντασιωθησομένῳ καὶ πάσας νοήσεις νοήσοντι. τὸ γὰρ εὐαλλοίωτον ἐν ἅπασι τοῖς τοιούτοις ἔργοις τε καὶ παθήμασιν ἐπιτηδειότατον, εὐαλλοιωτότερόν τε ἀεὶ τὸ μαλακώτερον τοῦ σκληροτέρου. διὰ ταῦτα μὲν δὴ μαλακώτερος τῶν νεύρων ὁ ἐγκέφαλός ἐστι. τῷ δ᾽ εἶναι χρῆναι διττὴν ἐκείνων τὴν φύσιν, . . . διττὸς καὶ αὐτὸς ἐγένετο, μαλακώτερος μὲν ὁ πρόσθεν, σκληρότερος δ᾽ ὁ λοιπός, ὃν ἐγκράνιόν τε καὶ παρεγκεφαλίδα καλοῦσιν οἱ ἀνατομικοί. καὶ γὰρ οὖν καὶ διείργεσθον ἐνδιπλουμένῃ τῇ σκληρᾷ μήνιγγι καὶ κατὰ μόνον τὸν ὑπὸ τῇ κορυφῇ τῆς

Apart from differentiating between cerebrum and cerebellum, this citation reminds Galen's audience of the importance of the brain's substance as the origin of the nerves. The further division of nerves into hard and soft reflects their origin from specific parts of the brain. As Galen formulates it: "Always, the further back one goes, the harder the brain becomes; moreover, the parts at its base are harder than the rest."[19] This is one way to account for the different texture and consistency of cerebrum and cerebellum. Galen makes a further, teleological interpretation of the softer outer part of the brain (cortex) and the harder inner part (medulla). It is appropriate that the hard dural covering of the brain be apposed by a firm part of the brain substance, and Galen prefaces this by saying the apposition of a soft and a hard body is contrary to nature. This also allows Galen to explain the relative hardness of some of the cranial nerves, which take their origin from the base of the brain. At times it is a relatively subtle distinction. The optic nerves, whilst thicker than the brain, nonetheless do not *appear* to be that much harder.[20] Galen thereby stresses

---

κεφαλῆς κείμενον πόρον καὶ τὰ τοῦτον περιέχοντα σώματα συνάπτεσθον. ἐπειδὴ γὰρ ἐχρῆν μαλακώτερον εἶναι τὸν πρόσθιον, ἀρχὴν ἔσεσθαι μέλλοντα τῶν ἐπὶ τὰς αἰσθήσεις ἰόντων νεύρων τῶν μαλακῶν, σκληρότερον δὲ τὸν ὀπίσθιον, ἀρχὴν καὶ τοῦτον ἐσόμενον τῶν εἰς ὅλον τὸ σῶμα διασπαρησομένων νεύρων τῶν σκληρῶν, καὶ ἦν οὐ πάνυ τι σύμφωνος ἡ τοῦ μαλακοῦ πρὸς τὸ σκληρὸν ὁμιλία, διὰ τοῦθ' ἑκάτερον ἀφώρισεν ἰδίᾳ καὶ μέσην αὐτῶν ἔταξε τὴν σκληρὰν μήνιγγα, τὴν καὶ τὸν ὅλον ἐγκέφαλον τὸν ἐκ τῶν εἰρημένων δυοῖν μορίων συγκείμενον περιέξουσαν. ἀλλὰ καὶ αὐτοῦ τοῦ πρόσθεν ἐγκεφάλου τὰ μὲν ἁπτόμενα τοῦ σκεπάσματος, ὃ δὴ σκληρὰν καὶ παχεῖαν ὀνομάζουσι μήνιγγα, σκληρότερα καὶ αὐτὰ κατὰ λόγον ἐγένετο, τὰ δ' ὑπὸ τούτων περιεχόμενα τὰ μέσα μαλακώτερα. τὰ μὲν γὰρ ἔξωθεν εἴς τε δυσπάθειαν ἐχρῆν παρεσκευάσθαι καὶ εἰς τὴν τῶν σκληροτέρων νεύρων ἀπόφυσιν, τὰ δ' ἐν τῷ μέσῳ τὴν μὲν δυσπάθειαν ἐξ αὐτῆς εἶχε τῆς θέσεως, ἀρχὴ δ' ἐπιτήδειος ἐγένετο τοῖς νεύροις τοῖς μαλακοῖς. *UP* I, pp. 461,4–462,17 Helmreich; III. 636–638 K.

[19] ἀεὶ καὶ μᾶλλον αὐτὸν ἑαυτοῦ σκληρότερον εὑρήσομεν τὸν ἐγκέφαλον, ἀλλὰ καὶ τὰ κατὰ τὴν βάσιν αὐτοῦ μέρη τῶν ἄλλων ἐστὶ σκληρότερα. *UP* II, p. 29,18–21 Hemreich; III. 723 K. The hardest nerves are those which come off the spinal medulla (*UP* II, p. 30,24–26; III. 724 K). Galen states that it is not possible for soft nerves to grow from the cerebellum (*UP* I, p. 462,17–18; III. 638 K).

[20] *UP* I, p. 463,4–6 Helmreich; III. 639 K. The nerves to the ears (vestibulocochlear) are similarly qualified, being "not that soft but yet not hard." (ἐπὶ τὰ ὦτα μαλακὸν μὲν οὐκέθ' ὁμοίως; οὐ μὴν ἤδη γέ πω σκληρόν. *UP* I, p. 464,5–6; III. 640 K). The fifth nerve (cranial nerves VII and VIII) is "precisely strong and hard" (ἰσχυρὸν ἀκριβῶς καὶ σκληρόν. *UP* I, p. 464,7–8; III. 640 K). In the fresh bovine brain, it is difficult to discern the exact consistency of the cranial nerves by visual examination alone, but their nature may be demonstrated by palpation; they are of firmer consistency than the brain (excluding the dural covering). In the fresh bovine brain, once the dural coverings have been removed, the cerebellum is found on palpation to be firmer in consistency than the cerebrum. On the problems of brain dissection in general see Meyer, 1954, 143–46.

the importance of the substance of the brain, especially its gradations into hard and soft parts, which serve respectively as the source of the hard and soft nerves. To understand the importance and distinctive role of the brain for Galen, its covering membranes and the vasculature they support must be examined. This will also highlight some of the techniques Galen employs in anatomical demonstration.[21]

3.3 *The cranial cavity and the anatomical exposure of the brain*

The instructions Galen gives for exposing the brain are framed to reveal the meninges in as undamaged a state as possible. For Galen, the meninges act as points of reference for important vascular structures, and in the case of the dura mater, comprise part of this vascular arrangement. Galen's comprehensive instructions for exposing the brain in *De anatomicis administrationibus* are prefaced with the following caveat:

> ... the first anatomy performed on the brain to be discussed will occur when the bones of the cranium have been removed, preserving intact the meninx which surrounds the brain.[22]

The exposure of the bovine brain for study is both complex and time consuming. As shall be seen below, Galen employs a special method and a specialised set of tools for this purpose.[23] To consider how Galen exposes the brain for observation, it is necessary first to examine the bones which Galen states make up the cranial cavity. Here and in the following accounts, drawn principally from the Greek and Arabic parts of Book IX of *De anatomicis administrationibus*, and Books VIII and IX of *De usu partium*, most of Galen's osteological descriptions are based on the primate and human skull.[24] According

---

[21] The best general survey of Galen's anatomical skills is Ullrich, 1919. In what follows in this chapter and for the rest of this book, my dissections are based as far as possible on Galen's descriptions and instructions. Any errors of interpretation or fact are mine alone.

[22] προτέραν ἐγχείρησιν ἀνατομῆς ἐγκεφάλου διηγήσομαι τὴν ἐξῃρημένων τῶν ὀστῶν τοῦ κρανίου γιγνομένην, σωζομένης ἐν κύκλῳ τῆς μήνιγγος· II. 708 K.

[23] In the sheep for example, as the instructions for one such dissection are prefaced, the skull is noted to be "very thick and strongly buttressed about the brain. It is therefore necessary rather to take the skull from the brain than the brain from the skull, and a special method and special tools become necessary." Wilkie, 1937, 1. Cf. Chauveau, 1891, 765–6.

[24] Although a detailed description of the bones of the human or primate skull is

to Galen, six bones make up the skull: the paired parietal and temporal bones, and the unpaired frontal and occipital bones.[25] The parietal bones meet at the sagittal suture (Galen's median suture), which runs from the *bregma* (the apex of the skull), to the *lambda*, the beginning of the lambdoid suture, where the occipital and parietal bones meet. The bregma was long considered an important landmark for incisions or deeper interventions to the brain, and was first formally acknowledged in the Hippocratic text *On wounds to the head* (*VC*).[26] Galen refers to the two parietal bones as the "bones of the bregma" (ὀστᾶ τοῦ βρέγματος), and these are described as *porous* (χαῦνος), a reference to the fact that, as will shortly be noted, the bones of the skull have been constructed to allow waste products from the brain to diffuse through.[27] Galen's description of the cranial bones is accurate as far as primate or human osteology is concerned. Galen uses human skull suture lines as reference points in his descriptions of the falx, tentorium, and venous sinuses of the bovine brain. For obvious reasons, there is no evidence that he used bovine skull sutures in a similar way.[28]

In what immediately follows from *AA* II. 708 K above, Galen lists the various names for the outer and inner meningeal layers of the brain (*dura mater* and *pia mater* respectively). Galen also reminds his students that it is of no great importance whether the outer layer is called *thick* (παχύς)—Galen's description here—or *hard* (σκληρός) or *skin-like* (δερματώδης).[29] Nor does it matter whether the inner meninx

---

given in *De ossibus*, this treatise does not, in my view, deal exclusively with human or primate osteology. The first three chapters furnish an accurate description of human or primate cranial osteology, but the fourth chapter makes better osteological sense when viewed from the perspective of the bovine cranium.

[25] In *De ossibus* II. 744 K, six bones make up the head, excluding the sphenoid: καὶ δὴ καὶ τούτων οὕτως ἐχοντων, ἓξ ὀστᾶ γίγνεται τῆς ὅλης κεφαλῆς τὰ πάντα χωρὶς τοῦ σφηνοειδοῦς· δύο μὲν κατὰ τὸ βρέγμα, κοινὴν ἔχοντα ῥαφὴν τὴν κατὰ τὸ μῆκος εὐθεῖαν· ἄλλα δὲ δύο τούτων κάτωθεν, ἕν καθ' ἑκάτερον οὖς· καὶ πέμπτον ἐπὶ τούτοις, τὸ κατ' ἰνίον· καὶ ἕκτον τὸ κατὰ μέτωπον.

[26] The brain is noted as lying more to the front of the head, and thus is under the bregma, where the bone is also at its thinnest and weakest (p. 64,9–12 Hanson; III. 188 L).

[27] *UP* II, p. 179,23–26 Helmreich; III. 935 K. Here and on the following section in Galen's description, the remaining bones of the skull and upper jaw are enumerated. The bregma is the area where the anterior fontanelle is situated, which may well account for the description which Galen has given here (cf. *Foet. Form.* IV. 673 K). Pliny alludes to the anterior fontanelle in the following: *uni homini in infantia palpitant, nec corroboratur ante primum sermonis exordium. HN* 11.49, 134; p. 71 Ernout.

[28] Cf. Sisson, 1930, 134–6.

[29] Cf. Greenhill, 1870, 550.

is called *thin* (λεπτός), *soft* (μαλακός), or *membrane-like* (ὑμενώδης).[30] The true nature of these structures is revealed only during dissection. As Galen reiterates, "the advantage obtained by anatomy is knowledge of the nature of the parts, not their appellation."[31] Galen continues with his instructions on exposing the brain. The following describes its preparation for dissection:

> In the large cities, ox brains, already prepared and stripped of most of the cranial parts, are usually for sale. If you believe more bone than necessary adheres to them, order its removal by the butcher who sells them.[32] If he is not present, do it yourself, using strong chisels for excision or carpenters' axes, such as you see I keep on hand.[33] These types of instruments are best made of hard iron,[34] for those made of soft iron will become useless after repeated blows. But I would not wish it that the cranium be violently and repeatedly struck, for this vigorously shakes the soft brain, separating and fragmenting it. You must prepare it for observation without any sort of accident . . .[35]

---

[30] *AA* II. 708 K.

[31] ἐν γὰρ τῇ γνώσει τῆς φύσεως τῶν μορίων ἡ ὠφέλεια τοῖς ἀνατέμνουσιν, οὐκ ἐν ταῖς προσηγορίαις. *AA* II. 708 K.

[32] How much bone should be removed is moot. It could not have been all the bone which forms the skull case, as the brain would collapse entirely. It is more likely that Galen means enough should be removed to expose the brain, but that the bones making up the base of the skull must remain intact. In this way, the brain is available in a good state for dissection.

[33] Cf. Gurlt, 1898, I, 428–474, Tables II & III; Bliquez, 1994; Jackson, 1986 and 1990; Künzl, 1982 and 1996; Møller-Christensen, 1938, 112–164. The *chisel* (ἐκκοπεύς) was mostly used as an osteotome or as a bone elevator in a depressed skull fracture (Cf. Milne, 1907, 122–3; Jackson, 1986, 143–4). These chisels were well made and robust, suitable for cutting through tough hide, muscle and bone. For the various types of *knife* (σμίλη), or the more precise μάχαιρα or μαξαίριον (a scalpel in the recognisably modern configuration of a thin, tapering blade, with a single cutting edge) see Milne, 1907, 24–9, and Plate V, figures 1 and 2. Cf. Bliquez, 1994, 113–121; Jackson, 122, fig. 1.

[34] For Galen the best surgical steel was from Noricum, the Roman Alpine province south of the Danube (*AA* II. 682 K). On the working of iron and steel, see Forbes, 1964, 198–213; Milne, 1907, 10–13. See also Healy, 1978, 233; Scarborough, 1969, 86–7. For an illustration of a scalpel from Noricum, see Künzl, 1982, 114. The *probe* (μήλη or κοπάριον), *forceps* (τριχολαβίς, πολυποξύστης, μύδιον, σταφυλάγρα, ἀκανθοβόλος) and trepanation devices (τρύπανον, τρύπανον ἀβάπτιστον, χοινικίς) which have survived impress by their practical design and craftsmanship. Cf. Künzl, 1982, 15–17 (scalpel); 17–18 (forceps); 27–8 (sound).

[35] ἕτοιμοι δὲ τοὐπίπαν ἐν ταῖς μεγάλαις πόλεσιν ἐγκέφαλοι βόειοι πιπράσκονται τῶν πλείστων τοῦ κρανίου μερῶν γυμνοί. καὶ εἰ προσκεῖσθαί σοί ποτε δόξειεν ἐν τοῖς πλαγίοις μέρεσι πλείω τοῦ δέοντος ὀστᾶ, κελεύσεις τῷ πιπράσκοντι μαγείρῳ περιελεῖν αὐτά· μὴ παρόντος δ' ἐκείνου, πράξεις οὕτως αὐτός, ἐκκοπεῦσι τοῖς ἰσχυροῖς χρώμενος, ἢ τοῖς τῶν τεκτόνων σκεπάρνοις, ἅπερ ὁρᾶτε κἀμοὶ προπαρεσκευασμένα. πρὸ πάντων δ' ἐκ σιδήρου σκληροῦ κατεσκευάσθω τὰ τοιαῦτα· τὰ γὰρ ἐξ ἀσθενοῦς ἐπὶ πολλαῖς πληγαῖς ἀνύσει τι σμικρόν. ἡμεῖς δὲ πλήττεσθαι σφοδρῶς πολλάκις οὐ βουλόμεθα τὸ κρανίον·

The fresh bovine brain is a delicate structure, and although the tough dural coat provides protection, once this is torn, damage to the underlying brain is usually inescapable. It is likely that at this stage of the dissection Galen has left the bones of the cranial fossae intact and adherent to the base of the brain in order to best preserve the integrity of the brain for examination.[36] After the brain has been exposed, the thick, hard, outer meninx is visible. Galen next places this meninx in a topographic relationship to two principal sutures of the skull, the median and the lambdoid:

> When the part is properly prepared, you will observe that the hard meninx does appear much thicker indeed in the midline, and dips down a little (sc. into the brain), especially in the region of the median suture of the cranium. In the same way you will observe that under the lambdoid suture the thick meninx is also doubled, and as well dips a little into the brain.[37]

In *De ossibus*, Galen defines a suture as a type of *union* (συνάρθρωσις) of bones between which there is no apparent motion.[38] A suture is therefore not an example of an *epiphysis* (ἐπίφυσις), which is an outgrowth between two bones (usually large like those of the limbs) for the sake of articulation.[39] For Galen, sutures serve three purposes:

ἁπαλὸν γὰρ ὄντα τὸν ἐγκέφαλον αἱ τοιαῦται πληγαὶ σείουσι σφοδρῶς, διαλύουσί τε καὶ διασπῶσι. χρὴ δὲ οὐδὲν τούτων αὐτὸν πεπονθότα παρεσκευάσθαι πρὸς τὴν θέαν. *AA* II. 708–709 K; p. 553 Garofalo, 2000. In attempting to replicate this method in its essentials, the following was performed using intact heads of young adult cows. A circular saw incised a groove in the coronal plane to a depth of several millimetres, in order to introduce a chisel. In this way, the skull cap was able to be removed in one piece, and the dura intact, except in places where it had adhered to the inner table of the skull. Even with the technical innovation of an electric saw, the entire process was tedious and lengthy, and one reflected with some envy on the ox brains available to Galen already stripped of most of their bones.

[36] The dural covering of the cranial fossae is tough and firmly adherent to the underlying bone.

[37] παρασκευασθέντος οὖν ἐπιτηδείως τοῦ μορίου, θεάσῃ τὴν σκληρὰν μήνιγγα κατὰ τὸ μῆκος αὐτοῦ μέσον ἑαυτῆς πολὺ παχυτέραν γε φαινομένην, ἐγκαταβαίνουσάν τε μέχρι τινὸς οὕτω, καθ' ὃ μάλιστα μέρος ἐστὶν ἐν τῷ κρανίῳ τῶν ῥαφῶν ἡ μέση. κατὰ δὲ τὸν αὐτὸν τρόπον ὑπὸ τῇ λαμβδοειδεῖ ῥαφῇ τὴν παχεῖαν μήνιγγα θεάσῃ, διπλουμένην τε ἅμα καὶ μέχρι τινὸς ἐγκαταβαίνουσαν τῷ ἐγκεφάλῳ. *AA* II. 709–710 K.

[38] II. 735 K. Cf. *UP* II, pp. 49,9–54,16 Helmreich; III. 750–759 K, for a detailed discussion of the sutures.

[39] Cf. *Oss.* II. 733 K. These terms are specific instances of a developing anatomical terminology. Singer, 1956, xviii, states: "The Greeks, unlike ourselves, had no classical language from which to draw scientific terms . . . Until modern times, and until the revival of the practice of dissection and the introduction of representational art, the anatomical works of Galen were almost incomprehensible." Judged in this way, the contemporary names of all parts of the brain are equally obtuse.

they are formed for the sake of transpiration of waste products, they allow passage of certain blood vessels, and they generate the *perikranion* (περικράνιον).[40] The skull sutures are not impermeable, for Galen interprets the cranial sutures in such a way that they act not only as strong linkages to the cranial bones,[41] but also serve as passages for the *vapour-like* (ἀτμώδη) or *smoke-like* (καπνώδη) residues or *waste-products* (περιττώματα), which pass up from the brain itself.[42] Like the roof space of a house, the brain needs adequate ventilation, and the cranial sutures provide appropriate sites of evacuation for the vaporous residues from the brain.[43] The sutures have another purpose; they anchor the *bands* (δεσμοί) composed of *thin membranes* (λεπτοὶ ὑμένες) that project from the dura to the sutures of the skull. These bands attach the dura mater to the inner table of the skull at the suture lines. These bands unite and give rise to a continuous membrane, the *perikranion*, which according to Galen attaches the dura to the cranium.[44] The perikranion is a noteworthy feature around the sagittal and lambdoid sutures.[45] Galen observes that the form and number of sutures vary depending on the shape of the skull.[46] But in a typical skull (one "according to nature", κατὰ φύσιν), the description

---

Terms such as *hippocampus, substantia nigra, septum pellucidum, gyrus, sulcus, thalamus, hypothalamus,* refer either to their shape, essential external characteristics or relationship to other parts of the brain, which is all that Galen's anatomical vocabulary sought to achieve.

[40] *UP* II, p. 170,11–21 Helmreich; III. 922 K. The context is a summary of the nature of the *articulations* (συνθέσεις) between the bones.

[41] Galen makes an analogy with the "peg-shaped extrusions" (γόμφοις) which are used in carpentry to firmly bind wooden joints. The sutures are also likened to the teeth of two saws set against each other (*UP* II, p. 5,5–16 Helmreich; III. 689–690 K).

[42] *UP* II, pp. 2,17–3,4 Helmreich; III. 686–687 K. These residues are also described as *vapour-like,* λιγνυώδη (*UP* II, p. 4,2; III. 688 K). Galen further differentiates the vapour-like residues into thin and thick types, with the *thicker* (παχύτερα) being evacuated, according to Galen, only via the sutures; the thinner residues can pass through the cranial bones (*UP* II, p. 6,14–19; III. 691 K). The heavier *water-like* (ὑδατώδη) or *slime-like* (ἰλυώδη) residues are expelled, via the ventricular system, through the perforations of the ethmoid and palatine bones. See also chapter 4.4.

[43] ὑπέρκειται γὰρ ἁπάντων τῶν ἐν τῷ σώματι μελῶν ἡ κεφαλὴ καθάπερ τις ὄροφος οἴκου θερμοῦ· πάντ' οὖν ὅσα λιγνυώδη τε καὶ ἀτμώδη περιττώματα τῶν ὑποκειμένων ἀναθεῖ, δεχομένη ταῦτα δαψιλεστέρας χρῄζει τῆς κενώσεως. *UP* II, p. 4,5–9 Helmreich; III. 688 K.

[44] *UP* I, p. 480,7–17 Helmreich; III. 662 K. Cf. *AA* X.5; p. 51 Duckworth. See also *AA* IX.11; pp. 15–16 Duckworth.

[45] *AA* IX.10; p. 11 Duckworth.

[46] In a conical or *pointed* (φοξός) skull, the posterior prominence is absent, as is the lambdoid suture. If the anterior prominence is missing, so is the coronal suture. Cf. *Oss.* II. 740–1 K.

Galen gives of the cranial sutures is virtually identical to that of the human cranial vault.[47] These sutures are the *lambdoid* (λαβδοειδής), *coronal* (στεφανιαία), the *parietal* or *sagittal* (μῆκος εὐθεῖα),[48] and, running parallel to the sagittal, two others which are described as *squamiform* (λεπιδοειδής).[49] In addition, the interior of the cranial cavity is distinguished by grooves for the venous sinuses and blood vessels. These, together with the sutures, form a series of fixed landmarks for Galen's dissection of the deeper parts of the brain. The sagittal and lambdoid sutures mark respectively the location of the dural folds (*falx* and *tentorium cerebelli*), and Galen, as will now be discussed, uses the attachment of these foldings to the sutures (as well as a dural attachment to the coronal suture), partly to account for the integrity of the ventricles of the brain.

3.4 *The* meninges *(i): the* dura mater *and the dural foldings*

By Galen's time, knowledge of the *dura mater* and the *pia mater* appears to have been detailed.[50] Galen states that according to Marinus, *all* membranes, not just those of the brain, were referred to generically by "the ancients" (οἱ παλαιοί) as *meninges*.[51] Galen describes a "double fold" (δίπλωσις) of the dura mater which separates the cerebrum

---

[47] *Oss.* II. 740 K.

[48] Literally the "straight-length"; a more explicit description for Galen than "the middle" (ἡ μέση).

[49] *Oss.* II. 742 K. These last two are most likely the squamosal and sphenofrontal sutures. Galen does not refer to them as sutures but as "squamous agglutinations" (λεπιδοειδῆ προσκολλήματα). Cf. Hyrtl, 1880, §354. These are said to be near the ear, running longitudinally, and parallel to the midline of the head (*UP* II, pp. 52,27–53,8 Helmreich; III. 755–756 K). May, 1968, I, 460, refers to them as squamosal and sphenoparietal. The latter label is dubious as there is no defined sphenoparietal suture (although a suture line can be noted at the region of the *pterion*, where the frontal, parietal, squamous part of the temporal and greater wing of the sphenoid meet, and where the sphenoidal fontanelle is located). There is a well marked sphenosquamosal suture, but this runs in the *coronal* plane.

[50] The origins of the word "mater" in the context of the meninges reflects the continuing development of anatomical terminology. According to Strohmaier, 1970, 204, the term "mater" is a neologism of Arabic origin, and has its origins in kinship relations. The brain is described as "Mutter der Schädelnähte" (mother of the sutures). Cf. 205–207.

[51] ὠνόμαζον γὰρ οὐ τούτους μόνους τοὺς ὑμένας, ἀλλὰ καὶ πάντας οἱ παλαιοὶ μήνιγγας, ὡς ἔνεστί σοι μαθεῖν ἔκ τ' ἄλλων πολλῶν συγγραμμάτων, ἃ γέγραπται τοῖς ἀνδράσιν ἐκείνοις, ἀτὰρ οὐχ ἥκιστα καὶ τῶν Ἱπποκράτους τε καὶ Διοκλέους, ὧν καὶ Μαρῖνος ἐμνημόνευσεν ἐν τῇ τῶν ἀνατομῶν πραγματείᾳ. *AA* II. 716 K; p. 561 Garofalo, 2000.

from the cerebellum; this is the *tentorium cerebelli*.[52] Another fold of dura (the *falx cerebri*) separates the two cerebral hemispheres. Within the falx run blood vessels that supply the brain.[53] An occipital prolongation of the *falx cerebri* (*falx cerebelli*) divides the cerebellar hemispheres.[54] The descriptions of the dural foldings that make up the falx and tentorium are given by Galen in *De anatomicis administrationibus*:

> Insert a hook into the dura mater and draw it upwards.[55] Then first cut through the piece of it that has been raised, so that it may not make contact with the part of the brain lying beneath it. Next insert two hooks, one on each side of the first hook, from the ends of the margin of the incision; also with these two hooks draw upon the dura mater found above, and cut through the whole raised portion of it; without touching any of the underlying parts of the brain. If you do this well, you can also introduce your fingers upwards beneath the part of the dura mater which you have incised, and you can split it until you have uncovered the whole of the brain lying beneath it. For the meninx . . . is in all its parts separable from it (sc. the brain), as I have shown, except at the sides of the sutural lines which I indicated, where the meninx becomes folded into two layers as described, and sinks downwards penetrating a considerable distance further into the parts of the brain lying beneath it [*falx cerebri and tentorium cerebelli*]. I said earlier that there in that region lies the parietal bone one on either side,[56] and that in the space at the back of the suture resembling the letter Λ of the Greek script is the whole of that portion of the brain found in those parts, which one calls the hind-brain or occipital brain [*cerebellum*]. Now you can cut away these three parts of the meninx [*dura mater*] and thus expose the underlying portions of the brain.[57] Two of these three parts lie in the region beneath the parietal bone, the third is the portion overlying the hind brain.[58]

There is also a fourth dural covering, overlying the cribriform plate of the ethmoid bone, in the anterior cranial fossa. Galen does not spe-

---

[52] *AA* II. 710 K; cf. *UP* I., p. 483,12–15 Helmreich; III. 666 K; T 77a *Herophilus*.

[53] *AA* II. 714 K.

[54] Cf. Frandson, 1965, 77.

[55] Galen here is probably employing one of several types of "sharp hooks" (ἄγκιστρον, ἀγκυρουήλη). Cf. Milne, 1907, 85–7, and Plate XXIV. For "blunt hooks" (τυφλάγκιστρον), see idem, 87–88.

[56] At the bregma (cf. Garofalo, 1991, II, 845), where the sagittal and coronal sutures meet.

[57] Duckworth, 1962, 18 n. 1, states that this should be a reference to the parts of the dural meninx (*falx* and *tentorium*) which divide the brain into two cerebral hemispheres and cerebellum. He points out that these three parts have been interpreted by Simon (1906, II, 16, and n. 61) to mean the three parts of the brain, but only if the cerebral hemispheres are counted separately.

[58] IX.12; pp. 17–18 Duckworth. Square brackets Duckworth; my gloss in round brackets.

cifically list it as a fourth infolding, but describes its nature, for, unlike the other dural foldings, it is variable, being quite thin over the cribriform part of the ethmoid and is "much perforated" (πολύτρητος):

> And when I detached and cut away from the bone the portion of the dura mater found at this spot [i.e. overlying the cribriform plate] and then with my hands stretched it out towards the sunlight, I saw clearly that in certain parts of this membrane the sunlight penetrated to a greater extent, as if there were in them fine perforations, whereas in other parts it shone through to a lesser extent.[59]

Galen again emphasises the importance of the right conditions necessary to carry out such a dissection: a well-lit room, for example, is mandatory.[60]

According to Galen, this thinner, apparently perforated part of the dural membrane facilitates the removal of heavier waste products from the brain, and Galen therefore interprets the relative variation in thickness of the dura according to his physiological requirements, as he does with the cranial sutures.[61] To Galen, there is another important part of the dural covering layer in this region, posterior to but contiguous with that covering the ethmoid. This covers three structures that will be discussed in the following chapters: the *hypophyseal gland* (ἀδήν), the *retiform plexus* (δικτυοειδὲς πλέγμα), and the *perforation* (διάτρησις) over the palate.[62] This dural membrane effectively separates these structures from the rest of the brain. The dural coverings will next be examined in relation to their contents, the venous sinuses.

---

[59] *AA* IX.7; p. 5 Duckworth. Square brackets Duckworth.

[60] This part of the dissection was duplicated in a room lit by bright sunlight. Holding this part of the dura to the light enables one to see clearly the pattern Galen has noted. This pattern is also noticeable when the dura of the falx and tentorium are dissected and stretched out in the light. Where did Galen perform such work? As noted in the General Introduction, in Rome Galen gave his public demonstrations in at least one "of the great auditoria" (τῶν μεγάλων ἀκουστηρίων, *Lib. Propr.* p. 101,4–5 Mueller; XIX. 21 K). But in regard to the bulk of his work, the opinion of Walsh, 1926, 179, seems reasonable: "Three possibilities occur to us: the stable of his own farmhouse on the road to the seaport of Elaea, an apartment associated with the quarters for animals destined for sacrifice within the precincts of the temple of Aesculapius, or for the animals destined for the arena connected with the amphitheater. After consideration, I believe he worked in all three, assisted by his slaves on the farm and by students or priests or gladiators in the other two places." Galen may also have had a special room linked to his residence in Rome.

[61] See n. 42, above.

[62] ταῦτα μὲν ἐντὸς τῆς σκληρᾶς μήνιγγος. οἷς δ' ἐπιβέβηκεν αὐτή τε καὶ ὁ κατ' αὐτὴν ἐγκέφαλος ὅ τ' ἀδήν ἐστι καὶ τὸ δικτυοειδὲς πλέγμα καὶ ἡ εἰς τὴν ὑπερῴαν διάτρησις. *UP* II, p. 22,9–12 Helmreich; III. 713 K.

3.5 *The* meninges *(ii): the dural venous sinuses*

The sutures orient the Galenic dissector to the dura and to that layer's complex system of double infoldings which make up the dural venous sinuses. For Galen, the dura is supportive as well as protective to several key deeper structures of the brain. The sutures and their attached dural membrane (the *perikranion*) are regarded by Galen as a set of anchoring points to which is fastened the dural foldings of the falx and tentorium. These structures are attached in turn to the deeper parts of the brain at critical points, maintaining the necessary tension to preserve the integrity of the ventricles. This crucial support is summarised in *De usu partium*:

> And indeed, . . . since the encephalon had to be double, (Nature) made use [of the thick membrane] for this purpose as well, in extending a part of it (falx) out to the forehead in order to divide the brain. But before this, the part [of the thick membrane, the *tentorium*] which is between the gland (the *pineal*) and the *torcular* lies perpendicular both to the passage (sc. the aqueduct) which unites the encephalon with the parencephalis, and to the worm-like outgrowth[63] lying upon the canal, so that by holding up the adjacent bodies toward itself it prevents them from bearing down heavily on the epiphysis of the passage . . . In this same way the meninx at the lambdoid suture holds up the bodies lying on the posterior ventricle (sc. the *tentorium cerebelli*). And the third suture, the one we call coronal, passes transversely along the middle of the anterior ventricles and raises up the very large part of the encephalon that lies between (the suture and the ventricles), removing the pressure from them, for they would be quite pressed, weighed down, and stenosed, if [the suture] had not been placed in this part of the head. Indeed, although the ventricles of the heart, due to the hardness of its body, remain uncompressed, and need no outside assistance to keep them so, the ventricles of the encephalon cannot without external assistance avoid being compressed, because the encephalon is of a soft consistency.[64]

---

[63] The cerebellar vermis. Here ἐπίφυσις is used in the more general sense of "outgrowth" or "excrescence" (cf. Durling, 1993, 164–5). The anatomy of the vermis is discussed in chapter 4.6, together with its alleged role in regulating the passage of psychic pneuma in the posterior (or fourth) ventricle.

[64] καὶ μέν γε καὶ δέον, . . . διφυῆ γενέσθαι τὸν ἐγκέφαλον, καὶ πρὸς τοῦτ' αὐτῇ κατεχρήσατο προσεπεκτείνασά τινα μοῖραν αὐτῆς ἕως τοῦ μετώπου διορίζουσαν τὸν ἐγκέφαλον. ἀλλὰ τό γε πρὸ τούτου μέρος αὐτῆς, τὸ μεταξὺ τοῦ τ' ἀδένος καὶ τῆς ληνοῦ, κατὰ κάθετον ἐπίκειται τῷ τε συνάπτοντι τὸν ἐγκέφαλον τῇ παρεγκεφαλίδι πόρῳ καὶ τῇ κατ' αὐτὸν ἐπιφύσει τῇ σκωληκοειδεῖ, ὥστε ἀνατεῖνον ἐπ' αὐτὴν τὰ συνεχῆ σώματα κωλύει βαρύνεσθαι πρὸς αὐτῶν τὴν ἐπίφυσιν τοῦ πόρου. . . . οὕτω δὲ καὶ ἡ κατὰ τὴν λαβδοειδῆ ῥαφὴν μῆνιγξ ἀνατείνει τὰ κατὰ τῆς ὄπισθεν κοιλίας ἐπικείμενα σώματα. καὶ μὲν δὴ καὶ ἡ τρίτη τῶν ῥαφῶν, ἡ στεφανιαία καλουμένη, κατὰ μέσον τῶν προσθίων

The dural foldings, the veins, and the surrounding structures which they support, are presented as part of a contiguous and logical whole that are critically important for the structural and functional integrity of the ventricles. Therefore, the key to understanding how Galen envisages the internal anatomy of the brain lies in understanding the role and function of the meninges and their infoldings. As noted in the previous section, the dura is thicker in the midline, and dips into the brain at the area of the overlying median or sagittal suture.[65] This marks the origin of the *falx cerebri*.[66] The blood in the falx supplies the cerebral hemispheres.[67] The lambdoid suture directs the dissector's attention to another doubling of the dura, in which two veins are noted (the *transverse sinuses*). These are key landmarks for Galen of the vasculature of the brain. Galen also describes veins that pass from the lambdoid suture, which have the same *tunic* (χιτών) proper to veins elsewhere.[68] For Galen, this venous complex supplies the cerebellum.[69] The lambdoid suture is another reference point for veins that traverse this region.[70] The transverse venous sinuses conjoin in a part of the brain that appears more elevated—the confluence of the sinuses or *torcular Herophili*, and this is used by Galen as a guide to the

---

κοιλιῶν ἐγκαρσία φερομένη, μετεωρίζουσα τὴν ἐν τῷ μεταξὺ χώραν ἐγκεφάλου παμπόλλην οὖσαν, ἀθλίπτους ἀπεργάζεται τὰς κοιλίας, πάντως ἂν θλιβείσας τε καὶ βαρυνθείσας καὶ στενοχωρηθείσας, εἰ μὴ κατὰ τοῦτο τῆς κεφαλῆς ἐτέτακτο τὸ μέρος. οὐ γάρ, ὥσπερ αἱ τῆς καρδίας διὰ τὴν σκληρότητα τοῦ σώματος αὐτῆς ἄθλιπτοι μένουσιν οὐδεμιᾶς εἰς τοῦτο βοηθείας ἔξωθεν δεόμεναι, οὕτως οἷόν τ' ἦν καὶ τὰς κατὰ τὸν ἐγκέφαλον μαλακὸν ἱκανῶς ὑπάρχοντα μένειν ἀθλίπτους ἄνευ τινὸς ἔξωθεν ἐπικουρίας. *UP* II, pp. 20,24–21,24 Helmreich; III. 711–712 K.

65 ἐν τῷ κρανίῳ τῶν ῥαφῶν ἡ μέση. *AA* II. 709 K.

66 The anatomy in ruminants is as follows: "The ventral longitudinal or sagittal sinus . . . runs backward on the corpus callosum in the concave edge of the falx cerebri, and joins the great cerebral vein to form the straight sinus." Sisson, 1930, 688. Cf. Dyce, Sack, Wensing, 1996, 305–6; Woollam, 1958, 15.

67 εἰς αὐτὸν δὲ τὸν ἐγκέφαλον, ὃν ἔνιοι πρόσθιον ἐγκέφαλον ὀνομάζουσι, διὰ τῆς παχείας μήνιγγος ἀφικνεῖται τὸ αἷμα, μέσης ἀκριβῶς τεταγμένης καὶ τεμνούσης δίχα τὸν ἐγκέφαλον ἴσοις τμήμασιν. ἐκνέμονται δὲ καὶ ταύτης φλέβες ἀκριβεῖς πάμπολλαι παρ' ὅλον τὸ μῆκος εἰς ἑκάτερον τῶν ἐγκεφάλου μερῶν, εὐώνυμόν τε καὶ δεξιόν, *AA* II. 714 K; p. 559 Garofalo, 2000.

68 According to Herophilus, the arterial wall is six times thicker than that of a vein (*UP* I, p. 325,8–10 Helmreich; III. 445 K). Cf. Lonie, 1973, 6.

69 ὀπίσθιον δὲ λέγειν ἐγκέφαλον, ἢ ἐγκράνιον, ἢ παρεγκεφαλίδα, διαφέρει οὐδέν. εἰς τοῦτο μὲν οὖν ἐμφυόμεναι φλέβες ἀπό τε τῶν ἀναφερομένων εἰς τὴν ληνὸν κατὰ τὰς πλευρὰς καὶ τῆς λαμβδοειδοῦς ῥαφῆς, ἔτι τε πρὸς αὐταῖς ἀπ' αὐτῆς τῆς ληνοῦ, τὸν ἀκριβῆ τῆς φλεβὸς ἔχουσι χιτῶνα, τοιοῦτον μάλιστα τὴν ἰδέαν, οἷός περ ἁπάσαις ταῖς καθ' ὅλον τὸ ζῷόν ἐστιν. *AA* II. 714 K.

70 Garofalo, 1991, II, 803, labels these vessels "internal jugular veins", which is not correct. However, according to Galen, the sinuses do arise from the internal jugular veins.

anatomy of the double folding of the dura (the *tentorium cerebelli*).[71] In this elevated region, Galen next notes a third vein, running forward in the sagittal plane (travelling "along the length", κατὰ τὸ μῆκος),[72] the straight sinus.[73] All these dural sinuses contain blood.[74] This can also be observed in incising or perforating the skull in cases of skull *fracture* (κάταγμα).[75] In a dead animal, clotted blood is noted in these sinuses.[76] Galen has now established that the dural foldings contain blood which he states is conveyed to the brain. The *torcular Herophili* comprise a significant part of these meningeal vascular invaginations.

3.6 *The* meninges *(iii): the venous sinuses and the* torcular Herophili

Between the two layers of the dura run the venous sinuses. With the exception of the *torcular Herophili*, Galen does not specifically name any of the venous sinuses of the brain.[77] As noted in the last section, according to Galen, once a vein enters the dura it gains thereby a venous tunic that all other veins in the body possess.[78] In this respect the venous sinuses and the veins of the brain are not an otherwise special set of vessels. In common with all other veins, they originate from the liver, and according to Galen, approach the brain via a branch of the external or superficial jugular vein, as well as from the internal jugular vein, which for Galen is the chief supplier of venous blood to the brain.[79] Galen's depiction of the venous supply of the

[71] ἔνθα δ' ἀλλήλαις εἰς ταὐτὸν ἥκουσιν αὗται, σχεδὸν μέντοι τοῦτο τὸ χωρίον ὑψηλότατον φαίνεται τῶν πέριξ· . . . ἐπὶ δὲ τὸ μετεωρότατον τοῦτο καὶ ἡ ἑτέρα διπλόη παραγίγνεται τῆς παχείας μήνιγγος, ὥστε φαίνεσθαι τετραπλασίαν τῷ πάχει τῶν ἄλλων ἁπάντων μερῶν ἑαυτῆς, ὅσα γε περιλαμβάνει τὸν ἐγκέφαλον ἐν κύκλῳ. *AA* II. 710 K; p. 553 Garofalo, 2000.

[72] *AA* II. 710 K.

[73] Sisson, 1930, 688, notes that in ruminants this sinus: "proceeds from the junction of the ventral longitudinal sinus with the great cerebral vein behind the corpus callosum. It ascends in the falx cerebri, inclining backward, and joins the dorsal longitudinal sinus at the confluens sinuum." From Galen's viewpoint, which examines these sinuses from the perspective of looking down into the brain substance, the straight sinus appears to run forward.

[74] *AA* II. 710 K.

[75] See chapter 5.3.

[76] ζῶντος μὲν οὖν τοῦ ζῴου, γυμνωθέντος οὕτως τοῦ ἐγκεφάλου, καθάπερ ἐπὶ τῶν κάταγμα ἐχόντων εἰώθαμεν κατατιτρᾶν τὸ κρανίον, αἷμα θεάσῃ περιεχόμενον ἐν ταῖς εἰρημέναις κοιλότησι, τεθνεῶτος δὲ θρόμβον, *AA* II. 710–711 K.

[77] May, 1968, 436 n. 17, states her own labels for the sinuses are "tentative". Cf. Guthrie, 1921, 94–95.

[78] *AA* II. 711 K.

[79] XIII.7; pp. 165–167 Duckworth. Cf. Sisson, 1930, 686–7, 719–720. It would

brain is not fully compatible with ungulate anatomy; the description of the course of the veins and their distribution in the neck is drawn from that of the primate. It should also be noted that Galen's descriptions carry the assumption that the veins are moving *towards* the head, in order to supply it.[80] As far as the jugular vein is concerned, Galen observes that it enters the skull via its eponymous foramen to supply the brain with nutriment. The distribution of the arteries to the brain is simpler, with the exception of the retiform plexus.[81]

The veins of the brain also form a complex anatomical interrelationship with the dural venous sinuses. Galen notes the presence of small veins between the inner and outer tables of the skull, this space being referred to as the *diploe* (διπλόη).[82] These eponymous veins are numerous, and anastomose freely in a complex and variable pattern. Some open into dural venous sinuses. Given the importance of the relationship Galen sets between the sinuses and the sutures, the anatomical relationship of the diploic veins allows him to regard the sutures as entry (and exit) points for these veins. Although Galen at times speaks of arteries and veins traversing the diploe, the diploic veins are usually unaccompanied.[83] Galen instructs the dissector to examine these dural layers in the following way:

> If you wish to observe this clearly (a sinus cavity), have available a long and slender instrument, known as a sound,[84] made either of boxwood or of similar solid wood. Introduce this into the cavities of the meninx, pushing it where there is no resistance and cutting through the meninx over it to meet the wood.[85] If it is not possible to do this,[86] insert into

---

seem from Galen's account that the external jugular provides only a few branches which enter via the sutures and emissary foramina. Galen maintains that most of the venous blood comes from the internal jugular veins. See further below.

[80] Galen also differentiates veins from arteries on what they supply. As well as blood, arteries carry pneuma to the retiform and choroid vascular plexuses in the brain. The singular exception in his system is the reversed roles of pulmonary artery and vein. Cf. *AA* II. 600–602 K.

[81] The arteries of the brain are discussed with the retiform plexus in chapter 6.2.

[82] Cf. *AA* II. 710, 712, 715, 725–726 K; *Meth. Med.* X. 445 K; *UP* II, pp. 18,19–19,3 Helmreich; III. 708 K.

[83] *UP* II, p. 18,19–24 Helmreich; III. 708 K.

[84] διπύρηνον. Milne, 1907, 56, defines it as a "slender sound with slight olivary enlargement at either end."

[85] The sound functions as a guide and protector to enable the dissector to cut away the outer layer of the dural sinus, so as not to damage the inner layer.

[86] At first reading, the text seems to indicate that if the διπύρηνον cannot by introduced, then one should try instead either a simple *probe* (μήλη, *specillum*. Cf. Milne, 1907, 27. 51–6) or a *spatula probe* (σπαθομήλη). Singer, 1956, 228, translates διπύρηνον ἢ μήλην ἢ σπαθομήλην (II. 711 K), as, "a *dipyrenon*, a *smile,* (sic) or a *spathomele*." A small emendation may be offered, given that, in the above cited

> the cavity of the meninx either a double-headed sound or a spatula probe,[87] using it by its non-olivary end;[88] push it forward into the cavity of the meninx and cut sideways, slanting the instrument so introduced from side to side in order to not break the scalpel, and thereby reach the cavity of the meninx.[89]

This passage is a good illustration of the sophistication of Galen's anatomical dissection and delineates some of the types of instruments he employed. Galen next deals with what he regards as the most noteworthy of the brain's venous sinuses, the ληνός:

> In the process of stripping the surrounding bones from the folds of the meninx, some part of it is frequently torn and separated. This will be the point for you to insert one of the instruments alluded to into the cavity containing blood. But if (the meninx) is not separated, make an incision with a sharp scalpel on either side of the fold of the meninx in the lower parts where it first reaches the skull,[90] and then, inserting the scalpel through the incision, attempt to force it up to the top where the two veins (*transverse sinuses*) meet; this is the area Herophilus refers to as a treading floor. But while the area to which he gives this name lies somewhat deeper, there is above it a different junction of

passage, Galen has just referred to a type of sound known as a διπύρηνον. It makes more sense for Galen to add that if this is unsuccessful then the σπαθομήλη should be employed, rather than use the term μήλη to refer to the employment of a simple probe. Thus the text could now read: διπύρηνον μήλην ἢ σπαθομήλην, as I have translated above. See below n. 89.

[87] Singer, 1956, 228, n. 172, inexplicably states: "the exact form of these instruments can hardly be recovered."

[88] Galen employs the spatulate end of this probe as a blunt dissector. This instrument consisted of an olivary end and a spatulate end, and was "a pharmaceutical rather than a strictly surgical instrument." Milne, 1907, 58. They were probably made of either wood or metal, depending on what sort of use they were put. Galen uses one as a protector or guard during excision of the ribs (II. 686 K), where it would be unlikely to be made of wood. Extant probes, "are mostly made of bronze. A few are overlaid with gold and silver, and a few are solid gold or solid silver. We read, however, of specilla of lead, tin, copper, and wood, and of the use of a boar's bristle or a stalk of garlic for searching fistulae." Milne, 1907, 56.

[89] ὅπως οὖν τοῦτ᾽ ἀκριβῶς ἴδῃς, ἔστω σοί τι παρεσκευασμένον σῶμα λεπτὸν καὶ πρόμηκες, οἷα τὰ διπύρηνα καλούμενα· τὴν δ᾽ ὕλην ἐχέτω ξύλον πύξινον, ἤ τι τῶν οὕτω πυκνῶν. καὶ τοῦτο καθιεὶς εἰς τὰς τῆς μήνιγγος κοιλότητας ὠθεῖν τε πρόσω πειρῶ, καθ᾽ ἅπερ ἂν ὑπείκῃ, καὶ τέμνειν ἐπ᾽ αὐτῷ τὸ σῶμα τῆς μήνιγγος, ἄρχι περ ἂν ἐντύχῃς τῷ ξύλῳ. εἰ δὲ μὴ παρείη τοῦτο, διπύρηνον μήλην ἢ σπαθομήλην κατὰ τὸ ἕτερον πέρας, ἐν ᾧ τὸν πυρῆνα κυκλούμενον ἔχουσι, καθιεὶς εἰς τὴν κοιλότητα τῆς μήνιγγος ὠθεῖν τε πρόσω καὶ τέμνειν αὐτὴν κατὰ τὸ πλάγιον, ἑκατέρωθεν ἐπὶ τὸ ἕτερον μέρος ἐκκλίνων τὸ καθιέμενον ὄργανον, ὅπως μήτε τὴν σμίλην θραύσῃς ἐντυχὼν αὐτῇ, καὶ τῆς κοιλότητος ἐπιτύχῃς. *AA* II. 711–712 K. Reading διπύρηνον μήλην: Rocca, for διπύρηνον ἢ μήλην: Kühn; Garofalo, 1991, II, p. 798; idem, 2000, p. 555.

[90] This region is probably the lowest part of the attached margin of the falx cerebri, where the superior sagittal or longitudinal sinus runs.

> small veins which rest on it, having arisen in a similar manner from the thick meninx. The narrow calibre (of this confluence) will not accept the olivary part of the spatula probe and therefore in small brains[91] it is either indistinct or does not appear. In this case try and introduce into this confluence one of the finer sounds, or ear-probes,[92] making an incision alongside it.[93]

According to Finlayson: "Nothing has served to preserve the memory of Herophilus so much as the name by which the conflux of the sinuses at the occipital protuberance is still known—the *torcular Herophili*."[94] In *De usu partium*, Galen gives further details of this structure and its relationship to other venous sinuses:

> At the crown of the head[95] the folds of the meninx that conduct the blood come together into a common space like a cistern, and for this very reason it was the custom of Herophilus to refer to it as a treading floor: thence, as from some acropolis, each (of the sinuses) send canals to all the parts which lie below; it is not possible to tell how many are their number, since the magnitude of the parts to be nourished by them cannot be counted. Some flow out from this central space itself into the entire cerebellum, dividing as well as splitting up in a way most like (those that supply) a garden bed; others, from which the confluence has received (a part) are carried forward, as, you could say, an aqueduct of blood which has been well fabricated from the thick meninx. For when these parts which convey the blood had been joined together at the confluence and (a branch) sent from there to the underlying bodies, not yet did (Nature) entrust the remaining (blood) to a single vein,

---

91 A reference which may apply either to different animals or to younger animals of the same class.

92 The ear probe "consists of a small narrow scoop at one end and a simple probe without olivary enlargement at the other." Milne, 1907, 63–4 (at 67, Milne quotes this part of *AA* but gives II. 581 K as the reference).

93 ἐν αὐτῷ μὲν οὖν τῷ γυμνοῦσθαι τῶν περικειμένων ὀστῶν τὰς διπλόας τῆς μήνιγγος πολλάκις ἀπορρήνυταί τι καὶ διασπᾶται. καὶ τοῦτ' ἀρχή σοι γενήσεται τοῦ καθιέναι τι τῶν εἰρημένων ὀργάνων εἰς τὴν κοιλίαν τοῦ αἵματος. εἰ δὲ καὶ μὴ διασπασθείη, τέμνων ὀξείᾳ σμίλῃ τὴν πλευρὰν ἑκατέραν τῆς διπλῆς μήνιγγος ἐν τοῖς κάτω μέρεσιν, ἔνθα πρῶτον ἐμπίπτει τῷ κρανίῳ, κἄπειτα καὶ ἐκεῖ ἐμβαλὼν διὰ τῆς τομῆς τὴν σμίλην, ἄνω βιάζεσθαι πειρῶ μέχρι τῆς κορυφῆς, ἔνθα συμβάλλουσιν ἀλλήλαις αἱ δύο φλέβες, ἥντινα χώραν Ἡρόφιλος ὀνομάζει ληνόν. ἔστι δ' αὕτη μέν, ἣν ἐκεῖνος οὕτως ὀνομάζει, διὰ βάθους μᾶλλον, ἐπιπολῆς δ' ἑτέρα συμβολὴ φλεβῶν μικρῶν ἐπικειμένων τῇ ληνῷ, κατὰ τὴν παχεῖαν ὡσαύτως μήνιγγα γεγενημένη. δέχεται δ' οὐκέτ' αὐτὴ σπαθομήλης πυρῆνα διὰ στενότητα· καὶ διὰ τοῦτ' ἐπὶ τῶν μικρῶν ἐγκεφάλων ἤτοι γ' ἀμυδρῶς, ἢ οὐδ' ὅλως φαίνεται. τῶν λεπτῶν οὖν τι <δι> πυρήνων ἢ μηλωτίδων διεμβάλλειν πειρῶ, καὶ τέμνειν παρ' αὐτήν, *AA* II. 712–713 K; T 122a,b *Herophilus* (part).

94 1893, 21.

95 This is the area of the *lambda*, where the lambdoid and sagittal sutures meet. Here, Galen stresses the importance of the sutures as meningeal reference points.

> but through the parts of the thick meninx which extend anteriorly, created the aqueduct. And not until this was done did (Nature) make the very many outflows along the entire course [of the aqueduct].[96]

Galen draws his analogies from structures and systems that deal with the containment or distribution of fluids. The analogy of the outflow tracts of the confluence of the venous sinuses to those which supply a garden bed is instructive. It is as effective as the reference to the treading floor of a winery in conveying the manner in which the meningeal folds direct the passage of venous blood. Galen's use of χώρα (*field*) for the area from whence these venous passages flow picks up the theme of an area being watered or supplied by nutriment.

Kühn's Latin translation renders ληνός as *torcular*, and this incorrect translation is still regarded in some quarters as valid, preventing a better understanding of Galen's anatomy of this region.[97] A torcular is a wine (or oil) squeeze or press.[98] The Greek is πιεστήρ or πιεστῆρος;[99] πιεστήρ 'ελαιουργεῖον is an oil press.[100] For ληνός, the correct Latin term is *calcatorium*, treading floor.[101] This exactly ren-

---

[96] συμβάλλουσι δὲ κατὰ τὴν κορυφὴν τῆς κεφαλῆς αἱ παράγουσαι τὸ αἷμα διπλώσεις τῆς μήνιγγος εἰς χώραν τινὰ κοινὴν οἷον δεξαμενήν, ἣν δὴ καὶ δι' αὐτὸ τοῦτο προσαγορεύειν ἔθος ἐστὶν Ἡροφιλῳ ληνόν· ἐντεῦθεν δ' οἷον ἐξ ἀκροπόλεώς τινος ἅπασι τοῖς ὑποκειμένοις μορίοις ὀχετοὺς ἐπιπέμπουσιν· οὐδ' ἔστιν ἔτι τὸν ἀριθμὸν εἰπεῖν τῶν ἐκροῶν, ὅτι μηδὲ τὸ πλῆθος τῶν τρεφομένων ἀριθμῆσαι μορίων. ἐκρέουσι δὲ τινὲς μὲν ἐξ αὐτῆς τῆς μέσης χώρας εἰς τὴν παρεγκεφαλίδα πᾶσαν ἐσχισμένοι τε καὶ κατατετμημένοι τρόπον ὁμοιότατον τοῖς κατὰ τὰς πρασιάς, τινὲς δ' ἐκ τοῦ πρόσω φερομένου τοῦ τὴν ληνὸν ἐκδεχομένου φαίης ἂν ἀγωγοῦ τινος αἵματος, ὃν καὶ αὐτὸν ἐκ τῆς παχείας μήνιγγος εὐμηχάνως ἐτεχνήσατο. συναψάντων γὰρ αὐτῆς κατὰ τὴν ληνὸν τῶν ἀγαγόντων τὸ αἷμα μορίων καί τινος ἐπιπεμφθέντος αὐτόθεν τοῖς ὑποκειμένοις σώμασιν οὐδέπω φλεβὶ μιᾷ τὸ λοιπὸν ἐπέτρεψεν, ἀλλ' ἔτι διὰ τῶν τῆς παχείας μήνιγγος μερῶν ἀποταθέντων πρόσω τὸν ἀγωγὸν ἐδημιουργήσατο. καὶ πρῶτον μὲν ἐξ αὐτοῦ παμπόλλας ἐκροὰς παρ' ὅλην τὴν ὁδὸν ἐποιήσατο. *UP* II p. 19,3–23 Helmreich; III. 708–709; T 123 *Herophilus* (part).

[97] *quam regionem Herophilus nominat lenon (torcular)*. II. 712 K. Cf. Dobson, 1925, 20. Longrigg, 1993, 191, refers to the article by Finlayson, where the matter is resolved, but fails to note that "torcular" is not a good translation for ληνός. Von Staden, 1989, 158, states that the "torcular" is "a Latin translation of Herophilus' label, *lenos*, 'wine vat' or 'wine press'." Hyrtl, 1880, §376, noted the error in assigning the word torcular as a translation of ληνός. Exactly when this particular error arose in the translation (or transcription) process is not clear. Daremberg 1841, 19 n. 2, offers the most plausible account. Cf. White, 1975, 149.

[98] The analogy with an oil press is just as valid since "there is no theoretical difference between the oil press and the wine press." Drachmann, 1932, 50; cf. figs. 14–17. The torcular may also refer to the press-beam. Cf. White, 1975, 113.

[99] Galen uses πιεστῆρος in the correct sense of the process of squeezing out a medicament (*Comp. Med. Gen.* XIII. 1044 K).

[100] Cf. Aristotle, *Pol.* 1259a13. Cf. Frankel, 1993, 109, fig. 2. See also Amouretti and Brun (eds), 1993, 588; Forbes, 1955, 135–38.

[101] It is correctly defined in LSJ, 1045, where it is described as "the meeting

ders the Greek ληνός, and thereby precisely qualifies the cistern as a *confluence* of the dural venous sinuses.[102] Rightly or wrongly, the term *torcular* has been institutionalised as the Latin translation of ληνός. Its use cannot be summarily dismissed, but it distorts the accuracy of both Galen's and Herophilus' descriptions of a complex area of vascular anatomy. For Galen, the torcular is the anatomical confluence of the four cranial venous sinuses, from which blood is distributed to the brain.[103] Von Staden points out that "Herophilus apparently identified only the two transverse or lateral sinuses."[104] Whether Herophilus based this discovery on human or ungulate neuroanatomy is not easy to determine, as will now be discussed.

In the passage from *De anatomicis administrationibus* (II. 712–713 K), quoted above, Galen refers to a venous confluence, located at the junction where the two veins meet, about the lambdoid suture. Von Staden states that this confluence "seems compatible with human anatomy."[105] However, the description of the confluence can also apply to bovine anatomy.[106] Given this potential discrepancy, the anatomy of the confluence in humans should be examined. This may be summarised as follows:

> The posterior (lower) end of the superior sagittal sinus is known as the confluence of the sinuses, where there is communication with the straight and occipital sinuses and the transverse sinuses of both sides.[107]

This description matches what Galen offers as Herophilus' contribution to the anatomy of the venous sinuses; namely, that the true ληνός is a deeper structure.[108] According to Von Staden, "it has not

---

point of the sinuses of the dura mater, still called *torcular Herophili*." Cf. White, 1975, 131: "*Calcatorium* is evidently the standard term in later Latin for the treading-vat, equivalent to the Greek ληνός and to the earlier Latin terms 'forus' and 'linter'." See also idem, 1970, 425.

[102] Potter, 1976, 48, renders *torcular Herophili* as "confluens sinuum."

[103] For Galen, veins, like arteries, distributed blood *qua* nutriment to an organ. The concept of venous return was unnecessary as venous and arterial blood were utilised by the tissues of the body.

[104] 1989, 179. Greenhill, 1843, 111, is also cautious: "Whether the common point to which the sinuses of the dura mater converge, corresponding with the internal occipital protuberance, to which we now give the name of "Torcular Herophili", can be the exact spot which Herophilus himself meant to designate by the term ληνός, it is almost impossible to decide."

[105] 1989, 179.

[106] Cf. Woollam, 1958, 5.

[107] Mc Minn, Hutchings and Logan, 1981, 169.

[108] This is not to exclude the possibility of any comparative work that Herophilus might have performed on the anatomy of this region.

been shown conclusively that his (Herophilus') neuroanatomy is based only on the discussion of the ox-brain."[109] The reason he gives is that the bovine sagittal sinus divides "equally" to form the lateral sinuses (that is, the transverse sinuses), and that in the human, this division is "unequal".[110] Yet, to quote from a text on veterinary anatomy, this division is said to occur "in a variable manner."[111] This variability is also a noticeable feature of the dural venous sinuses in man.[112] It is probably safer to conclude that the nature of the supposed division between bovine and human sagittal sinuses is insufficient *of itself* to establish conclusively whether Herophilus was using human brains in his description of the confluence. Finally, as Galen has noted in the context of the veins that run from the confluence: "it is not possible to tell how many are their number, since the magnitude of the parts to be nourished by them (sc. the brain substance and its contiguous parts) cannot be counted."[113] This variability exists in the human *sinus confluens* as well, but this does not mean Galen had access to a human brain nor was slavishly copying Herophilus *without* performing dissections of his own on the bovine brain.

Galen's account is entirely compatible with bovine anatomy. The above cited material may usefully be read as Galen *correcting* the account of Herophilus by shifting the site of the torcular to its true, deeper anatomical location. There is no doubt that Galen describes a confluence of four major venous sinuses. Herophilus may have used the expression ληνός to refer only to two of the four sinuses making up this confluence—the deeper or "true" one, according to Galen. If correct, this is not just a semantic distinction. A full differentiation of these superficial and deeper sinuses enables a dissector better to understand the relationships of these veins and their distribution. In the case of the smaller, superficial confluence, a mixture of veins of varying sizes (some, according to Galen, no larger in diameter than a hair) pass from it and are distributed to the sur-

---

[109] 1989, 158.

[110] Ibid., 158 n. 59.

[111] Dyce, Sack, Wensing, 1996, 305. Cf. Sisson, 1930, 687–8.

[112] Two fundamental studies are Gibbs and Gibbs and Browning. According to Gibbs and Gibbs, 1934, 419, "the torcular is not usually a single chamber . . . but a complex system of anastomosing channels." Browning, 1953, 313, notes that "no two sinus systems were completely identical."

[113] οὐδ' ἔστιν ἔτι τὸν ἀριθμὸν εἰπεῖν τῶν ἐκροῶν, ὅτι μηδὲ τὸ πλῆθος τῶν τρεφομένων ἀριθμῆσαι μορίων. *UP* II, p. 19,9–11 Helmreich; III. 708–709 K. What Galen discerned as "variable" might however be interpreted by others quite differently.

face of the brain. Those from the larger, deeper confluence are distributed to the cerebellum.[114] As noted in the previous section, the cerebellum also receives veins from the smaller confluence, as well as from the sinus that runs near the lambdoid suture. The supply of venous blood to the cerebrum is via the blood contained in the falx, from which vessels are distributed to both sides of the anterior part of the brain. All but two of these veins are small. One, arising from the ληνός, extends deep into the anterior part of the brain.[115] As will be shown in the following section, this is an important landmark vein, for its smaller branches extend into the brain to the *corpus callosum*. The second, larger vein is situated approximately in the middle of the brain (in contemporary anatomy the V. *cerebri magna*, the great cerebral vein of Galen). It runs down into the brain, where, shortly after leaving the meninx, it "forms branches of varying types" (σχίζεται πολυειδῶς).[116] This larger vein, like the first, is a guide to a deeper brain dissection. From the venous confluence represented by Galen's ληνός, veins are distributed to the brain, taking the dura with them as their covering layer. The ληνός serves therefore as a landmark in the brain for dissection of deeper structures. It is also important in its own right as the chief supplier of venous blood to the brain. Finally, Galen's adoption of the term ληνός and his partial differentiation of it from Herophilus' apparent deployment of the word, highlights one instance in the development of a more specific form of anatomical terminology in an area where precise dissection is essential.

---

[114] Ἑξῆς δὲ τὰς ἀποφυομένας ἑκατέρας τῶν ληνῶν θεάσῃ φλέβας λεπτάς· ἐνίας μὲν ἰσχνὰς πάνυ καὶ τριχὸς μόνης δεχομένας κάθεσιν, ἐνίας δὲ μείζους. ὄψει δὲ τὰς μὲν ἀπὸ τῆς ἐπιπολῆς ληνοῦ τῆς μικρᾶς εἰς τὰ πλησίον ἐγκεφάλου μόρια τὰ ἐπιπολῆς διασπειρομένας, τὰς δ' ἀπὸ τῆς διὰ βάθους τῆς μεγάλης εἴς τε τοὐπίσω μέρος ὅλον ἐγκεφάλου, τὸ καλούμενον ὑπό τινων ἐγκράνιον, εἴς τε τὸ πρόσω κατασχιζομένας, ἃς καὶ πρὶν εἰς αὐτὸ τὸ σῶμα καταδύεσθαι τοῦ ἐγκεφάλου, θεάσῃ σαφῶς ἐκπιπτούσας τῶν κατὰ τὴν μήνιγγα φλεβῶν, ἐάν γε μὴ τύχῃς αὐτὰς διεσπακώς. *AA* II. 713–714 K.

[115] ἅπασαι μικραὶ πλὴν δυοῖν, μιᾶς μὲν τῆς κατὰ τὴν ληνόν, ἀποφυομένης εἰς τὸ πρόσω διὰ βάθους κατὰ τὸ μῆκος ὅλης τῆς κεφαλῆς· *AA* II. 714–5 K. Cf. Sisson, 1930, 688.

[116] δευτέρας δὲ πολὺ μείζονος, οὔτ' ἐγγὺς πάνυ τῶν κατὰ τὴν ληνὸν χωρίων, οὔτε πόρρω, κατὰ μέσον μάλιστα τοῦ παντὸς ἐγκεφάλου· . . . δύεται δ' ἡ φλὲψ αὕτη κατάντης εἰς τὸ βάθος, ἔνθεν καὶ σχίζεται πολυειδῶς. οὐ μὴν αὐτίκα τοῦτο πάσχει μετὰ τὴν ἐκ τῆς μήνιγγος ἔκφυσιν, ἀλλὰ προελθοῦσα μέχρι τινὸς οὐ πολλοῦ. *AA* II. 715 K. Cf. Sisson, 1930, 689; Woollam, 1958, 15.

3.7 *The* meninges *(iv): the* pia mater

In *De anatomicis administrationibus*, Galen ends the discussion of the confluence of the sinuses and their venous distribution by again noting that all these observations have taken place without the brain being opened. The dural folds have been probed and dissected, but the substance of the brain remains intact:

> All these things will be observed before the brain is incised and only the thick meninx is exposed. The dura is exposed in three places, since it divides the brain in three by its double folds (*falx cerebri* and *tentorium cerebelli*). Stretch the folds out, incising them, and use your fingers to separate the right and left parts which cover the front part of the brain, doing the same by peeling back the layer which covers the rear of the brain. In this way you will see the outgrowths of all the veins which are dispersed into the three parts of the brain. Some are on the surface, so that their divisions are apparent; some dive into the depths. The fine membrane binds together the veins that embrace the outside of the brain, and goes with the veins into the cavity of the brain itself.[117]

If the dura mater separates the chief parts of the brain substance and provides channels for venous blood flow, then the inner membranous layer, the *pia mater*, acts for Galen as a supporting network for delicate blood vessels. The distinction between these two meninges may be placed in a slightly different perspective by stating that whereas the dura has supportive, protective, and distributive functions, the pia not only supports blood vessels but is itself an entirely vascular construct which also forms part of the brain itself. The brain beneath the dura is described by Galen as consisting of arteries and veins woven together and further bound by the enveloping thin meninx.[118] As it happens, this is a very good description of the appear-

[117] ταῦτα μὲν οὖν ἅπαντα θεάσῃ, πρὶν διελεῖν κατά τι τὸν ἐγκέφαλον, ἐπὶ μόνῃ τῇ παχείᾳ μήνιγγι γυμνωθείσῃ. γυμνώσας δ' αὐτὴν κατὰ τρεῖς τόπους, ἐπειδὴ καὶ τρίχα τέτμηκε ταῖς διπλόαις τὸν ὅλον ἐγκέφαλον, εἶτ' ἀνατείνας τὰ κατὰ τὰς τομὰς τοῖς σαυτοῦ δακτύλοις, ἰδίᾳ δὲ τὸ δεξιόν, ἰδίᾳ μὲν τὸ ἀριστερὸν μέρος αὐτῆς, οἷς τὸν πρόσθιον ἐγκέφαλον ἔσκεπεν, ἰδίᾳ δ' αὖ πάλιν τὸ λοιπόν, ᾧ τὸν ὀπίσθιον, ἁπάσας τε τῶν φλεβῶν ὄψει τὰς ἐκφύσεις εἰς τὰ τρία μέρη τοῦ παντὸς ἐγκεφάλου διασπειρομένας, ἐνίας μὲν οὕτως ἐπιπολῆς, ὡς φαίνεσθαι τὰς σχίσεις αὐτῶν, ἐνίας δὲ καταδυομένας εἰς τὸ βάθος, ἅπαντά τε τὸν λεπτὸν ὑμένα, τὸν συνδοῦντα τὰς εἰρημένας φλέβας, ἔξωθέν τε περικείμενον τῷ ἐγκεφάλῳ, μετά τε τῶν φλεβῶν εἰς τὴν κοιλίαν αὐτοῦ κατιόντα. *AA* II. 715–716 K.

[118] οὕτω γάρ τοι καὶ τὰ τῶν ἐγκεφάλων ἐκ φλεβῶν καὶ ἀρτηριῶν πέπλεκται, συνδουμένων τῇ λεπτῇ μήνιγγι, τὴν αὐτὴν οὐσίαν ἐχούσῃ τοῖς ἄλλοις ὑμέσι τοῖς λεπτοῖς, *AA* II. 720 K. According to Galen the substance of the brain is also made up of veins and arteries. Galen is trying to emphasise the essentially vascular nature of the brain substance, and this might constitute part of his reply (albeit implicit) to Aristotle's contention that the brain is an avascular structure.

ance of the unpreserved brain once it has been stripped of the dura. Galen then describes the thin meninx as having the same *substance* (οὐσία) as other fine membranes, listing as examples the foetal *chorion* (χόριον) and those which line the thoracic cavity, peritoneum and other areas.[119] This use of the pia emphasises the essential role the blood vessels play in making up part of the substance of the brain; in effect they constitute the ὑποκείμενον of the brain. In contrast, the substance that makes up the deeper parts of the brain is not referred to by Galen as vascular in nature. The more superficial parts of the brain are thoroughly invaginated by the thin meninx and its vessels.[120] In a fresh bovine brain, even in good light, the pia is not easy to observe. It is firmly adherent to the surface of the brain and requires great care in its removal to avoid tearing the underlying brain substance. This intimate relationship is perfectly captured by Galen when he describes the pia as *embracing* (περιπεφυκυῖα) the brain.[121] It is testimony to his skill as an anatomist that he was able to identify this inner meningeal layer to the degree in which he did. This does not mean that Galen was the first to discover this layer.[122] But at least part of Galen's audience would have been familiar with the history of the meningeal layers of the brain.[123] Nevertheless, despite Galen's lucid anatomical exposition, he also takes pains to confirm the distinctive character of both pia and dura by means of a relatively straightforward experiment. In contrasting the relationship of the dura and pia of the brain with that of the meningeal coverings of the spinal cord, Galen states that while both are similar, those surrounding the spinal cord are in closer apposition. According to Galen, the reason is that the brain needs the potential space between the dura and pia to move during its respiration.[124] For Galen, the

[119] τὴν αὐτὴν οὐσίαν ἐχούσῃ τοῖς ἄλλοις ὑμέσι τοῖς λεπτοῖς, αὐτοῖς τε τοῖς κατὰ τὸ χορίον, ὑπεζωκότι τε καὶ περιτοναίῳ, καὶ ὅσοι τοιοῦτοι. *AA* II. 720 K.

[120] *UP* I, pp. 448,22–449,2 Helmreich; III. 619 K. The arachnoid mater—the vascular layer between the dura and pia—is not mentioned by Galen. It is quite difficult to see in its own right, and lies well apposed to the pia. Cf. Rubinstein, 1953, 29.

[121] *AA* II. 716 K. Cf. 720, where, as noted above, the pia also binds together the superficial parts of the brain substance.

[122] According to Galen, part of the seventh book of Marinus' *Anatomy* dealt with "the dura mater and other membranes". *Lib.Propr.* p. 106,2–3 Mueller; XIX. 27 K. It is likely that these other membranes included the pia.

[123] Cf. Strohmaier, 1970.

[124] The brain, according to Galen, has its own cycle of inspiration and expiration to bring in air for use in the manufacture of psychic pneuma and to remove residues and waste products. See chapter 6.5.

spinal cord does not move, and hence does not require a potential space between the meningeal membranes.[125] After making a small opening in the dura at one of the places where its folds divide the brain (where the dural folds are a little looser and easier to pierce), Galen introduces a "small tube" (σαλπίγγιον), into this opening, and blows air through it, inflating, as he interprets it, the space between dura and pia.[126]

Galen's meticulous anatomical account of the meninges emphasises their importance as supporting structures for the vascular network of the brain, and in the case of the pia, as part of the substance of the brain. Both meningeal layers also function as guides to the underlying parts of the brain.

3.8 *Ventricular protection and support: the brain's deeper bodies*

For Galen the deeper bodies of the brain are a well-defined anatomical group crucially important for the protection and support of the ventricles; they are paler in appearance than the surrounding brain and firmer in consistency. These deeper structures are accessible following the separation and incision of the two cerebral hemispheres in the sagittal plane, and by following the path of the superior sagittal sinus.[127] After this has been done, two structures in particular

---

[125] *UP* II, p. 263,19–23 Helmreich; IV. 113 K.

[126] ἡ μὲν δὴ λεπτὴ μῆνιγξ ἔξωθέν σοι φαίνεται περιπεφυκυῖα τῷ ἐγκεφάλῳ, ἔν τε τῷ βάθει κατὰ τὸν αὐτὸν τρόπον ὁμιλοῦσα, τὴν παχεῖαν δὲ πλεῖστον ἀφεστῶσαν αὐτοῦ θεάσῃ. καὶ πόσον γε τοῦτ' ἔστι, δυνήσῃ μαθεῖν, εἰ καθ' ἕν τι τῶν τριῶν μερῶν, εἰς ἃ τέτμηκεν ὅλον τὸν ἐγκέφαλον, ὀπὴν ἐργασάμενος βραχεῖαν ἐμβάλλοις εἰς αὐτὴν ἀρχὴν σαλπιγγίου τινός, οἷά περ ἐμοὶ παρεσκευασμένα βλέπετε, παραπλήσια τοῖς τῶν χρυσοχόων φυσητῆρσιν. ὀνομάζουσι δ' οὕτως, ὡς ἴστε, δι' ὧν φυσῶσιν αὐτοὶ τὸ πῦρ ἀνάπτοντες. ἐὰν οὖν ἐμβαλὼν τῷ διῃρημένῳ τὴν ἀρχὴν τοῦ σαλπιγγίου καὶ περισφίγξας ἐν κύκλῳ τὴν μήνιγγα φυσήσῃς δι' αὐτοῦ, θεάσῃ πληρουμένην ἀέρος οὐκ ὀλίγου τὴν ὑπ' αὐτῇ χώραν. αὕτη μὲν γὰρ ἡ μῆνιγξ ἡ παχεῖα τὸ κρανίον ὑπέζωκεν· ὁ δ' ἐγκέφαλος ἐν τῷ διαστέλλεσθαί τε καὶ συστέλλεσθαι προσέρχεταί τε καὶ ἀποχωρεῖ κατὰ τὴν μεταξὺ χώραν τὴν κενήν. *AA* II. 716–717 K. Galen probably inflated the space between the dural foldings at this point (as I did in recreating this experiment, by blowing air into this space through a fine plastic tube). Galen notes that he keeps several such "small tubes" for just this purpose (Milne, 1907, does not mention this instrument, which may mean the σαλπίγγιον is probably a non-surgical tool which Galen has adapted for this use). Cf. LSJ, 1582.

[127] The instructions for dissection are as follows: ἀρχὴν τῆς ἀνατομῆς αὐτοῦ ποιησάμενον ἀπὸ τῆς διχοτομούσης μήνιγγος τὸ πρόσω μέρος, ἧς ἀποτεμὼν ἢ ἀποσπάσας τὰς εἰς τὸ πλάγιον ἐκφύσεις τῶν φλεβῶν, ἀρξάμενοω ἀπὸ τοῦ πρόσω πέρατος, ἀνάτεινε τοῖς σαυτοῦ δακτύλοις, ἄχρι περ ἂν ἔλθῃς ἐπὶ τὴν ἀποφυομένην αὐτῆς μεγάλην φλέβα,

stand out: the *callus-like* (τυλοειδής) body, the *corpus callosum*, and the *arch-like* (ψαλιδοειδής) body, the *fornix* (also called by Galen the καμάριον), which lies under the corpus callosum. Together, they literally support the ventricles at critical points and help maintain their patency. In addition, another deep structure, the *pineal body* (σῶμα κωνοειδές or κωνάριον), which for Galen is a supporting structure for certain blood vessels, and acts as a landmark for the third ventricle. Although the pineal will be discussed in detail in the following chapter, it is mentioned here in order to emphasise once more that for Galen the deeper parts of the brain have a purpose and collectively represent a comprehensive functional entity on which the ventricles depend for their identification, support, and integrity. Perhaps more than his exposition of the dural venous sinuses, Galen's descriptions of these parts of the brain testify to the comprehensiveness of his methodology of dissection, and the relentless single-mindedness of his teleological vision.

The vein of the inferior sagittal sinus is a guidepost to an area of the brain described by Galen as "in substance like a callus".[128] This pale structure lies between the two parts of the cerebrum in the sagittal plane.[129] The *corpus callosum* is different from the brain substance which surrounds it.[130] Galen also notes that the corpus is associated with a natural *hollow* (κοιλότης), remarking that the function of this hollow is to receive the *residues* (περιττώματα) of incomplete coction from the surrounding brain substance.[131] Only after the roof

---

τὴν εἰς τὸ βάθος εἰρημένην φέρεσθαι κατάντη. κἀνταῦθα πάλιν ἀνατείνας ὑψηλὴν ἑτέρῳ δὸς κρατεῖν, εἶτ' αὐτὸς διάλυε κατὰ μῆκος ἑκάτερον τῶν μερῶν τοῦ ἐγκεφάλου χωρίζων ἀπ' ἀλλήλων ἀτρέμα τοῖς ἑαυτοῦ δακτύλοις, ἄχρι περ ἂν ἔλθῃς ἐπὶ τὴν τεταμένην φλέβα κατὰ τὸ μῆκος ἐνταῦθα, περὶ ἧς ἔμπροσθεν εἶπον, ὡς ἔστιν ἀξιόλογος, ὑπὲρ δὲ τῆς καταντους. *AA* II. 717–718 K.

[128] αὐτὴν μὲν τὴν φλέβα τῶν ὑποκειμένων σωμάτων ἀφελών, ἤτοι γ' ἔκκοπτε πᾶσαν <ἢ> μέχρι τῶν κατὰ τὴν ληνὸν χωρίων, ᾗπερ ἐκπέφυκεν ἀνατείνας ἐπ' αὐτῶν ἐκείνων κατάθες. σὺ δὲ τὴν χώραν, ἣν ἐγύμνωσας, ἐπιμελῶς κατάσκεψαι, τυλώδη πως οὖσαν, *AA* II. 718 K.

[129] Clarke and O'Malley, 1968, 577, note that Galen's concept of the corpus callosum referred to "a more extensive structure than is implied today . . . It could include not only the midline connection of the white matter but also the white matter in the cerebral hemispheres that it connects."

[130] εὐθείαις τομαῖς ἑκατέρωθεν τῆς μέσης χώρας ἕως τοῦ βάθους τέμνοντι, μέχρι περ ἂν ἐντύχῃς ταῖς κοιλίαις. γνωρίσεις δ' αὐτάς, τὸ τετυλωμένον θεασάμενος ἐνταῦθα σαφεστάτην ἔχον τὴν διαφορὰν τῶν τετμημένων ἐγκεφάλου μερῶν. *AA* II. 719 K.

[131] ὡς φαίνεσθαί τινα φυσικὴν ἐνταῦθα ὑπάρχειν κοιλότητα, δεχομένην τῶν ὑπερκειμένων τε καὶ περικειμένων σωμάτων τὰ μὴ κατεργασθέντα τελέως τῆς τροφῆς, ἃ καλοῦσιν ἰδίως περιττώματα· *AA* II. 718–719 K. Although at this point in the text Galen has not said he has opened the corpus callosum—the ventricular spaces lie

of the corpus callosum has been incised does examination reveal that it partly overlies each of the "anterior ventricles" (προσθίαι κοιλίας) of the brain. Galen notes that by continuing the dissection, two "fine passages" may be observed leading from the anterior ventricles to the "middle ventricle" (μέση κοιλία). Galen also notes, obviously enough, that this dissection must be undertaken with care; one's hand must be steady and "without trembling" (ἀτρέμας), because of the presence of a septum separating the anterior ventricles.[132] In a fresh brain, this region is extremely friable. The dissection therefore must be precise, skilful, and meticulous.[133] As in other places where Galen exhibits his skill at dissection, the anatomy described and the structures defined speak of many hours of work and of a broad and detailed knowledge. As will be discussed further below, these two "fine passages" in the anterior ventricles are the interventricular foramina, by which the two anterior ventricles communicate with the middle or "third ventricle" (τρίτη κοιλία). The structure separating the anterior ventricles is the *septum pellucidum* or *septum lucidum* (διαφράττον). Within each anterior ventricle are found the complex of veins and arteries known as the choroid plexus. Although this vascular structure will be examined in chapter 6, it is cited here because, together with the corpus callosum they serve—and Galen states this explicitly—as visible markers or "reference points" (γνωρίσματα) for the first incision into the ventricles.[134] The exposure and dissection of the corpus callosum reveals the septum lucidum.[135] The upper

---

under this structure. To reach this area however, Galen has already incised the brain in the median plane, following the anatomical landmark of the sagittal sinus. It is likely Galen is getting a little ahead of himself by describing what is to be found once the corpus callosum is incised. Alternatively, if Galen has not yet opened the corpus callosum, two possibilities are suggested. Galen is either describing the sulcus of the corpus callosum or its midline longitudinal stria. The former is more likely as Galen notes the corpus callosum to have a cavity-like configuration, and the incised groove of the stria does not fit such a description. Cf. Daremberg, 1841, 24.

[132] προσαναδέρων δὲ ἀτρέμα καὶ πόρων λεπτῶν ἐμφάσεις εὑρήσεις, ἄχρι τῆς μέσης κοιλίας τῶν κατὰ τὸν ἐγκέφαλον ἡκόντων. ἀτρέμα δ', εἶπον, ἀναδέρειν χρὴ κατὰ χωρίον τοῦτο διὰ τὴν κορυφὴν τοῦ διαφράγματος ἐνταῦθ' ἀνήκουσαν καὶ διαφράττουσαν τὰς προσθίας ἐγκεφάλου κοιλίας. *AA* II. 719 K.

[133] Galen emphasises that the dissection at this stage must be gentle in two successive lines (*AA* II. 719.4,5 K).

[134] ἐπὶ δὲ τῇ πρώτῃ τομῇ τῶν κοιλιῶν ἑκατέρας ἥ τε τῶν χορ<ι>οιεδῶν πλεγμάτων ὄψις ἔστω σοι γνώρισμα καὶ τὸ τυλῶδες τοῦ σώματος. *AA* II. 720–721 K.

[135] οὐ γὰρ δὴ στήριγμά γε δυνατὸν εἶναι τῶν ὑπερκειμένων σωμάτων τὸ διάφραγμα, καθάπερ τινὰ τοῖχον, ἐσχάτως ἁπαλόν τε ἅμα καὶ λεπτὸν ὑπάρχον· ὧν εἰ καὶ θάτερον ἦν αὐτῷ μόνον, οὐκ ἂν ἐβάσταζε τῶν ὑπερκειμένων ἐγκεφάλου μερῶν οὐδὲ τὸ βραχύτατον. ἀλλὰ τούτου μὲν ἡ χρεία κατὰ τοὔνομα· διαφράττει γὰρ ἀπ' ἀλλήλων τὰς προσθίας κοιλίας, οὐ στηρίζει τὰ κατ' αὐτῶν ἐπικείμενα. *AA* II. 726 K.

border of this septum is attached to the inferior aspect of the corpus callosum. The delicate nature of this membrane is emphasised when Galen says its substance is very like that of the brain itself, and is easily damaged if stretched too far. Galen likens the nature of the septum to the thinly cut translucent stones which are used as windows. This is only appreciated in *bright* (φωτεινός) conditions, for, according to Galen, it must be stretched for its translucent nature to be made evident.[136] The septum is Galen's key to the dissection of the anterior ventricles, for once removed it allows an examination of their internal anatomy.

For Galen the following structure also has a supportive function. This is the *vault* (καμάρα), *arch* (ψαλίς), or "arch-like body" (ψαλιδοειδὲς σῶμα), known as the *fornix*. The function of the corpus callosum is to support the overlying brain and to anchor the septum pellucidum in place, providing it with the necessary tension to separate the anterior ventricles. Galen's elucidation of the corpus callosum is the key to understanding the relationship of the fornix to the surrounding brain. The fornix lies below the corpus callosum, extending like a broad arch from the interventricular foramen of the anterior ventricles and roofing over the middle or third ventricle.[137] Galen compares the fornix to the vault or arch of a roof, with the same load-bearing and protective function.[138] Architects, according

---

[136] ἐπὶ δὲ τῇ πρώτῃ τομῇ τῶν κοιλιῶν ἑκατέρας ἥ τε τῶν χορ<ι>οειδῶν πλεγμάτων ὄψις ἔστω σοι γνώρισμα καὶ τὸ τυλῶδες τοῦ σώματος. εὐθὺς δὲ πειρῶ τὸ διαφράττον σῶμα τὴν δεξιὰν κοιλίαν ἀπὸ τῆς ἀριστερᾶς θεάσασθαι, τὴν αὐτὴν ἔχον οὐσίαν ὅλῳ τῷ ἐγκεφάλῳ, καὶ διὰ τοῦτο ἀπορρηγνύμενον ῥᾳδίως, εἰ ἐκταθείη σφοδρότερον. ἔστι δ' οὕτω λεπτόν, ὥσθ' ὅταν ἐν φωτὶ λαμπρῷ ποιῆταί τις τὴν ἀνατομήν, ὥσπερ οὖν καὶ προσήκει, διαλάμπειν αὐτοῦ τὴν αὐγὴν ὁμοίως τοῖς διαφανέσι τούτοις λίθοις, οὓς τέμνοντας εἰς λεπτὰ ταῖς θυρίσιν ἐντιθέασιν. οὔτ' οὖν σφοδρῶς ἀνατείνειν χρὴ τοῦτο, διασπᾶται γάρ, οὔτ' ἰδεῖν ἐγχωρεῖ σαφῶς αὐτὸ μὴ ἀνατείναντα. συμφυῶν δ' ὄντων αὐτοῦ τῶν ἄνω περάτων τοῖς τετμημένοις σώμασι, μᾶλλον δ' οὐ συμφυῶν χρὴ λέγειν, ἀλλ' ἡνωμένων αὐτῶν τοῖς τετμημένοις, χρὴ λαβόμενον ἀνακλᾶν ἀτρέμα αὐτὰ πρὸς τὴν ἑτέραν κοιλίαν, ἐπιβάλλοντα τῇ κορυφῇ τοῦ διαφράγματος. οὕτω γὰρ ἡ γεγυμνωμένη κοιλία φανερωτέρα γενήσεται, καὶ τὸ διάφραγμα μετρίαν ἕξει τὴν ἀνάτασιν, ἧς μάλιστα δεόμεθα. πρὸ μὲν γὰρ τοῦ ὅλως ἀναταθῆναι, χαλαρόν τέ ἐστι καὶ ῥυσόν, οὔτε διαλάμπειν δυνάμενον, οὔτε τὴν αὐτοῦ συνέχειαν ἐνδεικνύμενον ἐναργῶς. ἀνελκυσθὲν δὲ εἰς τοσοῦτον, ὡς τετάσθαι μὲν ἅπαν, οὐ μὴν ἤδη γέ πω διεσπάσθαι, φανεῖται σαφέστατα. *AA* II. 720–722 K. Cf. Daremberg, 1841, 25.

[137] Ὅπως οὖν χρὴ τοῦτο γυμνοῦν, ἤδη μοι πρόσεχε τὸν νοῦν. οὐ γὰρ δὴ τὸ τυχόν γέ ἐστι μέρος ἐγκεφάλου τὸ σκεπόμενον ὑπὸ τούτου τοῦ σώματος, ἀλλὰ καί τις ἄλλη κοιλία τρίτη παρὰ τὰς εἰρημένας ὀλίγον ἔμπροσθεν, ἃς ἐχώριζέ τε καὶ διώριζεν ἀλλήλων τὸ διάφραγμα. *AA* II. 724 K.

[138] γυμνώσεις οὖν αὐτὸ κατ' ἐκεῖνα μάλιστα τὰ χωρία, δι' ὧν αἱ φλέβες, ὥσπερ ὀπῶν τινων ἐξιοῦσαι, ταῖς προσθίαις ἐμπίπτουσι κοιλίαις· κατ' αὐτὰς γάρ τοι τὰς ὀπὰς ἡ μέση χώρα συντέτρηται ταῖς προσθίαις κοιλίαις. καὶ χρὴ πυρῆνα μήλης ἢ τὸ πλατὺ μέρος τῶν καλουμένων σπαθίων ἢ καὶ σπαθομήλης ταὐτὸ τοῦτο κατὰ τὰς ὀπὰς ἀμφοτέρας

to Galen, employ the same term.[139] The curved shape of the fornix has the correct configuration to bear loads evenly.[140] The fornix provides more immediate support to the anterior and middle ventricles. This tension or elevation of the fornix is only appreciated "with the animal still living" (ἔτι ζῶντος τοῦ ζῴου). Here, Galen notes that whilst all the dura is attached to the skull, the dura is only attached to the brain at the points of the double folding (the falx and tentorium); their tent-like support collapses at death. In a similar way, one would not *disbelieve* (ἀπιστεῖν) that the upper part of the fornix is also kept elevated, producing the large cavity beneath it. The patency of the anterior ventricles is maintained in a similar fashion.[141] This structural integrity is necessarily destroyed in a dissection.[142]

ἀτρέμας ὑποβάλλοντα μετέωρον ἐξαίρειν ἄνω τὸ ταῖς φλεψὶν ἐπικείμενον σῶμα. τοῦτο γάρ σοι πράξαντι καθ' ἑκατέραν τὴν ὀπὴν ἀπαντήσουσιν ἀλλήλαις αἱ μῆλαι, καὶ φαίνεται τὸ σῶμα τουτὶ τὸ ταῖς φλεψὶν ἐπικείμενον ὃ διεξέρχονται κρυπτόμεναι, καθάπερ τις ψαλὶς οἰκοδομήματος σφαιροειδοῦς. ὀνομάζουσι δ' οὐ ψαλίδα, ἀλλὰ καμάρας οἱ πολλοὶ τὰ τοιαῦτα. καὶ τοίνυν καὶ κέκληται τοῦτο τὸ σῶμα ψαλιδοειδὲς ὑπὸ τῶν γνόντων αὐτό. τῶν δ' ἀγνοησάντων ἔνιοι μὲν οὐδ' ὅλως φασὶν εἶναι τὸ ψαλιδοειδὲς τοῦτο κατὰ τὸν ἐγκέφαλον, ἔνιοι δὲ παρακούσαντες οἴονται, τὸ κατὰ τοῦ διαφράγματος ἐπικείμενον οὕτως ὀνομάζεσθαι. ἀλλ' οὔτ' ἐκεῖνο καλεῖται ψαλιδοειδές, ἔστι δὲ τοῦτο ψαλιδοειδὲς ὄντως, ὥσπερ ὠνόμασται. *AA* II. 724–725 K; p. 571 Garofalo, 2000. Willis, 1681, states that the: "*Fornix* so called, or arched Vault, as it were a string or ligament, which arising before, where the brain is hanging to it, is carried to its hinder border, to which it is united as it were with two stretched out arms, and so it holds together the whole frame of the brain." Feindel, 1966, 60. The Latin *fornix* is a direct translation of the Greek. Although Willis gave the term a more general currency, reinforcing both the anatomical accuracy as well as the appropriateness of its name, it remains, in all probability, Galen's discovery. Cf. Rocca, 1998a.

[139] τὸ δ' ὑπὲρ τὴν κοινὴν κοιλότητα μόριον ἐγκεφάλου, καθάπερ οἰκίας τις ὄροφος, εἰς κοίλης σφαίρας ἐπιφάνειαν περιαγόμενον, οὐκ ἂν ἀλόγως δόξειεν ὠνομάσθαι καμάριόν τε καὶ ψαλιδοειδές, ὅτι καὶ τὰ τοιαῦτα τῶν οἰκοδομημάτων ἔθος ἐστὶ τοῖς ἀρχιτεκτονικοῖς καμάρας τε καὶ ψαλίδας ὀνομάζειν . . . τὴν δὴ τοῦ ψαλιδοειδοῦς ἐκείνου σώματος χρείαν οὐκ ἄλλην τινὰ εἶναι τῆς τῶν ψαλίδων αὐτῶν τῶν ἐν τοῖς οἰκοδομήμασιν ὑποληπτέον. *UP* I, p. 484,6–25 Helmreich; III. 667–668 K. Although the reference to architects is a general one, it is not unlikely that Galen was aware of the term *fornicationes* in Vitruvius, even though Galen nowhere uses the Latin expression itself (cf. *De architectura* VI, 8, p. 152,11–13 Rose – Müller-Strübing).

[140] ὡς γὰρ κἀκεῖναι βαστάζειν τὸ ἐπικείμενον ἄχθος ἐπιτηδειότεραι παντὸς ἄλλου σχήματος, οὕτω καὶ τοῦτο τὴν ὑπερκειμένην ἐγκεφάλου μοῖραν ἅπασαν ἀλύπως ὀχεῖ. *UP* I, p. 485,1–4 Helmreich; III. 668 K.

[141] εἰ δ' ἐννοήσεις, ὅπως, ἔτι ζῶντος τοῦ ζῴου, τῷ μὲν κρανίῳ προ<σ>είχετο πάντα τῆς παχείας μήνιγγος τὰ μέρη, τῷ δ' ἐγκεφάλῳ μόνα τὰ κατὰ τὰς διπλόας αὐτῆς, οὐκ ἀπιστήσεις, τὴν κορυφὴν τοῦ ψαλιδοειδοῦς ἀνατεταμένην φυλάττεσθαι, κοιλίαν ὑφ' ἑαυτῇ μεγάλην ἐργαζομένην. οὕτω δὲ καὶ τῶν προσθίων κοιλιῶν ἔτι μειζόνων οὐσῶν, ὅλην τὴν κεφαλὴν τοῦ διαφράγματος ἀναγκαῖόν ἐστιν ἀνατετάσθαι πρὸς τὸ μετέωρον ἅμα τοῖς συνεχέσι σώμασιν. *AA* II. 725–726. Daremberg, 1841, 27, finds it difficult to reconcile the description of the role of the fornix described here with that in *De usu partium*. But Daremberg perhaps has failed to appreciate fully the analogous role Galen gives here to the dural foldings.

[142] στήριγμα δ' οὐδὲν οὐδὲ ταύταις ἐστὶ ταῖς κοιλίαις, οὔτε τῇ μετὰ ταύτας, ἀλλὰ

The supportive role of the fornix is also appreciated by examining the interventricular foramina, which allow the anterior ventricles to communicate with the middle or third ventricle. Once these passages on either side of the septum pellucidum have been exposed, Galen instructs the dissector to insert a probe into each communication, in such a way that they meet each other in the midline. In so doing, the probes are now positioned underneath the anterior arch of the fornix.[143] The appropriateness of its curved shape also fulfils an important, didactic role. For Galen, the fornix describes part of the curvature of a sphere, which he regards as the best geometric form for any organ or structural component with a load-bearing capacity. A sphere allows an even distribution of weight, and Galen extends the analogy to account for the shape of vessels, passages, canals, and ventricles; that is, for every structure in the body that holds some substance.[144] Galen takes the analogy a step further: the ventricles, considered as a whole and stripped of their surrounding adnexae, can be visualised as "precisely sphere-like" (σφαιροειδὲς ἀκριβῶς).[145] In this way, Galen invites the comparison of these anatomical structures not only with the geometrical perfection of the sphere, but with the axiomatic properties of spherical geometry.[146] The deeper structures of the brain provide Galen with another example of how geometric demonstration may be extrapolated to an empirical methodology.

It is also likely that the fornix is Galen's discovery. He states that some (unnamed) anatomists either deny that it exists, or else label the corpus callosum as the fornix.[147] The fornix has the same consistency

---

τῷ κρέμασθαι μετέωρα τὰ ὑπερκείμενα σώματα τὴν ἐντὸς εὐρυχωρίαν ἐργάζεται τῶν τριῶν κοιλιῶν, ἣν ἀπόλλυσθαι κατὰ τὰς ἀνατομὰς ἀναγκαῖόν ἐστι, καταπιπτόντων τῶν ὑπερκειμένων σωμάτων, *AA* II. 726–727.

[143] This is the pillar of the fornix, above the interventricular foramen (of Monro). In the ungulate, "the lower and anterior limits of the foramen of Monro are formed by the pillar of the fornix." Wilkie, 1937, 28; cf. 52–4. See also Sharp, 1961, who illustrates this part of the fornix to good effect.

[144] πάντη τε γὰρ ὁμοιότατον ἑαυτῷ τὸ κυκλοτερές ἐστι καὶ διὰ τοῦτο πάντων σχημάτων δυσπαθέστατον, καὶ μέντοι καὶ μέγιστον ἁπάντων τῶν ἴσην ἐχόντων περίμετρον. ἔστι δ' οὐδὲ τοῦτο σμικρὸν ἀγαθὸν ἀγγείοις καὶ πόροις καὶ κοιλίαις καὶ πᾶσιν, ὅσων ἡ γένεσις ἕνεκα τοῦ δέξασθαί τινας οὐσίας· ἄριστα γὰρ ἐν αὐτοῖς, ὅσα πλείστην ὑποδέχεται, σμικρότατα τοῖς τοῦ σώματος ὄγκοις ὑπάρχοντα. *UP* I, p. 485,4–11 Helmreich; III. 668 K.

[145] οὕτω τοι καὶ αὐτῶν τῶν προσθίων κοιλιῶν ἐὰν ἀφῃρημένην νοήσῃς τήν τε τῆς κοίλης καὶ μέσης χώρας ψαλίδα καὶ τάς γ' ἐπὶ τὰς ῥῖνας καταφερομένας ἀποφύσεις, ἔτι τε τὰς εἰς τὰ πλάγιά τε ἅμα καὶ κάτω . . . σφαιροειδὲς ἀκριβῶς εὑρήσεις τὸ λοιπόν. *UP* I, p. 486,2–7 Helmreich; III. 669–670 K.

[146] Cf. Lloyd, 1987b, 192–194, 231f.

[147] The description of the fornix in *AA* complements that of *UP*. In the latter description however, Galen also uses the term καμάριον, the diminutive of καμάρα, a vaulted chamber, and a more general term, according to Galen, than ψαλιδοειδές.

as the corpus callosum, which might account for some confusion regarding its separate existence. As discussed in chapter 1, Galen mentions the fornix together with Herophilus' contribution to the importance of the ventricles. However, there is no indication from Galen that Herophilus either cited or employed the fornix in a way similar to his own.[148] Galen does not deny Herophilus priority of discovery regarding other parts of the brain—the *torcular Herophili* and retiform plexus being two examples. Although this priority dispute cannot be entirely resolved, it would seem unusual of Galen to appropriate the anatomy and function of the fornix without citing Herophilus as its discoverer, if that were indeed the case.

3.9 *Conclusion*

The corpus callosum and fornix support the paired anterior ventricles and the middle ventricle. The critical nature of this support is revealed when it is realised that these ventricles represent only a potential space in a dissection. Here, the role of the dural foldings that make up the tentorium and falx should be recalled. These also act to support the underlying brain. As the dural foldings are seen in the living animal to elevate the underlying brain substance (thereby maintaining its integrity), so too, according to Galen, the fornix and corpus callosum maintain the structural integrity of the anterior and middle ventricles. Without the support of the meninges and the underlying parts of the brain, the ventricles cannot be envisaged as distinct structures. If this support fails, their physiological role is extinguished. The anatomy of the corpus callosum and fornix provide Galen with the means to identify as well as to construct a patent ventricular system.

[148] καὶ οἶς γε τετάρτη τις αὕτη κοιλία νενόμισται, κυριωτάτην εἶναί φασιν αὐτὴν ἁπασῶν τῶν καθ' ὅλον τὸν ἐγκέφαλον. Ἡρόφιλος μὴν οὐ ταύτην, ἀλλὰ τὴν ἐν τῇ παρεγκεφαλίδι κυριωτέραν ἔοικεν ὑπολαμβάνειν. *UP* I, p. 484, 11–15 Helmreich; III. 667 K; T 138 *Herophilus*. Galen's point in this citation is to emphasise Herophilus' understanding of the importance of the fourth ventricle. It is not to be taken as evidence that Herophilus himself employed the term καμάριον or otherwise cites this structure. In *AA* II. 725 K, there is a reference to some unnamed anatomists, who, not having seen the fornix, deny its existence. Whether this is an allusion to Herophilus cannot be determined.

CHAPTER FOUR

THE ANATOMY OF THE VENTRICULAR SYSTEM

And now we shall first examine the ventricles (of the brain), their size and also the position of each, and their form and communications with each other, and how many there are, and after that the parts (of the brain) which lie upon or are adjacent to them.

De usu partium.[1]

4.1 *Introduction*

For Galen, the *hegemonikon* of the rational soul functions because of four communicating cavities lying deep within the brain substance known as the "ventricles of the brain" (κοιλίαι τοῦ ἐγκεφάλου). Their anatomical description and physiological function represent one of Galen's chief contributions to the history of anatomical science. Galen based his decision to accord the ventricles the functional locus of the brain's activities strictly on empirical grounds: his knowledge of anatomy, his vivisectional experiments, and partly by his observations of brain-injured patients. Although Herophilus and Erasistratus were the first to state formally the importance of the ventricles (or at least some of them), it is Galen who makes "the study of the anatomical structures of the ventricular system one of his special interests."[2] Galen does not claim originality for his discovery of the ventricles, but he brings an especial rigour to this aspect of the anatomy of the brain, creating a detailed anatomical and physiological portrait of the ventricles that goes far beyond that laid down by the two Alexandrians.[3] This chapter will examine the distinctive

[1] 'Αλλὰ νῦν γε πρῶτα μὲν ὑπὲρ τῶν κοιλιῶν αὐτοῦ διεξίτω, μεγέθους τε πέρι καὶ θέσεως ἑκάστης καὶ σχήματος καὶ τῶν πρὸς ἀλλήλας συντρήσεων καὶ τοῦ παντὸς αὐτῶν ἀριθμοῦ, μετὰ ταῦτα δὲ καὶ περὶ τῶν ἐπικειμένων τε καὶ παρακειμένων αὐταῖς μορίων. I, p. 481,1–6 Helmreich; III. 663 K.

[2] Meyer, 1971, 7.

[3] What is extant does not permit the conclusion that the direction of their work necessarily paralleled that of Galen in all respects.

anatomy of the ventricles of the brain, providing the necessary groundwork for the remaining chapters, which deal with Galen's ventricular physiology. The discussion of the corpus callosum and fornix in the preceding chapter began the discussion of the structural organisation of the ventricles, noting the bilaterally symmetrical anterior ventricles and the single, centrally spaced, middle or third ventricle. This account will be augmented here, and will show how these ventricles, together with the posterior or fourth, are considered by Galen to be an integrated whole. Although an important part of the anatomy of the ventricular system—the choroid plexus—will be cited, the details of its anatomy and physiology will be discussed together with the retiform plexus in chapter six.

4.2 *Ventricular symmetry: a constructed harmony*

In the Arabic part of Book IX of *De anatomicis administrationibus*, Galen asks us to study carefully the "reciprocal symmetry" or "reciprocal harmony" of the four ventricles in the brain.[4] Galen believes this to be necessary so that their anatomy may better be understood.[5] In his discussion of the importance of the symmetry of the body, Galen cites the Greek sculptor Polykleitos, who exhibited the correct symmetry of the surface of the body.[6] Galen exploits this analogy with sculpture as follows. Properly conveying the external symmetry of the body is difficult enough for sculptors.[7] By implication, only a careful anatomical demonstration by a trained physician can reveal useful information about the internal structures of the body. This example from the technical world is of some importance to Galen because the symmetry of parts of the body is easy neither to grasp nor to defend.[8]

---

[4] IX.7; p. 2 Duckworth. Cf. Simon, 1906, II, 3.

[5] Plato notes that if the component proportions of a structure are incorrect, then that which is constituted from such elements will destroy them as well as itself. *Phlb.* 64e; cf. *R.* 530a, *Sph.* 228c. For Aristotle, symmetry is commensurability (*Metaph.* 1004b.11, 1061b1). Cf. Lloyd, 1966, 342–5. Galen's concept of symmetry is that of an ordered pattern imposed by Nature (cf. Weyl, 1952, 3–38). Symmetry is also associated quantitatively with καιρός, right proportion or measure, and it is this sense of correct proportion that seems to be at work in Galen's description of ventricular symmetry. Cf. Pigeuad, 1993, 92 n. 18; Wilson, 1980, 202–3.

[6] *PHP* p. 308,20–25 De Lacy; V. 428 K. According to Galen, Polykleitos stressed the "precise symmetry" (ἀκριβῆ συμμετρία) of each part to the other (*Temp.* I. 566 K). See also Gourevitch, 1987, 275ff.; Pigeaud, 1993a.

[7] Cf. *UP* II, pp. 441,17–442,9 Helmreich; IV. 352–353 K.

[8] An example of this is the topographic relationship of certain abdominal organs.

An idea of the ventricles as three-dimensional *symmetrical* structures aids Galen's target audience in understanding their anatomy and topographic relationships. This is brought out in the following:

> When you have cut away these bones (sc. of the top of the skull), then pay particular attention to that spot where the two meninges alone lie over the end of the posterior ventricle, since this is one of the most useful points for you in connection with the dissection which you intend to carry out in the living animal body. Pay attention also to the middle ventricle which the cupola or vault-like structure roofs over (sc. the fornix).[9] This ventricle is situated in the upper regions of the head. Observe further, very thoroughly and carefully, the two anterior ventricles at the sides of the longitudinal (sc. sagittal) suture,[10] and search out and remember well the position of each of the two as you see it. Then, when you have also seen quite clearly how the two optic nerves mount upwards towards the two anterior ventricles, and how this takes place at each side of them, preserve this also in your recollection. For in the course of this operation which I have described for you, you will have to investigate what takes place in the body of the living animal when one compresses or severs all these single structures.[11]

Galen asks for the ventricles to be considered as three-dimensional entities which, both in their description and their relation to other structures in the brain, also possess a symmetric order. The above citation invites the dissector to reflect upon the anatomy of the ventricles, and thereby preserving this information "in your *recollection*" (ἀνάμνησις). A symmetrical relationship is necessary in Galen's eyes for the proper function of each constituent part of the body.[12] The idea of the ventricles as symmetrical structures is a deliberate construct designed to allow for better systematisation and comprehension of their anatomy.

For example, each kidney may be considered as reciprocally symmetrical with the other. Nevertheless, Galen holds that the right kidney is higher than the left (*AA* II. 579 K). Cf. Scarborough, 1976, 173ff.; Triolo, 1966, 113–4, 119. This seeming imbalance is necessary in order for all the abdominal organs to be in correct equilibrium with each other, which allows for their proper function. Cf. *AA* XI.1; pp. 70–71 Duckworth. In *PHP* pp. 594,30–596,4 De Lacy; V. 789 K, all internal organs are said to be symmetrical. There the context is that even single structures are symmetrical by virtue of their constituent parts (cf. n. 5 above).

[9] Cf. Simon, 1906, II, 13.

[10] Cf. Garofalo, 1991 II, 842.

[11] IX.10; p. 14 Duckworth. My glosses in round brackets.

[12] The correct constitution of our bodies, which arises from the right proportion of the elements that constitute them, is stressed in *PHP* pp. 584,38–586,6 De Lacy; V. 776–777 K.

4.3 *An overview of Galen's ventricular system*

Galen's account of the ventricular system testifies to a meticulous knowledge of the subject.[13] It is also a complex record. For this reason, the following prefaces a more detailed analysis of Galen's account, and those areas noted as problematic by later commentators are highlighted for subsequent discussion. Galen records his anatomical demonstrations on the ventricles chiefly in *De anatomicis administrationibus* and *De usu partium.*[14]

It is remarkable that anything constructive in an anatomical sense can be made with a fresh brain. In its unpreserved state, the brain is "an amorphous gruel", the ventricles forming its "distinguishing features."[15] Determining their exact anatomical relationships is no easy task. For example, their intercommunications are delicate passages readily obliterated unless great care be taken.[16] A brain maintains its shape due to the cushioning effect of the cerebrospinal fluid.[17] Once removed from the skull it collapses under its own weight, and is well compared to a *blancmange.*[18] Galen is aware of this potential loss of structural integrity, which forms, as already noted in chapter three, one of the reasons why he stresses the importance of the structural support of the dural folds. It is also why Galen emphasises that much of the work of detailed dissection take place whilst the brain remains partially confined by the bones of the base of the skull. This will not prevent the brain from partly collapsing over time, but provides some measure of stability. What must not be underestimated is Galen's achievement in constructing a coherent account of a set of cavities that are not immediately discernible on first inspection of a fresh brain.

---

[13] Kier, 1977, 2787–2914, provides a comprehensive comparative study.

[14] Although the fifteen anatomical books were revised and expanded some twenty-five years after *UP*, which may partly account for some of the differences in anatomical descriptions between the two works, this has no direct bearing on the overall consistency of the anatomy of the brain recorded in these texts. Duckworth, 1962, xii, gives a date of sometime after 192 AD for the entire, revised text of *AA*. Cf. Simon, 1906, I, ix.

[15] Woollam, 1958, 14.

[16] The corollary—that false passages may thereby be created—will be examined in the course of this chapter.

[17] It is likely that what Galen described as "residues" (περιττώματα) in the ventricles were traces of cerebrospinal fluid (cf. *UP* II, p. 2,17–19 Helmreich; III. 686 K). According to Siegel, 1968, 124, Galen, "did not recognise that cerebrospinal fluid was lost by his faulty autopsy technique." There is no basis for this statement.

[18] Cf. King, 1987, 22.

The nineteenth century comparative anatomist Richard Owen observed that in man and the higher vertebrates, each "primary encephalic division has its cavity or cavities called 'ventricles.'"[19] They consist of two paired lateral ventricles, which are Galen's anterior (or first) ventricles, deep within each cerebral hemisphere, and which communicate with each other and with the third (or middle) ventricle across the midline via the interventricular foramen (of Monro).[20] The third ventricle in the ox brain is noticeably smaller in comparison with the lateral ventricles, consisting of "a ring-like space."[21] It communicates with the fourth (Galen's posterior ventricle) via the cerebral or mesencephalic aqueduct (of Sylvius). This aqueduct is relatively large and quite long in the ox, progressively widening as it approaches the fourth ventricle.[22] When Galen says that some anatomists have referred to this passage as a ventricle, it may have been because this distal dilatation resembles a cavity in its own right. There is some controversy in the secondary literature on whether Galen described the aqueduct proper or a dissection artefact. In section 6, it shall be argued that Galen's descriptions are such that it is highly likely he correctly depicted the nature of the communication between the third and fourth ventricle. The roof of the fourth ventricle is dominated by the mass of the cerebellum. The floor of the fourth ventricle narrows into the central canal of the spinal cord, which Galen interprets as a passage through which psychic pneuma can access the nerves.

The most important features of the ventricular anatomy of the brain of higher mammals is the large size of the lateral ventricles, the presence of the interventricular foramen, the cerebral aqueduct and the proximal part of the fourth ventricle with which the aqueduct merges.[23] Galen's ventricular system is a continuous series of

---

[19] 1868, 80. Cf. Chauveau, 1891, 763–803; Solly, 1847, 120f. The fundamental comparative study is Ariëns Kappers, Huber and Crosby, 1936.

[20] Alexander Monro *Secundus* (1733–1817). But Monro is not the "discoverer" of this passage, and was mistaken in regard to the actual site of the communication. Cf. Last and Tompsett, 1953, 535; Sharp, 1961.

[21] Fitzgerald, 1961, 41.

[22] Here the anatomy in the ungulate is distinctive: "In large ruminants the fourth ventricle is an elongated almost quadrangular cavity . . . (it) communicates through the mesencephalic aqueduct with the third ventricle." Sisson and Grossman, 1975, 1073.

[23] "The true relationship between the ventricles is most accurately envisaged by means of casts of their cavities." Sharp, 1961, 88. Leonardo Da Vinci first performed castings (in wax). Cf. O'Malley and Saunders, 1952. Wax casting is a difficult technical procedures. Cf. Kier, 1977, 2788; Last and Tompsett, 1953, 525.

symmetrical chambers, linked by passages or canals, and communicating with the brain substance, the spinal cord and the cranial nerves. The following encapsulates Galen's notion of a series of communicating cavities deep within the brain:

> What you should carefully study is the reciprocal symmetry of the four ventricles in the brain. For if you proceed carefully, you find that the part which the anatomists have compared to the sharpened end of the writing reed is fashioned similarly to an outflow which discharges into the spinal marrow. Further you see how, above this part, a passage opens out from the posterior ventricle, which extends itself to the middle ventricle. Then you see how the two anterior [*R. and L. lateral*] ventricles open themselves, discharging into the middle ventricle, as I have described above [Book IX, chap. 4].[24] And you see how the anterior end of each of the two anterior ventricles [*rostral horn*][25] goes to each one of the two nasal cavities like a hollow horn, wide at its commencement from its upper part and then steadily narrowing itself. You must now detach and remove all that surrounds these two horns, and go on uninterruptedly along them until you arrive at their termination. Now this is something that it is most desirable for you to see in the brain of a starved and emaciated animal, such a one as starvation and exhaustion have afflicted so that it has become lean, being emaciated by disease. For when you set about this examination, it is necessary that the substance of the brain be exceptionally hard and desiccated. Previously also, if you will, inspect thoroughly the discharging effluents on the two sides [R. and L.] of the anterior ventricles, and remove completely all the parts lying around them, so as to get a fair view of the duct which comes out from the end of each of the two ventricles, noticing how for a wide stretch it descends in the same way as the commencement of the spinal marrow [*central canal*]. However, the top of that duct does not resemble the point of the calamus scriptorius. For it has in this spot no sort of vaulting; on the contrary each of the two ventricles steadily narrows and diminishes, so that the duct comes into being thereout.[26]

---

[24] That is, at the interventricular foramen.

[25] Duckworth glosses this as the "inferior horn". Garofalo, 1991, II, 829, refers to the "terminazione inferiore" of the anterior ventricle. Simon, 1906, II, 3, makes no parenthetical comment. But, as will be seen below, Galen is describing that part of the anterior ventricle which terminates at the olfactory tract. This is the superior (rostral) horn. However, this interpretation depends on how Galen has oriented the brain in the dissection. If he has turned the brain over and is examining it from its base, then the anterior horn is "inferior." But Galen nowhere in the above description cites any manipulation of the brain in this way, and it should be assumed that he is proceeding from the surface of the brain downwards, with the brain still *in situ*. It is better then to regard this part of the anterior ventricle which he has described as the superior horn.

[26] IX.7; pp. 2–3 Duckworth. Square brackets Duckworth.

Galen stresses the inter-communicability of the ventricles, an absolute necessity for the passage of his physiological agent of the nerves, psychic pneuma. His more usual reference to the third ventricle as the middle better reflects its anatomical relationship with the anterior and posterior ventricles.[27] Galen sometimes varies his nomenclature of the ventricular system. In *De usu partium*, for example, he describes the large communication between the anterior ventricles of the brain and the ventricle of the cerebellum (the fourth ventricle). The third ventricle is not specified separately here, but in this instance Galen is stressing the necessity of a communication for psychic pneuma between cerebrum and cerebellum. An expression used in *De usu partium*, ἔμπροσθεν κοιλιῶν, refers to both the anterior and middle ventricles.[28] Galen employs this term not only to stress the communication between cerebrum and cerebellum, but to help visualise the ventricles as an unbroken intercommunicating set of passages from the anterior part of the brain to the spinal cord. In view of the wide-ranging physiological claims Galen makes on each of the ventricles of the brain, their individual anatomical characteristics should be thoroughly understood.

4.4 *The anterior ventricles*

As discussed in chapter 3, Galen states that the anterior and middle ventricles are so shaped because of the support they receive from the *corpus callosum* and *fornix*. It is *inevitable* (ἀναγκαῖος) that this sense of ventricular space will "be destroyed by dissection" (ἀπόλλυσθαι κατὰ τὰς ἀνατομάς).[29] To comprehend ventricular internal anatomy and to visualise them in three dimensions, require from Galen anatomical precision of a high order. The description of the anterior ventricles cited in the preceding section conveys something of their complicated nature. Galen himself is moved to remark that: "Often, indeed, when I have wished to attain to a complete knowledge of the nature of this region, have I met with no slight uncertainty."[30] Only by building an image of the anatomy under examination through repeated dissection is Galen saved from this particular ἀπορία.[31] Of

---

[27] Cf. May, 1968, 414 n. 66; Meyerhof, 1928, 17.5–8.
[28] *UP* I, p. 482,18–22 Helmreich; III. 665 K.
[29] *AA* II. 726 K.
[30] *AA* IX.7; p. 4 Duckworth.
[31] Cf. Garofalo, 1991, II, 830 n. 42. See also chapter 2, n. 39.

course, this kind of statement can be seen as self-serving—designed to reinforce Galen's claim as an anatomist without peer. Yet Galen's uncertainty here revolves around one particular anatomical point—the nature of certain of the passages of the anterior ventricles, and as will be discussed shortly, there is no doubt that their elucidation represents a challenge to Galen's skill as an anatomist.

In the ox, the anterior ventricles are larger than all the remaining ventricles combined. Each consists of a broad, C-shaped structure whose upper part (superior or rostral horn) consists of a long process (the olfactory stem) which terminates as the olfactory bulb. For Galen, whilst the superior horn terminates in a passage for the olfactory tract (not the olfactory *nerve*, as will be noted below), the inferior ends at a supposed communication with the optic nerve which Galen refers to as the *thalamus* (θαλάμη) of the anterior ventricle. The body of each anterior ventricle communicates with its fellow and with the third ventricle at the interventricular foramen. The olfactory and optic tracts are two of the most important features of the anterior ventricles. To consider the olfactory apparatus first. According to Galen, the nature of the communication between the termination of the olfactory tract (the olfactory bulb) and the ethmoid fossa of the eponymous bone is partly *inferred.*[32] Galen regards it as "visible through reason" (λόγῳ θεωρητός).[33] Galen intends that the nature of the passage between the inferior horn of the anterior ventricle and the optic tract be apprehended in a similar way. It is therefore not for rhetorical effect alone that Galen enjoins the dissector to "proceed carefully" and to "inspect thoroughly" these areas. Properly executed, a dissection of this part of the brain should reveal the following:

> Most of the anatomists are aware of the lower (perforation) at the eyes, but I would say that the upper origin of these nerves has escaped their notice;[34] it is at the point where the anterior ventricles turn sideways. Each of these (sc. the anterior ventricles) has a lower end that is stenosed and prolonged, where the nasal passages have their root, and an upper end that widens out as it comes to the middle ventricle; and

[32] To Galen, the olfactory tract or stem is an *outgrowth* (ἔκφυσις) of the brain; the olfactory bulb is the terminal portion of this tract.

[33] According to Galen, the older term was αἰσθήσει ἄδηλος, "imperceptible by sense" (*Hipp. Epid VI.* p. 233,1 Wenkebach; XVIIB. 193 K), where Galen uses it to describe the invisible pores in the skin (cf. *Hipp. Aph.* XVIIB. 421, 521 K). See also Greenhill, 1870, 556–557.

[34] One of a number of references in Galen to other anatomists. These anonymous physicians may also indicate an ongoing debate concerning the nature and function of the parts of the body.

> from this point there is a slanting curve which inclines to some extent toward the lower region and the base (sc. of the brain). The anterior ventricles do not reach as far as the base, however, but narrow gradually and terminate like a horn[35] that is not precisely straight but has a gentle recurvature. The beginning of the optic nerves, which has a channel that is difficult to see, extends to this end of the ventricles.[36]

Here Galen notes the curved nature of the anterior ventricles, recalling his description of the shape of the fornix, which overlies part of the anterior ventricles. Galen observes that the anterior ventricles end in the olfactory tracts at the region of the ethmoid bone. This anatomical arrangement allows outside air to enter the ventricles and *residues* or *waste products* (περιττώματα) to be expelled.[37] But for Galen the nature of the communication between tract and ventricle is more than a question of proximity; whilst there is no such direct passage between the olfactory system and the anterior ventricles in man, in higher mammals such as the ox there is a visible communication between olfactory tract and the anterior ventricles.[38] This is noted in *De*

---

[35] Or, "resembling in form a cone" (κῶνος σχήματι παραπλησίως), as stated in the similar description of this part of the anterior ventricles in *Nerv. Diss.* II. 832 K.

[36] τὸ μὲν οὖν κάτω καὶ πρὸς τοῖς ὀφθαλμοῖς οἱ πλεῖστοι τῶν ἀνατομικῶν ἐγνώκασιν, ἡ δ' ἄνωθεν ἀρχὴ τῶνδε τῶν νεύρων σχεδὸν ἅπασιν αὐτοῖς ἠγνόηται κατὰ τὴν εἰς τὸ πλάγιον ἐπιστροφὴν οὖσα τῶν προσθίων κοιλιῶν. ἑκατέρας γὰρ αὐτῶν στενὴν καὶ προμήκη τὴν κάτω τελευτὴν ἐχούσης, ἔνθα τῶν τῆς ῥινὸς πόρων ἐστὶν ἡ ῥίζα, τὸ δὲ ἄνω πέρας, ὃ πρὸς τὴν μέσην ἀνήκει κοιλίαν, εὐρυνόμενον, ἐπιστροφή τις ἐντεῦθεν εἰς τὸ πλάγιον γίγνεται ῥέπουσα βραχὺ πρός τε τὸ κάτω καὶ τὴν βάσιν. οὐ μὴν ἐξικνοῦνταί γ' ἐπ' ἐκείνην, ἀλλὰ κατ' ὀλίγον στενωθεῖσαι παύονται κέρατος δίκην οὐκ ἀκριβῶς εὐθέος ἀλλ' ἀτρέμα πως ἀποστρεφομένου. πρὸς τοῦτ' οὖν τὸ πέρας τῶν κοιλιῶν ἀνήκει τῶν ὀπτικῶν νεύρων ἡ ἀρχὴ τρῆμα δυσθεώρητον ἔχουσα. *PHP* pp. 448,29–450,3 De Lacy; V. 613 K.

[37] εἰσὶ γὰρ αὐτῷ[ν] καὶ δύο κατὰ τοῦτο ἀποφύσεις προμήκεις τε καὶ κοῖλαι, τὴν ἀρχὴν μὲν ἐκ τῶν προσθίων ἔχουσαι κοιλιῶν, καθήκουσαι δ' ἐπ' ἐκεῖνο τοῦ κρανίου τὸ μέρος, ὅθεν ἡ ῥὶς ἄρχεται. κατὰ τοῦτο καὶ ἡ τῶν ἠθμοειδῶν ὀστῶν ἐστι θέσις, ὧν τὴν χρείαν ἱκανὸν ἐνδείξασθαι τοὔνομα· καὶ ἥ γε μῆνιγξ ἡ παχεῖα, καθ' ὃ ψαύει τῶνδε τῶν ὀστῶν, ὀπαῖς λεπταῖς τέτρηται. διὰ ταύτης γέ τοι πρώτης διηθεῖται τὰ παχύτερα τῶν τοῦ ἐγκεφάλου περιττῶν. *Inst. Od* p. 36,4–10 Kollesch; II. 859 K. Siegel, 1970, 61, 63, provides two illustrations of this area which are useful for a general orientation, albeit transposing Galen's account to an outline of a human brain.

[38] "In most Mammals which derive so important a share of their ideas through the olfactory sense, the 'lateral ventricle' is continued into the 'rhinencephalon'." Owen, 1868, 80–81. Cf. Solly, 1847, 121. In higher mammals, the superior horn projects forward as the olfactory stem. It is well demarcated in the ox, sheep, pig, dog, and horse. It has been stated that in man: "The posterior horns are not constantly present, and a unilateral absence of a posterior horn can be seen in normal cases." Torkildsen, 1934, 491. On the importance of the human posterior horn as an epistemological tool in the context of the often acrimonious disputes concerning human evolution in the nineteenth century, see Cosans, 1994. This debate focussed on the best way to interpret anatomical findings, a point of relevance to the polemic in Galen's time.

anatomicis administrationibus, where Galen instructs the dissector on how to follow the course of the superior horn of the anterior ventricles:

> . . . now examine carefully in this spot how the outgrowths of the two anterior ventricles stretch themselves out towards the nasal cavities [*olfactory bulbs*] and seek to ascertain that fact from the ventricles themselves, as I will describe for you. Betake yourself to the part where each of the two ventricles, stretching towards the front, is narrowed. Into this part insert the head end of a thick sound.[39] For the part itself is wide at its commencement. When you have done that, then point the sound a little upwards, cut or tear through the part stretching itself upwards, and again go gradually forwards until you come to the first part of the nose where the bone lies which the anatomists name the 'cribriform' (ἠθμοειδὲς ὀστοῦν).[40] When you have fully set free all this offshoot, in such a manner as described, then there reveals itself clearly the rigid quality of the [interior] surface of the duct. Also you see that in this region the brain is softer, more like the nature of mucus.[41]

Galen's instructions for this dissection are as follows. First, he states that "when you set about this examination, it is necessary that the substance of the brain be exceptionally hard and desiccated."[42] In the fresh brain, most structures, including the ventricles, are relatively soft and friable.[43] In the passage from *AA* IX.7; p. 3 Duckworth, cited above, Galen has opened the roof of the anterior ventricle and inserted a probe or sound along the olfactory stem. By advancing the probe, Galen next encounters a bony obstruction, the ethmoidal fossa, where the olfactory bulb ends. The importance of this bony region for olfaction is expressed in *De usu partium*:

---

[39] This is probably a *double-headed sound* (διπύρηνος μήλη or ἀμφίσμιλος).

[40] "From the brain two long horn-like processes spring out, which reach to the two nasal fossae. Besides being lengthy, these two processes are hollowed out internally by an excavation which resembles that of a tubular flute, and their substance is exactly like that of the brain." ibid., XIV.2; p. 186 Duckworth.

[41] IX.7; p. 4 Duckworth. My gloss in round brackets; square brackets Duckworth.

[42] IX.7; p. 3 Duckworth. It is unlikely that starvation alone will produce the required desiccation, and it may be that Galen relied on animals whose brains had been subject to a degree of desiccation from some form of disease process (although there is also a degree of shrinkage of the brain with age).

[43] The olfactory passages, being for Galen prolongations of the brain itself, are of the same substance. It is the only instrument of sensation to be formed within the brain. *UP* I, p. 472,1–4 Helmreich; III. 650–651 K. In contrast, the optic tract is a nerve whose "substance is dense and tough, and consequently a nerve is of firmer consistency than the brain." *AA* XIV.2; p. 186 Duckworth. Cf. Simon, 1906, II, 170, n. 583.

Since the covering for this (sc. the olfactory sense)[44] had to be full of holes[45] and loose textured so as to transmit air promptly to the brain for respiration[46] and vapour for distinguishing odours, and to expel large quantities of waste products instantly, should the need arise. Since such a construction would necessarily make the covering itself very vulnerable and result in great damage (to the brain), that most dominant viscus, Nature therefore put near it a bone intricately perforated like a sponge, to keep hard bodies from inflicting external blows and impure cold from directly entering the ventricles of the brain when we inhale . . . These bones overlying the meninges,[47] the intricately perforated, porous bones anatomists call ethmoid, were made to guard against such damage. It would be better, however, not to refer to them as sieve-like, but sponge-like, which was Hippocrates' own comparison,[48] since their perforations are as intricate as those in a sponge, and not drilled through in straight lines like those in a sieve.[49]

Even though, as Galen states immediately following this citation, the dura is perforated like a colander or *sieve* (ἠθμός), those of the ethmoid are more intricate, convoluted and winding, exactly like a sponge.[50] This offers better protection for the ventricles of the brain.

---

[44] τῶν αἰσθήσεων ἡ ὀσφρητική. Cf. Aristotle, *Sens.* 445a5: ἡ αἴσθησις ἡ τοῦ ὀσφραίνεσθαι.

[45] Cf. Greenhill, 1871, 555.

[46] This is the most important function of the olfactory tract; it is necessary for life itself (*UP* I, p. 471,15 Helmreich; III. 650 K).

[47] According to Galen, perforations are present in both meningeal layers which line this bone Cf. *Inst. Od.* p. 60,24–27 Kollesch; II. 883–884 K.

[48] It is a bone likened to a sponge (cf. *UP* II, p. 146,7–10 Helmreich; III. 888 K). It is also "the bone they call *colander-like* (ἠθμοειδές)." *UP* II, p. 150,14; III. 894 K. Cf. Kollesch, 1964, 78; Simon, 1906, II, 4 n.15. Although Garofalo, 1991, II, 830 n. 40, ascribes knowledge of the cribriform plate of the ethmoid bone to Celsus, this is not found in the text of *De Medicina*.

[49] ἐπειδὴ γὰρ ἐχρῆν εἶναι τὸ σκέπασμα αὐτῆς πολύτρητόν τε ἅμα καὶ ἀραιόν, ἑτοίμως μὲν παραπέμψον εἰς ἐγκέφαλον ἀέρα μὲν ἀναπνοῆς ἕνεκεν, ἀτμὸν δὲ διαγνώσεως ὀσμῶν, κενῶσον δ' ἀθρόως, εἴ ποτ' ἄρα δεήσειε, πλῆθος περιττωμάτων, ἐξ ἀνάγκης δ' ἠκολούθει τῇ τοιαύτῃ κατασκευῇ μεγάλη μὲν αὐτοῦ τοῦ σκεπάσματος εὐπάθεια, μεγάλη δὲ καὶ ἡ τοῦ κυριωτάτου σπλάγχνου [ἐγκεφάλου] βλάβη, προστίθησιν αὐτῷ ἡ φύσις ὀστοῦν ποικίλως κατατετρημένον, ὥσπερ σπογγιάν, ὑπὲρ τοῦ μήτ' ἄλλο τι προσπίπτειν ἔξωθεν σκληρὸν σῶμα, μήτ' ἀκραιφνῆ τὴν ψῦξιν εἰσπνεόντων ἡμῶν εὐθὺ τῶν κατὰ τὸν ἐγκέφαλον ἰέναι κοιλιῶν . . . 'Αλλὰ τὰ προκείμενα τῶν μηνίγγων ὀστᾶ ταῦτα τὰ πολύτρητα καὶ σηραγγώδη, τὰ καλούμενα πρὸς τῶν ἀνατομικῶν ἠθμοειδῆ, τῆς τοιαύτης βλάβης ἀλεξήματα γέγονε. βέλτιον δ' ἦν οὐκ ἠθμοειδῆ καλεῖν αὐτὰ μᾶλλόν περ ἢ σπογγοειδῆ, καθάπερ 'Ιπποκράτης εἴκαζε. ποικίλα γοῦν ἐστι ταῖς κατατρήσεσιν, ὥσπερ αἱ σπογγιαί, καὶ οὐκ εὐθύτρητα, καθάπερ οἱ ἠθμοί. *UP* I, pp. 472,4–473,2 Helmreich; III. 651–652 K. Cf. Kollesch, 1964, 78; Sisson and Grossman, 1975, 779.

[50] αὐτὴ μὲν γὰρ ἡ σκληρὰ μῆνιγξ ἡ σκέπουσα τὸν ἐγκέφαλον διατέτρηται δίκην ἠθμοῦ, τὰ προκείμενα δ' αὐτῆς ὀστᾶ πολυειδέστερον ἔτι καὶ ὥσπερ αἱ σπογγιαί, μητ' ἐπ' εὐθείας ἀλλήλοις τῶν πόρων κειμένων μήθ' ὅλως εὐθέων ἁπάντων ὑπαρχόντων, ἀλλ' ἔστιν ὧν καὶ τοιούτων, τῶν πλείστων μέντοι σκολιῶν τε ἅμα καὶ περιφερῶν, ὥστ' ἄλην

But for Galen, it also permits air to enter the anterior ventricles, and as will be discussed in chapter 6, this is essential for the elaboration of the psychic pneuma. Furthermore, Galen's description of the soft, mucus-like nature of the brain (*AA* IX.7; p. 4 Duckworth) implies that there is a part of the brain between the cribriform bone and the end of this part of the anterior ventricles. Galen will use the particular quality of this part of the brain to elaborate a communication between the anterior ventricle and the ethmoid.

The chief problem regarding the anatomy of this region lies in the possibility of creating a false passage. The citation above (*AA* IX.7; p. 4 Duckworth), conveys Galen's awareness of this potential problem, and he expands on it as follows:

> Often indeed, when I have wished to attain to a complete knowledge of the nature of this region, have I met with no slight uncertainty. And so will you likewise feel uncertainty, and see grave difficulties confronting you. The reason for my difficulty was that it was not possible for me to ascertain by the eye whether both these two cavities come finally to a single orifice, as is the case with the cavity of the place named the 'pool' [*infundibular recess*], or whether the two come to an end, closed up and blocked, or whether fine apertures and effluents are present in the duct. The very nature of the thing makes it necessary for the ducts not to be closed up, since you often see that from the brain there flows down very abundant thick mucus into the duct opening into each of the two nasal cavities. And this mucus cannot be poured out explosively without an effluent orifice of a certain diameter.[51] Nevertheless it is not here my purpose to derive the knowledge of the nature of the things which I wish to understand from analogy; for this is not the aim of anatomy. Rather I am simply trying to give an account of those things which manifest themselves to the eyesight.[52]

Galen is forthright about his inability to observe the connection between the anterior ventricles and the nasal passages. Yet he seems capable of observing very fine structures, although the polemical advantage of such an ability should be kept in mind.[53] Although he introduces the analogous comparison of the single communication that the middle ventricle makes at the *infundibulum*, he dismisses it

τέ τινα μακρὰν καὶ περίοδον συχνὴν ἐκπεριελθεῖν χρὴ πρότερον, εἴ τι μέλλει δι' αὐτῶν ὁδοιπορῆσαν ἐπὶ τὸν ἐγκέφαλον ἰέναι. *UP* I, p. 473,3–11 Helmreich; III. 652 K.

[51] Galen might have had in mind the multiple diameter piping arrangement of an aqueduct (cf. Frontinus, *De aquis urbii Romae* I.23ff.).

[52] *AA* IX.7; p. 4 Duckworth. Square brackets Duckworth.

[53] In his last work, *De propriis placitis* p. 74,25–26 Nutton, Galen describes some inexperienced doctors cutting through a "hair-like nerve" (νεῦρον τριχοειδές) during venesection. Cf. *AA* II. 395–396 K.

as well as the existence of such a communication at this particular place, for the necessity of the removal of "very abundant thick mucus" from the brain into the ventricles dictates that there exist several apertures of a certain size. But these small apertures, as Galen also states, do not reveal themselves to the eye:

> And it is not possible for you to perceive with the eye how the canals leading from the brain to the nose terminate. For the matter is as I explained, that is to say, in this region the brain is soft. For the rest, I do not believe that the outlet of the two effluent ducts is a single one [i.e. common to both], because you see plainly that that substance which envelops those two hollow spaces is a brain structure of which the individual parts on both sides, right and left, are united to one another in one and the same manner. I have often investigated and explored this in well-lighted surroundings, and have continued until I have come close to observing clearly enough to allow me to state definitely that the ending of each of the two prolongations is pierced by numerous openings.[54]

Galen is not disputing the evidence he has already obtained which purports to show a patent duct or passage between the anterior end of the lateral ventricle and the olfactory bulb. Rather, it is the *nature* of the communication between the olfactory bulb and the ethmoid which Galen seeks to establish. For he has asserted that the communication between the olfactory end of the anterior ventricles and the ethmoid is pierced by "numerous openings." Why is Galen now so confident? According to Savage-Smith, "after attempting an unbiased description of the olfactory bulb Galen is forced by his *a priori* desire for a direct passage to conclude that each prolongation of the lateral ventricle is perforated by numerous openings."[55] But Galen states that he explored this area in "well-lighted surroundings". As discussed in the previous chapter, Galen stresses the necessity of this condition in his account of the anatomy of the meninges. As Galen makes clear in the continuation from the above citation, what surrounds the terminations or processes of the anterior ventricles is pia mater—the permeable nature of which the dissector should already be aware:

> That which surrounds these processes is the more delicate of the two meninges of the brain, and this meninx itself lies in the middle between the two processes, passes on beyond each of them, defines its limits, and separates it clearly from the other zone.[56]

---

[54] *AA* IX.7; pp. 4–5 Duckworth. Square brackets Duckworth.
[55] 1971, 83.
[56] *AA* IX.7, p. 5 Duckworth. Cf. Garofalo, 1991, II, 831 n. 43.

Since the pia invaginates the surface of the brain, Galen uses this to fashion a presumed communication between the anterior ventricles and the nasal passages. Galen has already forewarned the reader of the use of the pia in this way by the earlier reference to the mucus-like nature of the brain in the immediate region of the ethmoid. Galen now continues to exploit this "permeable" quality of the pia, but significantly also extends its permeable qualities to the dura mater:

> And the middle part of this delicate meninx ends opposite the portion to be found here of the thick meninx, which likewise intervenes between these two processes, together with a very thin bone [*crista galli*].[57] These are things which I have repeatedly seen quite clearly. And when I detached and cut away from the bone the portion of the dura mater found at this spot [i.e. overlying the cribriform plate] and then with my hands stretched it out towards the sunlight, I saw clearly that in certain parts of this membrane the sunlight penetrated to a greater extent, as if there were in them fine perforations, whereas in other parts it shone through to a lesser extent. These again are details from which a man can by surmise reach a correct solution, to the effect that the ends of the two hollow spaces are furnished with perforations, except that these two endings, as demonstrated, are not accessible to the eye.[58]

As noted earlier from *De usu partium* (I, pp. 472,4–473,2 Helmreich), for Galen the term "ethmoid" is not entirely accurate; it is better to regard this bone as "sponge-like" (σπογγοειδής), a word which accurately conveys the complex pattern of the perforations. For Galen,

---

[57] Savage-Smith, 1971, 82–3, raises an interesting point in relation to the comparative anatomy of the ethmoid crest or crista galli, pointing out that the description Galen gives of the superior horn of the anterior ventricles in *Nerv. Diss.* (II. 832 K) is "a remarkably accurate description of the olfactory bulbs in the Rhesus monkey where there is no ethmoid crest and the falx cerebri attaches itself directly to the cribriform plate." This description is "clear evidence" that Galen "had actually dissected this aspect of a monkey." This raises implications for Galen's use of animals for brain dissection. One might wonder why Galen would restrict himself to the use of an ape for this particular detailed dissection and not for other areas of the anatomy of the brain. However, Savage-Smith's evidence of the lack of an ethmoid crest in the Rhesus is based on "two samples". According to the authoritative text of Hill, 1974, 218, the Macaca species of ape (which include the Barbary ape and Rhesus monkey) possess a "low crista galli." And in the tailed apes there is a "well developed crista galli." Owen, 1866, II, 533. These tailed apes were known to the Rome of Galen's day (as Hill, 1966, 213, notes). As far as the evidence from *Nerv. Diss.* is concerned, it might also be that the subject used by Galen is a dog, a possibility raised by Savage-Smith, although regarded as "not as probable." Cf. Sisson and Grossman, 1975, 323, 343, 779.

[58] *AA* IX.7; p. 5 Duckworth. Square brackets Duckworth.

since the ends of the anterior ventricles are a part of the brain, they are surrounded by pia mater; this thin meninx is therefore a part of the terminal prolongation of the anterior ventricles, allowing the communication between anterior ventricle and nasal passage to occur.[59] As for the dura that overlies the cribriform plate of the ethmoid bone, this too, when examined in sunlight, is seen to be perforated. These observations are used by Galen to infer that the olfactory terminations of the anterior ventricles are also perforated, even though these endings, as he states above, "are not accessible to the eye."

The perforated nature of the ethmoid is not the only inference from anatomy Galen makes. He also maintains that some nerves—the optic for example—contain hollow passages, stating: "it is obvious that these [nerves] have in themselves perceptible pores."[60] To better understand this, Galen's notion of the alleged passages between the right and left ventricle of the heart should be noted. This type of communication, Galen states, also occurs through *perforations* (τρήματα) which are "not possible to observe".[61] This can of course be used to reinforce a picture of Galen inferring what he wants to suit his claims. In regard to the alleged lumen in the optic nerves, May states that Galen: "has persuaded himself that he has seen what he needed to see for the support of his conception of the physiology of the nervous system."[62] But Galen's explicit hesitancy in asserting that such passages cannot be seen with the eye, together with his awareness of how easy it would be to imagine, much less create a false passage, entitle Galen to infer a similar communication between the terminal olfactory prolongation of the anterior ventricle and the ethmoid fossae, and to compare—albeit implicitly—such communications with those of the heart. It would satisfy Galen's requirements if he simply *claimed* that such passages were present, but this would represent an abrogation of his chosen role of the methodically accurate dissector. Galen may with justification be accused of creating the conditions for imagining a false passage by his statement that one needs to probe the alleged passage in the optic nerve in order to detect it. But that one cannot see passages in the nerves and ducts

---

[59] Similarly, the lateral ventricles discharge into the hypophyseal recess which is also covered by pia mater. See further below.

[60] ἀτὰρ οὖν καὶ φαίνεται μόνα ταῦτα πόρους αἰσθητοὺς ἐν ἑαυτοῖς ἔχοντα, *UP* I, p. 463,9–10 Hemreich; III. 639 K.

[61] *Nat. Fac.* pp. 251,27–252,8 Helmreich; II. 207–208 K.

[62] 1968, 399 n. 42.

between the end of the anterior ventricle and the cribriform plate does not mean that such passages do not exist. Only after their beginning has been empirically determined are they perceptible λόγῳ θεωρητός. Such a combined method of dissection and reasoned inference is used by Galen in his discussion of the supposed communication between the optic tract and the inferior horn of the anterior ventricles, which will next be examined.

The nature of the relationship between the optic nerve and the anterior ventricle was sketched out in the *De placitis Hippocratis et Platonis* (pp. 448,29–450,3 De Lacy), cited above. According to Galen, "most anatomists" know of the apparent lumen or passage in the optic nerve at its upper commencement.[63] This information does not in any case require a highly sophisticated empirical methodology to determine it, other than the excision of the eye together with the part of the optic nerve behind the globe.[64] However, the origin of the optic nerve and its relationship with the ventricular system is different. In the citation from *PHP*, Galen claims priority for the discovery of the ventricular origin of the optic nerves.[65] And in *De anatomicis administrationibus*, Galen states that each optic tract, ". . . lies on one side of the two outgrowths which lead from the brain to the nose [*olfactory bulbs*]."[66] Here the anatomical orientation is the *base* of the brain, and Galen notes the close relationship between the optic and olfactory tracts in this region:

> The material substance of these two [*optic*] processes is also dense and compact . . . In addition, the material substance of this nerve is farther removed from that of the brain than is the substance of the rest of the nerves from that of this one. This pair of processes is the nerve known as the optic nerve. Its place is behind that of the two processes

---

[63] This "lumen" is due to the presence of the central retinal artery, vein and nerve. In humans: "The central retinal artery pierces the optic nerve about 1.25 cm behind the eyeball to enter the eyeball with the optic nerve and supplies the retinal area." Kaplan and Ford, 1966, 76. This account is similar to that in the ox and other higher mammals.

[64] As noted in chapter 1.2, this observation apparently goes back to Alcmaeon.

[65] The testimony does not reveal Herophilus linking the optic nerve to the anterior ventricle as Galen does. But given that Herophilus' physiology of the brain and nerves apparently utilised pneuma, this, together with the importance of the optic nerve for Herophilus, meant that it is likely he discussed the anterior ventricles, at least insofar as they relate to the optic tract.

[66] IX.9; p. 8 Duckworth. Cf. Simon, 1906, II, 244 n. 18. The optic nerves are the first pair of cranial nerves for Galen. Cf. Daremberg, 1841, 47–50; Savage-Smith, 1971, 85–87.

> which come to the nose on a similar course. You can obtain a good view of this nerve pair after you have detached the surrounding bones from the skull as we have indicated in the ninth[67] book of this work.[68]

Galen states that the substance of the olfactory tract and bulb are exactly like that of the brain, each being part of the anterior ventricles. The olfactory outlet is the only sense instrument created within the anterior ventricles of the brain, and which contains "vaporous pneuma" (ἀτμῶδες πνεῦμα).[69] In contrast, nerve substance "is dense and tough, and consequently a nerve is of firmer consistency than the brain."[70] Galen makes a further, significant qualification in the case of the optic nerve, remarking that "its density and compactness are less than those of the substance of all the other nerves."[71] With this in mind, the dissector can next observe the following about these key structures in the base of the brain:

> And you will find that the origin of each one of these two nerves extends upwards as far as the thalamic region, belonging to the two anterior ventricles of the brain, which come into existence near the oblique lateral deviation of the cerebral ventricle. Then the two processes [*optic tracts*] make their way from this region and come to the middle region (sc. of the base of the brain), where the one process meets and blends with the other [optic chiasm].[72]

The term "thalamic region" is appropriately chosen. Aristotle uses the term θαλάμη to refer to what he maintained was the middle ventricle of the heart, which allegedly communicates with the left and right chambers.[73] Although Galen's view of the heart is anatomically

---

[67] "Wohl Lapsus", Simon, 1906, II, 171 n. 585. The description of removing the cranial bones to expose the optic nerve is found in Book X (X.1; pp. 27–8 Duckworth). Cf. Garofalo, III, 1991, 1045 n. 49.

[68] XIV.2; p. 187 Duckworth. Square brackets Duckworth.

[69] τὸ τῶν ὀσμῶν αἰσθητήριον ἔνδον ἐγένετο τοῦ κρανίου, μόνον ἁπάντων ἐν αὐταῖς τοῦ ἐγκεφάλου ταῖς προσθίαις κοιλίαις ἀτμῶδές τι πνεῦμα περιεχούσαις. *UP* I, p. 469,14–17 Helmreich; III. 647 K.

[70] *AA* XIV.2; p. 186 Duckworth.

[71] Ibid., XIV.2; p. 187 Duckworth. This is an aid to the tactile sense of the dissector. In a fresh brain, the optic nerves are of a firmer consistency than the olfactory tracts. Galen compares the difference between the consistency of other nerves to that of the optic as between curdled milk and milk which has started to turn (ibid.).

[72] *AA* XIV.2; p. 187 Duckworth. Square brackets Duckworth; my gloss in round brackets.

[73] *Somn.Vig.* 458a17. Cf. LSJ, 781. Savage-Smith, 1971, 85 n. 34, points out that in the Arabic part of *De anatomicis administrationibus*, θάλαμος rather than θαλάμη was translated (cf. Garofalo, 1991, 1045 n. 50; Simon, 1906, 338 n. 586). Daremberg, 1841, 47, refers to θάλαμος, not θαλάμη, albeit rendering it as "lit".

far superior to that of Aristotle's, Galen takes up the notion of ventricular communicability using an already established expression. It is crucial for his physiology of the brain that the ventricles intercommunicate, so that physic pneuma may have access to the nerves. In the case of the optic nerve, Galen's largest and most important cranial nerve, pneuma is essential for the visual process.[74] This is achieved by the alleged relationship of the thalamic region to the inferior horn of each anterior ventricle.[75] Galen enlarges on this necessary proximity in the following:

> For the optic nerves grow out from the place where each of the two anterior ventricles terminates obliquely, and the thalamus, for instance, of the (anterior) ventricles was constructed for the sake of these nerves. The anatomists have not understood this marvellous work of Nature since they have not followed the ventricles to their extremities, nor inspected for what purpose they have been formed, nor seen that the outgrowths of the optic nerves[76] from above are connected to the ends of the ventricles. So for these reasons the optic nerves have been made hollow as well as very large and very soft, although the other senses also possess large and soft nerves.[77]

Galen lays claim to this particular anatomical discovery, implicitly maintaining, as he frequently does elsewhere, that the correct dissections not only provide the right answers, but are sufficiently awe-inspiring in themselves to reflect the "marvellous work" (ἔργον θαυμαστόν) of

[74] This notion of ventricular communicability is not explicitly repeated with the other cranial nerves, as they lie much further from the ventricles than the optic nerves, although clearly they also must have access to psychic pneuma. As noted above, Galen states that only the optic nerves "contain narrow perforations" (*AA* IX.8; p. 6 Duckworth). As for the question of πόροι in other nerves, there is also a limitation imposed by size. Galen does not believe that very small nerves have passages; if they did, they would be so very fine that they would risk being constantly occluded (cf. *PHP* p. 452,8–15 De Lacy; V. 616–617 K).

[75] This region should not to be confused with the *thalamus* of contemporary anatomy, which for Galen is an important, although unnamed part of the anatomy of the third ventricle. See the following section.

[76] Cf. *Nerv. Diss.* II. 832 K, where the optic nerve is described as an "outgrowth from the brain itself" (ἐκφυόμενον ἐξ αὐτοῦ τοῦ ἐγκεφάλου).

[77] ἔνθα γὰρ ἑκατέρα τῶν προσθίων κοιλιῶν τελευτᾷ πρὸς τὰ πλάγια, τῶν ὀπτικῶν νεύρων ἐστὶν ἡ ἔκφυσις, καὶ αὕτη γε τῶν κοιλιῶν ἡ οἷον θαλάμη δι' ἐκεῖνα τὰ νεῦρα γέγονεν. ἠγνόηται δὲ τοῖς ἀνατομικοῖς ἔργον τοῦτο θαυμαστὸν τῆς φύσεως οὔτ' ἀκολουθησάντων αὐτῶν τοῖς πέρασι τῶν κοιλιῶν οὔτ' ἐπισκεψαμένων, ὅτου χάριν οὕτω διεπλάσθησαν, οὔτε τῶν ὀπτικῶν νεύρων θεασαμένων τὰς ἄνωθεν ἐκφύσεις συνημμένας τοῖς πέρασι τῶν κοιλιῶν. διὰ ταύτας μὲν οὖν τὰς αἰτίας τοῖς ὀφθαλμοῖς τὰ νεῦρα κοῖλά τε ἅμα καὶ μέγιστα καὶ μαλακώτατα γέγονεν ἐχουσῶν καὶ τῶν ἄλλων αἰσθήσεων καὶ μεγάλα καὶ μαλακὰ νεῦρα. *UP* II, pp. 384,21–385,7 Helmreich; IV. 275–276 K.

Nature. In regard to Galen's descriptions of the thalamus, Savage-Smith notes: "Galen here is referring not to the thalamus of modern terminology, but to the recess (*inferior cornu*) in the lateral ventricles which he considered to be the source of the optic nerves."[78] But this recess is not perceptible on the surface of the base of the brain. In an examination of the base of the ox brain, two of the most prominent features are the olfactory and optic tracts. Each can be visualised in the way Galen has just described, for both appear to merge with the surrounding brain substance. Therefore, to fully delineate the relationship of each optic tract to the anterior ventricles, Galen must dissect the anterior ventricles from within. He has already performed this procedure in the case of the olfactory tract and bulb. And Galen will find the connections of the optic tract with the anterior ventricles to be as intimate as those described for the olfactory bulb.

In *De anatomicis administrationibus*, Galen directs the dissector's eye to observe that a specific (albeit unnamed) part of the brain marks the origin of the optic tract:

> You find here, at the side of this duct (end of inferior horn of anterior ventricle),[79] a part of the brain of a peculiar kind, different in boundaries and in circumference from the other parts of the brain and so singled out from the others (optic thalamus).[80] Direct your efforts to this part, and loosen away from it that which surrounds it and follow it in its course towards the front. There you find it firmly connected with and intimately united to the two nerves of sight [*optic nerves*]. For speedy discovery of its union and its contiguity, it will help you to draw these two nerves from below upwards towards their upper parts, their specific roots.[81]

---

[78] 1971, 85. In the only other instance in Galen's medical corpus which I have found, ἡ θαλάμη is used to refer to the bony recess of the temporal fossa (*UP* II. p. 119,4 Helmreich; III. 850 K).

[79] This duct is not specified, nor does the text immediately preceding this passage enable one to say with complete confidence that Galen is referring to the inferior horn; it is possible that he is referring to the optic tract. But in any case, Galen's intention is to direct the dissector's attention to the origin of the optic tract and its relationship with the brain.

[80] Galen does not name this part of the brain. Simon, 1906, II, 3, inserts *optic thalamus*, which makes better Galenic anatomical sense than Duckworth's gloss: *lateral geniculate body* (cf. Garofalo, 1991, II, 829). The optic tract emerges from this area at the base of the bovine brain; in appearance it is no different from the rest of the brain in this region.

[81] IX.7; p. 3 Duckworth. Cf. Briggs, 1939, 26. My glosses in round brackets; square brackets Duckworth.

This description is based on an examination of the interior of the anterior ventricles, especially the inferior horn. Galen notes that both the olfactory bulb and optic tract have a passage or channel which is "difficult to see"; but, as with the olfactory bulb, such a passage may be probed:

> Besides this cavity (sc. olfactory tract) which you see in the two processes that go to the nasal fossae, you will see no other cavity in the (sc. optic) nerve since, although admittedly a hollowing out is found in the two processes [*optic nerves*] making their way to the eyeballs from the brain, yet this [hollowing] is not so large as that other one (sc. in the olfactory tract), and also that it does not reach so far forwards as that. On the contrary, it is only a small hollow passage, difficult to expose to view. It can only be detected by the insertion of a fine probe, a hog's bristle, or something else of similar tenuity.[82]

To this may be added the following from *De placitis Hippocratis et Platonis*, which partly parallels the above:

> The beginning of the optic nerves has an aperture that is difficult to see, and reaches to this end of the ventricles. You will see it when you have achieved the following three conditions: first, that the animal is large; second, that it is dissected immediately after its death; third, that the surrounding air is bright. If with these provisions you sufficiently lay bare the end of the [inferior horn of the anterior] ventricle by removing all the bodies that lie upon it, without tearing or crushing the outgrowth of the nerve, you will see the aperture at its beginning.[83]

Galen seeks to demonstrate an aperture that is "difficult to see" (δυσθεώρητον). To do so, he adduces the above three conditions. That the animal be large is obvious, since one is dealing with a relatively small part of the brain. The necessity for dissection to be carried out *immediately* (παραχρῆμα) after death is also obvious when dealing with such a friable organ. The necessity of *bright* (φωτεινός) conditions is a Galenic *sine qua non* for a well organised dissection. Further, consider Galen's use of the word τρῆμα, *aperture*.[84] It may

[82] *AA* XIV.2; p. 187 Duckworth. My glosses in round brackets; square brackets Duckworth. Cf. Simon, 1906, II, 170 n. 583. See also May, 1968, 401 n. 42.

[83] πρὸς τοῦτ' οὖν τὸ πέρας τῶν κοιλιῶν ἀνήκει τῶν ὀπτικῶν νεύρων ἡ ἀρχὴ τρῆμα δυσθεώρητον ἔχουσα. θεάσῃ δὲ αὐτὸ τριῶν τούτων στοχασάμενος, ἑνὸς μὲν τοῦ μέγα τὸ ζῷον εἶναι, δευτέρου δὲ τοῦ παραχρῆμα μετὰ τὸν θάνατον ἀνατέμνεσθαι καὶ τρίτου τοῦ τὸν πέριξ ἀέρα φωτεινὸν εἶναι. ἐὰν γὰρ ἐπὶ τούτοις οὕτω παρεσκευασμένοις γυμνώσῃς ἐπιτηδείως τὸ πέρας τῆς κοιλίας ἀφελὼν ἅπαντα τὰ κατ' αὐτῆς ἐπικείμενα σώματα χωρὶς τοῦ διασπάσαι τε καὶ θλάσαι τὴν ἔκφυσιν τοῦ νεύρου, θεάσῃ τὸ κατὰ τὴν ἀρχὴν αὐτοῦ τρῆμα. *PHP* p. 450,1–9 De Lacy; V. 613–614 K.

[84] Aristotle uses the term for the tracheal (*HA* 495a29) and urethral (*HA* 497a25) orifices.

be large or small, and the term can be placed in association with another to indicate a blind passage or channel. In *De nervorum dissectione*, for example, Galen refers to the course of the two component parts which make up the fifth cranial nerve. One penetrates the foramen of the organ of hearing where an invagination of dura is also found. The other part of the nerve passes through a canal in the petrous part of the temporal bone, known "by the older anatomists" (οἱ παλαιοὶ τῶν ἀνατομικῶν) as a blind passage, for they were unable to probe its full extent owing to the helical nature of its course.[85] Galen does not say he is any more successful. However, since this part of the nerve is seen to emerge on the other side of the bone, then the channel through which it passes is only blind *apparently*. If Galen is not explicitly asking for this to be considered when the relationship between the optic tract and inferior horn of the anterior ventricle is discussed, then at the very least there is an established precedent testifying to the difficulties in properly recognising such apparently blind apertures, and inferring their patency.

Galen has dissected the structures surrounding the inferior horn of the anterior ventricle so as not to disturb the outgrowth of the optic tract from the base of the brain.[86] Is this sufficient to enable the passage at the beginning of the optic nerve to be apprehended? It has been claimed that Galen could not have seen such a communication. Daremberg states that his own dissections, both in man and other animals, could not demonstrate what Galen says he has done.[87] Yet, in following Galen's instructions, it is possible to visualise an aperture.[88] When the anterior ventricle has been completely opened, its narrow confines are seen to be filled with the arteries and veins of the choroid plexus. According to Galen, this plexus *undergirds*

---

[85] ἐν μὲν τοῖς πρόσω μᾶλλον ὃ καλοῦσιν ἀκουστικὸν, εἰς τὸ τρῆμα τῆς ἀκοῆς ἐκπίπτον ἅμα τῇ συνεκφυομένῃ μήνιγγι τῇ σκληρᾷ, μεθ' ἧς πλατυνθὲν ὑπαλείφει τὸν πόρον, ἐκ δὲ τῶν ὀπίσω θάτερον εἰς ἕτερόν τι τρῆμα τοῦ λιθοειδοῦς ἐκπίπτον ὀστοῦ τὸ καλούμενον τυφλόν· ὠνόμασαν γὰρ οὕτως οἱ παλαιοὶ τῶν ἀνατομικῶν αὐτὸ, μὴ δυνηθέντες ἀκριβῶς ἀνατρῆσαι τὴν ἕλικα, δι' ἧς ἐκπίπτει τὸ νεῦρον πρὸς τὸ ἐκτὸς ὀπίσω τῶν ὤτων. II. 837–838 K. Cf. *Oss.* II. 745 K. See also Durling, 1993, 316. This nerve is the vestibulo-cochlear, Galen's (and Marinus') fifth cranial nerve. Cf. Daremberg, 1841, 54–6; Savage-Smith, 1971, 181–3.

[86] In his instructions on preparing the ox brain for dissection, Galen cautions that this be carefully done in order to observe clearly the outgrowths of the cranial nerves, as well as arteries and veins. *AA*. II. 709 K.

[87] 1841, 48.

[88] May, 1968, 401 n. 42, provides a good account of a dissection of the anterior ventricle in the fresh bovine brain, which my dissections confirmed.

(ὑπέζωκε) the interior of the anterior ventricles.[89] Only after its removal can the floor of the inferior horn be examined.[90] The inferior horn extends forward somewhat, terminating in a groove which resembles an orifice.[91] If this is compared to what adjoins it at the base of the brain—the optic tract—then it lends weight to Galen's assertion of a communication between the two structures.[92] To verify the patency of this communication, Galen has also stated, as noted above, that it should be probed. Given the proximity of the two structures and their delicate nature, this can easily create a false passage. Galen is well aware of this possibility, but it does not absolve Galen from the charge of creating a false passage here, particularly when the distance between the two areas is so short. Galen's awareness of this possibility is confirmed when he states that a demonstrated communication does not extend the *full length* of the optic tract.[93] Galen therefore does not claim to have demonstrated the patency of this passage for the entire length of the optic tract,[94] helpful though this would be for confirming his thesis that psychic pneuma passes from the ventricles through the optic nerve to the eye.[95] Galen seems aware of the limitations of anatomical demonstration in this instance, even though it would benefit his pneumatic physiology were he to override those limitations by simply asserting that he has *demonstrated* the passage in the optic nerve throughout its entire length. The strength of this particular exhibition must rely on the type of analogy Galen seeks to show in his elucidation of the supposed communication between the superior horn of the anterior ventricle and the olfactory bulb; that is, by apprehension of something λόγῳ θεωρητός. With this form of justification, Galen attempts to overcome the empirical limitations of anatomical demonstration set by the sense of sight.

---

[89] *UP* I, p. 476,21 Helmreich; III. 657 K. The peritoneum, pleura, pia, and the periosteum are similarly organised (*AA* II. 591 K; cf. Garofalo, 1991, II, 623 n. 6).

[90] The removal of the choroid plexus is essential if the detailed nature of the underlying structure is to be appreciated. Cf. Sisson & Grossman, 1975, 221.

[91] Cf. May, 1968, 401 n. 42.

[92] This aperture lies at the terminal part of the floor of the anterior ventricle which overlies the choroid fissure; through this fissure the anterior choroid artery enters to supply the choroid plexus. It is just possible that Galen probed this artery. As already noted, to visualise this area, the choroid plexus must be removed; perhaps the stump of the choroid artery appeared to Galen to represent the means of communication between the ventricle and the optic tract.

[93] Cf. *AA* XIV.2; p. 187 Duckworth.

[94] The nature of the optic chiasm would negate any attempt to probe the tract throughout its full length.

[95] The optic nerve contains "luminous pneuma" (αὐγοειδὲς πνεῦμα). *UP* II, p. 384,16 Helmreich; IV. 275 K. Cf. *PHP* pp. 448,26–27; 460,32 De Lacy; V. 612–613, 627 K.

4.5 *The middle ventricle*

For Galen, the anterior ventricles encompass four functions: they elaborate psychic pneuma, they are the instruments of olfaction, they discharge residues, and ensure the passage of psychic pneuma to the eye via the optic tract.[96] The physiology of the middle or third ventricle is twofold.[97] First, through its communication with the anterior ventricles via the interventricular foramina, it allows psychic pneuma to pass from there to the channel between third and fourth ventricles. Second, by using its complex series of ducts or passages which lead ultimately to the pharynx via the base of the brain, it is responsible for the removal of the heavier waste-products of nutrition.[98]

Elevating the fornix exposes the interventricular foramina. By following the veins that pass through these foramina, the dissector enters what Galen refers to as "the middle space" (ἡ μέση χώρα), the middle ventricle.[99] These veins also supply the choroid plexus.[100] The

[96] αἱ μὲν δὴ πρόσθιοι δύο [sc. τῶν κοιλιῶν] τήν τ' εἰσπνοὴν καὶ τὴν ἐκπνοὴν καὶ τὴν ἐκφύσησιν ἐργάζονται τὴν ἐξ ἐγκεφάλου . . . ἀποδέδεικται δὲ καὶ ὅτι προκατεργάζονταί τε καὶ προπαρασκευάζουσιν αὐτῷ τὸ ψυχικὸν πνεῦμα. καὶ μὲν δὴ καὶ ὅτι τοῖς κάτω μέρεσι σφῶν αὐτῶν τοῖς πρὸς τὰς ῥῖνας ἅμα μὲν ὀσφρητικόν ἐστιν ὄργανον, ἅμα δ' οἷον ὀχετός τις εἰς περιττωμάτων ἐκροὴν ἐπιτήδειος, ὀλίγῳ πρόσθεν εἴρηται. *UP* I, p. 481,6–14 Helmreich; III. 663 K. This last function is unstated here. Cf. *PHP* p. 450,10–13 De Lacy; V. 614 K.

[97] The "middle" because it lies between the anterior and posterior ventricles: ἀλλὰ σύ γε καλῶς ἅπαντα γυμνώσας τὰ προκείμενα κατὰ τὸν λόγον μόρια θεάσῃ τὴν τρίτην κοιλίαν, μέσην κειμένην τῶν τε δύο τῶν προσθίων καὶ τῆς τετάρτης τῆς ὀπίσω. *AA* II. 728 K.

[98] καὶ γὰρ οὖν κἀνταῦθα τῶν κατὰ τὸν ἐγκέφαλον ὑπερκειμένων κοιλιῶν [καὶ] δεχομένων ἐξ ἀνάγκης πολλάκις ἐκ τῶν περιεχόντων αὐτὰς σωμάτων συρρέοντα περιττώματα, συνεχῶς ἂν ἀποπληξίαις ἡλίσκετο τὸ ζῷον, εἰ μή τινα κἀνταῦθ' ἡ φύσις ἐτέμετο πρὸς ἔκρουν ἐπιτηδείαν ὁδόν. ἀλλὰ μὴν οὐδ' ἐπινοῆσαι δυνατὸν ἑτέραν ἀμείνω τῆς εὐρείας θ' ἅμα καὶ κατάντους. ἔσωθεν μὲν οὖν ἔξω διὰ τῶν κατὰ τὰς ῥῖνας πόρων τὰ περιττώματα φέρεται, ἔξωθεν δ' ἔσω τὰ τῆς ὀσφρητικῆς δυνάμεως αἰσθητά, καὶ δύο ταύταις ταῖς χρείαις ἓν ὄργανον ὑπηρετεῖ, τῇ μὲν εἰς αὐτὸ τὸ ζῆν ἀναγκαίᾳ, τῇ δ' εἰς τὸ βέλτιον ζῆν. εἰσὶ μὴν καὶ ἄλλοι δύο τινὲς ὀχετοὶ κατάντεις, ἐξερευγόμενοι δι' ὑπερῴας εἰς τὸ στόμα τὰ τοῦ παντὸς ἐγκεφάλου περιττώματα· καὶ ὅταν γε τελέως εὐπραγῇ καὶ καλῶς περιγίγνηται τῆς τροφῆς, ἀρκοῦσιν οὗτοι μόνοι. ὥσθ' ἡ πρώτη χρεία τῶν εἰς τὰς ῥῖνας ἐγκεφάλου συντρήσεων, ἧς ἕνεκα μάλιστα γεγόνασιν, οὐχ ἡ τῶν περιττωμάτων ἐστὶν ἔκκρισις, ἀλλ' αὕτη μὲν ἐξ ἐπιμέτρου κακοπραγοῦντος ἐγκεφάλου βοήθημα, προτέρα δ' αὐτῆς ἡ τῶν ὀσμῶν διάγνωσις, καὶ ταύτης ἔτι πρεσβυτέρα καὶ πρὸς αὐτό γε τὸ ζῆν ἀναγκαία ἡ εἰς ἐγκέφαλον εἰσπνοή. *UP* I, p. 471,4–26 Helmreich; III. 649–650 K.

[99] γυμνώσεις οὖν αὐτὸ κατ' ἐκεῖνα μάλιστα τὰ χωρία, δι' ὧν αἱ φλέβες, ὥσπερ ὀπῶν τινων ἐξιοῦσαι, ταῖς προσθίαις ἐμπίπτουσι κοιλίαις· κατ' αὐτὰς γάρ τοι τὰς ὀπὰς ἡ μέση χώρα συντέτρηται ταῖς προσθίαις κοιλίαις. *AA* II. 724 K.

[100] In saying that the veins may be observed passing from the anterior ventricles to the middle ventricle via the interventricular foramina, Galen implicitly affirms the presence of the choroid plexus in the middle ventricle. Galen makes no mention of a choroid plexus in the posterior or fourth ventricle. See n. 225, below.

third ventricle begins at the interventricular foramina as a duct that passes directly down to the base of the brain at the *infundibulum*, which marks the anterior limit of the base of the third ventricle. The anatomy of the third ventricle is more complex than any of the other ventricles, a fact dictated by a structure known as the *interthalamic adhesion* or *massa intermedia*.[101] The anatomy of the third ventricle is really a series of spaces which are ducts or passages formed around the anterior, posterior, and inferior surfaces of this paired intermediate mass.[102] Their presence means that the third ventricle is "only a ring-like space."[103] And, as will be seen below, Galen does refer to part of the third ventricle's structure as forming a portion of an "exact circle" (κύκλος ἀκριβής). Galen is aware that some anatomists regard the middle ventricle only as a communicating duct or else consider it part of the fourth ventricle:

> Indeed the encephalon is separated from the cerebellum . . . by the doubling of the thick meninx (*tentorium cerebelli*), but needs to be attached even if only at one part for the sake of forming the canal just referred to, and it has made both anterior ventricles end first in one space, which some anatomists have counted as a fourth ventricle of the whole encephalon. There are some, however, who call it (sc. the third ventricle) the opening (σύντρησις)[104] between the two (sc. anterior and posterior) ventricles and do not grant that it is proper to regard it as another ventricle. I myself believe that it does nothing either to help or to hinder the progress of our discourse on this point if one wishes to interpret this region either as an outlet of the two ventricles or as a third ventricle in addition to them, but I do believe it worthwhile knowing the cause for the anterior ventricles uniting in one place. The reason the anterior ventricles are brought together is for the creation

---

[101] The thalamus is the largest structure in this region, lying on either side of the third ventricle, and is connected across the midline by the interthalamic adhesion. Cf. May, 1986, 687 n. 13; De Lahunta, 1983, 18. It is likely that Galen regarded part of the structure of the thalamus as contiguous with the corpus callosum (cf. Clarke and O'Malley, 1968, 577; Sisson, 1930, 836–837).

[102] "Indeed there is really no third ventricle as such in the ox but rather a series of short interconnected tubes, and it is precisely in these terms that Galen described the ventricle." Millen & Woollam, 1962, 8. Cf. Clarke and O'Malley, 1968, 713.

[103] Fitzgerald, 1961, 41. In prepared sagittal sections, the third ventricle looks larger than it does in a fresh brain, as the fixation process tends to shrink brain tissue and slightly augment ventricular size.

[104] Galen also uses σύντρησις in a more technical sense to describe the surgical creation of a passage or opening into the thoracic cavity (*AA* II. 633, 635, 636, 643 K; cf. *Loc. Aff.* VIII. 52, 54 K). Galen is trying to convey the slit-like nature of the communication between the anterior ventricles, and does so by use of a word which an anatomist would understand as commensurate with the shape made by an incision.

of the canal uniting them with the cerebellum. For from the beginning of the hollow space (sc. the third ventricle), the canal receives the pneuma held there, sending it to the cerebellum.[105]

The first interpretation offered by these anonymous anatomists implies that the area where the two anterior ventricles meet should be considered part of the fourth ventricle of the entire brain.[106] The second interpretation is more restrictive, and maintains that the third ventricle is only the passage between anterior and posterior ventricles. As will be shown below, Galen's interpretation falls between these two views. Apart from this, the fact that Galen sets down rival accounts indicates the complexity of the anatomy of this region, as well as fulfilling a perhaps more urgent desire to dispose of his rivals. Thus in *De anatomicis administrationibus*, what may be the same anonymous anatomists are described as "completely ignorant" (μὴ γνόντες ὅλως) of the presence of the middle ventricle.[107] Yet they have a valid point to make regarding the *function* of the ventricle, which Galen appears to acknowledge in *De usu partium*, above. The anterior ventricles must communicate with the remaining ventricles so that psychic pneuma may ultimately be disseminated to the nerves. In this respect, whether some anatomists refer to the opening between the two anterior ventricles as a third ventricle or simply as a communicating passage is immaterial to Galen's physiology of the brain.

Regardless of whether the middle ventricle is construed as a series of passages or as a ventricle in its own right, to Galen its function is clear. According to him, the reason the anterior ventricles come together at the middle ventricle (via the interventricular foramina) is for the sake of the creation of the canal (the cerebral aqueduct)

---

[105] διειργόμενος οὖν ἀπὸ τῆς παρεγκεφαλίδος ὁ ἐγκέφαλος, . . . τῇ τῆς παχείας μήνιγγος διπλώσει, δεόμενος δὲ κἂν καθ' ἕν τι συναφθῆναι μέρος ἕνεκα τῆς τοῦ προειρημένου πόρου γενέσεως, εἰς μίαν πρότερον χώραν τὰς κοιλίας ἀμφοτέρας ἐπεράτωσεν, ἣν δὴ τετάρτην ἔνιοι τῶν ἀνατομικῶν ἀριθμοῦσι τοῦ παντὸς ἐγκεφάλου κοιλίαν. εἰσὶ δ' οἳ σύντρησιν μὲν αὐτὸ δὴ τοῦτο καλοῦσι τῶν δύο κοιλιῶν, ἑτέραν δέ τινα χρῆναι νομίζειν οὐκ ἐπιτρέπουσιν. ἐγὼ δ', εἴτε κοινὴν ἀμφοτέρων εἴτε καὶ τρίτην τινὰ παρ' αὐτὰς ἑτέραν κοιλίαν ἐθέλοι τις ὑπολαμβάνειν αὐτήν, ἡγούμενος ἔκ γε τούτων οὔτ' ὠφελεῖσθαί τι τοῦ λόγου τὴν προκειμένην διέξοδον οὔτε βλάπτεσθαι, τὴν αἰτίαν τοῦ συνελθεῖν ἐς ταὐτὸ τὰς προσθίας κοιλίας ἀξιῶ γιγνώσκειν. ἔστι δ' αὕτη τις ἡ τοῦ συνάπτοντος αὐτὰς τῇ παρεγκεφαλίδι πόρου γένεσις. ἐκ ταύτης γὰρ ἐκεῖνος ὁρμώμενος τῆς κοιλότητος ἐκδεχόμενός τε τὸ περιεχόμενον ἐν αὐτῇ πνεῦμα, διαπέμπει τῇ παρεγκεφαλίδι. *UP* I, pp. 483,12–484,6 Helmreich; III. 666–667 K.

[106] Cf. May, 1968, 414; T 77a *Herophilus*; cf. Von Staden, 1989, 198; idem, 2000, 14.

[107] *AA* II. 727 K. Cf. II. 416 K, for similarly dismissive views on his contemporaries' inability to make full use of their alleged anatomical training.

which *joins* (συναπτός) the anterior ventricles with the fourth ventricle (the ventricle of the cerebellum). This connection is essential in enabling psychic pneuma to be sent from the anterior ventricles to the spinal cord.[108] Anatomy has shown Galen that all nerves of the body below the head take their origin from the cerebellum and spinal cord.[109] This being so, the ventricles must communicate with these two structures:

> Now since all the nerves of the body below the head grow either from the cerebellum or from the spinal cord, the ventricle of the cerebellum (sc. the fourth ventricle) had to be large and had to get a share of the psychic pneuma previously elaborated in the anterior ventricles. Just as there must of necessity be a passage from those (ventricles) to this ventricle. This ventricle appears large but the passage which enters it from the anterior ventricles is very large; it is by this passage alone that the union (ἡ σύμφυσις)[110] between the cerebellum and [the rest of] the brain exists.[111]

The existence of a third ventricle or its equivalent is assumed by Galen in the above citation, for the "large passage" which connects the anterior and middle ventricle to the fourth ventricle is the aqueduct, the posterior prolongation of the third ventricle. The duct system of the third ventricle is indispensable for the removal of waste products from the brain. Their anatomy is as follows. The beginning of the third ventricle is the conjoined interventricular foramina of each anterior ventricle. Posterior to this, and roughly at the same level, lies the *pineal* (κωνάριον) and its recess. The beginning of the aqueduct is behind the pineal. At the base of the third ventricle, and almost perpendicularly below the interventricular foramina, is the *infundibulum* (πύελος), and its recess leading to the underlying *gland* (ἀδήν) the

[108] *UP* I, p. 484,2–6 Helmreich; III. 667 K.

[109] As noted in chapter 3.2, the cranial nerves are a special case since they emerge from the underside of the brain itself. Cf. *AA* IX.9, pp. 8–10 Duckworth. See Savage-Smith, 1971. On the dissection of the spinal cord see *AA* II. 681–684 K; IX.13, pp. 20–26 Duckworth. Cf. Daremberg, 1841, 38–42.

[110] This term is perhaps used here in an Aristotelian sense of organic continuity (cf. *Metaph.* 1014b22), in order to emphasise the contiguous nature of cerebrum and cerebellum.

[111] Ἀλλ' ἐπεὶ πάντα τὰ κατὰ τὸ σῶμα νεῦρα τὰ κάτω τῆς κεφαλῆς ἢ ἐκ τῆς παρεγκεφαλίδος ἢ ἐκ τοῦ νωτιαιόυ πέφυκεν, ἐχρῆν καὶ τὴν ταύτης κοιλίαν ἀξιόλογόν τ' εἶναι τὸ μέγεθος καὶ τὸ προκατειργασμένον ἐν ταῖς προσθίοις ψυχικὸν πνεῦμα μεταλαμβάνειν. ὥστ' ἀναγκαῖον ἦν γενέσθαι τινὰ πόρον ἐξ ἐκείνων εἰς ταύτην. ἀτὰρ οὖν καὶ φαίνεται μεγάλη μὲν αὕτη, μέγιστος δὲ καὶ ὁ ἀπὸ τῶν ἔμπροσθεν κοιλιῶν εἰς αὐτὴν ἐμβάλλων πόρος. καὶ κατὰ τοῦτόν γε μόνον ἡ σύμφυσίς ἐστι τῇ παρεγκεφαλίδι πρὸς τὸν ἐγκέφαλον. *UP* 1, p. 482,14–23 Helmreich; III. 665 K.

hypophysis.[112] Each of these features represents a corner of a roughly quadrilateral space, with most of it taken by the massa intermedia. Before continuing with Galen's anatomy of the third ventricle, three further matters should be noted. First, Galen, to have made the following observations, has opened the middle ventricle from above. Second, Galen has also exposed the roof of the aqueduct, or at least its proximal part. Third is the overwhelming presence of the large massa intermedia.[113] Galen does not distinguish it by name, regarding it as part of the surrounding brain substance, or, as he puts it: "all that which lies above the oblique duct".[114] The massa intermedia then, dictates the nature of the discharge ducts as well as determining the manner of Galen's anatomical demonstration of the third ventricle, especially the conjoined ducts that make up its base. The key to understanding the structure and function of the third ventricle therefore, lies in tracing the ducts which lead to the infundibulum and underlying hypophysis.[115] In the Arabic part of Book IX of *De anatomicis administrationibus*, Galen examines these passages:

> As for what relates to the other two 'outlet' ducts, the ones at the middle ventricle, of those you can see that which flows out of the duct (sc. the aqueduct) connecting the middle with the posterior ventricle, of which I said that it [*inferior horn of lateral ventricle*][116] travels obliquely downwards and forwards.[117] First you can examine properly the union of the two,[118] then see how both are picked up by the single duct, which goes straight downwards, and into which both discharge. Then you can see afterwards how that delicate meninx covers over the duct externally.[119] Next see how this meninx also is perforated, in the place where the brain finally ends. You can see here how the dura mater[120] lies beneath its base [the lower aspect of the brain].[121]

---

[112] On the use of the term 'gland' in Galen see Marmelzat, 1991, 423.

[113] What Woollam, 1958, 18, describes as the "complicated bifid path" of the effluent ducts of the third ventricle is due to the presence of this structure.

[114] *AA* IX.7; p. 6 Duckworth.

[115] Its function, according to Galen, is to filter phlegmatous waste-products passed down to the infundibulum. Cf. May, 1968, 425 n. 2.

[116] Duckworth's gloss is not entirely accurate (cf. Simon, 1906, II, 5, n. 14). Better would be: "in the same direction as the inferior horn of the lateral ventricle."

[117] As will be explained below, this refers to the duct which will form the conjoined duct in the infundibular region. It cannot be a reference to the aqueduct since this structure projects downwards and backwards into the fourth ventricle.

[118] From the point where the passage from the outlet of the aqueduct bifurcates owing to the massa intermedia.

[119] Probably the conjoined hypophyseal and infundibular recesses.

[120] The dura mater lining the cranial fossa in the hypophyseal region, the base of the brain being of course *above* this (cf. Simon, 1906, II, 5, n. 23).

[121] IX.7; p. 5 Duckworth. My gloss in round brackets; square brackets Duckworth.

This is a complex anatomical narrative. Hunayn's use of the present continuous does not help in trying to determine whether Galen is discussing structures previously examined, or is referring to them by means of other brains either accessible at the time, or dissected previously (although in the Greek part of Book IX it is sometimes not clear either). But since we are dealing with a teaching text, it is likely there were several specimens in different stages of dissection.[122] But however Galen organised his material, the reader *qua* student is presented with an account that is intended to be definitive. Attention is drawn first to the "outlet" that arises from that duct which connects the middle and posterior ventricle (the aqueduct).[123] This "outlet" duct is said to run obliquely downwards and forwards, and therefore lies first behind and then below the massa intermedia. As noted earlier, the presence of this brain substance accounts for the obliquity of the bifid path to the infundibulum at the base of the brain.[124] Galen next gives instructions for the midline division of this mass. The following is a direct continuation of the above citation:

> Begin the examination of these things by introducing the sound from out of the oblique duct (i.e. anterior to the commencement of the aqueduct),[125] and carry it onwards to the duct descending downwards from the middle ventricle (sc. the descending duct from the interventricular foramina), and then, as you wish, strain it upwards as strongly as you may be able, so that all that which lies above the oblique duct may be cut through,[126] and it may be completely uncovered, as far as to the spot where it unites itself with the duct travelling directly downwards.[127]

Galen has introduced a sound from the proximal part of the aqueduct, which is then passed forwards through a passage that communicates with the lower end of the duct and is almost at right

The account here is paralleled in *UP* II, pp. 7,18–8,10 Helmreich; III. 693–694 K, but there the stress is placed upon the anatomical detail of the infundibular region.

[122] And if so, then another reason why Galen singles out the plentiful supply of ox brains in the major cities.

[123] Above this point lies the pineal, marking the transition point from third ventricle to aqueduct. See below.

[124] Cf. Millen & Woollam, 1962, 8; Woollam, 1958, 16–17.

[125] If "the oblique duct" is understood only as a reference to the interventricular foramina, then Galen's subsequent instructions for dissection become both superfluous and misleading.

[126] That is, the obstructing bulk of the massa intermedia is bisected by the probe. From Galen's description it becomes obvious that the base or floor of the third ventricle is exposed by this procedure. This is only possible if the sound is placed where Galen has indicated and which I have glossed accordingly.

[127] IX.7; pp. 5–6 Duckworth. My glosses in round brackets.

angles to it. The massa intermedia is always above the sound. The concept of a large space, interpreted by Galen as a passage under the massa intermedia, is well brought out in the extant Greek of Book IX of *De anatomicis administrationibus*:

> Lying below in the base of the third ventricle there is a large duct which, as I said a little earlier, receives, via the holes (sc. the interventricular foramina), the residues from the anterior ventricles as well as from the overlying bodies.[128]

This "large duct" is the floor of the third ventricle which lies under the massa intermedia and leads to the infundibular recess. It consists of the two conjoined ducts derived from the proximal part of the aqueduct and the interventricular foramina. Galen also stresses here the importance of this space as the drainage area not only of the anterior ventricles but of the surrounding brain itself, referred to here as the "overlying bodies". When the sound has been drawn firmly upwards, the massa intermedia is split apart, and the contents of the floor of the third ventricle are completely exposed:

> When you have executed this exposure, then it is clearly seen that the hardness of its inner surface (sc. of the base of the infundibulum) is not something which came into existence through our stretching of those bodies which we strained, but on the contrary it was in this state previously. Also, the duct into which both the two open out (sc. into the infundibular recess) will give you a better indication of its own nature, and thereupon convince you of what, in fact, the position is with regard to both of the two. That is, what surrounds it is the pia mater, and beyond, a pair of arteries, which ascend upwards from the basis cranii, lie upon the pia mater [*middle cerebral arteries*].[129]

In the ox brain, the region of the infundibulum at the base of the third ventricle is prolonged downwards into the hypophysis as a definite space, the infundibular recess.[130] To further orient the dissector, Galen has drawn attention to a pair of arteries which lie on

---

[128] καὶ μὲν δὴ καὶ πόρος τις ὑπόκειται μέγιστος ἐν τῇ βάσει τῆς τρίτης ταύτης κοιλίας, ἔκ τε τῶν προσθίων δεχόμενος τὰ περιττὰ κατὰ τὰς εἰρημένας ὀπάς, καὶ προσέτι τῶν ὑπερκειμένων σωμάτων, ὡς ὀλίγον ἔμπροσθεν ἐρρέθη. *AA* II. 727 K. Singer, 1956, 235, renders καὶ προσέτι τῶν ὑπερκειμένων σωμάτων as "and there is another [duct] from the tissues there also". His gloss, n. 176, states, "The first duct is the groove in the floor of the third ventricle leading to the infundibulum. The second is the aqueduct." But the subject of the sentence is the large duct in the base of the third ventricle.

[129] IX.7; p. 6 Duckworth. My glosses in round brackets; square brackets Duckworth.

[130] Cf. May, 1968, 429 n.6; Simon, 1906, II, 5, n. 20.

the pia mater, which, as will be discussed in chapter 6, are important landmarks to the retiform plexus. These two arteries demarcate the region of the infundibulum:

> Also you see that just as two arteries, one on each side, ascend from below and reach the brain and unite themselves and pass over the one into the other at the place called the 'pool' or 'cistern' [*infundibular recess*],[131] similarly at that place, still other numerous structures lie partly on the anterior, partly on the inferior subdivisions, partly at the sides. That is to say, the two passages that lead to the eyes . . . at the place where they are preparing to pierce the cranial bones, are to be found conjoined at a point near the anterior region of the pool or cistern.[132]

The two ducts join at the infundibulum of the third ventricle, whose recess is visible to the naked-eye. The infundibulum is the stalk of the *hypophysis* (*pituitary gland*). The *infundibular recess* is the cavity in the upper part of the stalk which Galen believed led directly to the gland. The above two accounts from *AA* may usefully be compared with the following from *De usu partium*:[133]

> Of the two other (outlet) ducts which extend down to the palate, one arises at the base of the middle ventricle of the brain, passing steeply down; the other begins at the canal[134] which connects the encephalon to the cerebellum and slants downwards to the first duct. When they first come together, both are received into a common space, hollow and (sloping) downwards, the lip of which above is a perfect circle.[135] Thence, becoming ever more narrow, it forms itself beneath into a gland that is like a flattened sphere, and which has a perceptible cavity. It is followed by a bone which resembles a sieve (sc. the ethmoid) and which finishes at the palate, and this is the path of the thick residues.[136]

---

131 *Infundibulum*: Simon, 1906 II, 6. But Duckworth's gloss rightly stresses that this is a reference to the recess proper.

132 IX.8; p. 6 Duckworth. Square brackets Duckworth.

133 All things being equal, one would expect the account in *AA* to augment that of *UP* and not the other way around, as seems the case here. This may or may not indicate a hiatus in this part of *AA*, but Galen often uses *UP* to augment a particular anatomical point, the retiform plexus being one example.

134 May, 1968, 428 n. 5, states that this canal is "not the aqueduct." In fact, it is its proximal part. See section 4.6, below.

135 The curved lower part of the massa intermedia.

136 τῶν δ' ἄλλων δυοῖν τῶν εἰς τὴν ὑπερῴαν καθηκόντων ὁ μὲν ἐκ τοῦ πυθμένος τῆς μέσης κατὰ τὸν ἐγκέφαλον κοιλίας ὁρμηθεὶς τὸ κάταντες φέρεται, ὁ δ' ὑπόλοιπος ἄρχεται μὲν ἀπὸ τοῦ συνάπτοντος πόρου τὸν ἐγκέφαλον τῇ παρεγκεφαλίδι, φέρεται δ' ἐπὶ τὸν ἕτερον εἰς τὸ κάτω λοξός. ἡνίκα δ' ἂν πρῶτον εἰς ταὐτὸν ἀλλήλοις ἀφίκωνται, δέχεταί τις ἀμφοτέρους χώρα κοινὴ κοίλη καὶ κατάντης. κύκλος μὲν ἀκριβής ἐστιν αὐτῆς τὸ ἄνω χεῖλος· ἐντεῦθεν δ' ἀεὶ καὶ μᾶλλον στενουμένη καταφύεται εἰς ὑποκείμενον ἀδένα

Galen's "perceptible cavity" (αἰσθητὴ κοιλία), is the *infundibular recess*, and connects the infundibulum to the underlying gland (hypophysis). It is a notable feature of the ruminant brain.[137] In domestic animals: "The funnel-shaped infundibular recess extends the ventricle vertically into the infundibulum, which is the stalk of the hypophysis."[138] This is well brought out by Galen:

> The upper part of the cavity (the infundibulum) which receives the channels and which some name basin on account of its shape, and others funnel on account of its usefulness, serves as it were the purpose of a cistern; its lower part is likened to a funnel, as its name betokens; for it is pierced by a perceptible channel (sc. the infundibular recess) which leads down as far as the cavity in the gland.[139]

The *infundibulum* (πύελος, χοάνη, or δεξαμένη) is prolonged as a recess which does not communicate with the hypophysis. However, if the end of the infundibulum is torn when the brain is removed from the skull, then Galen could legitimately have deduced that the infundibular recess communicated with the hypophysis. The cavity in the gland to which Galen refers is especially noticeable on the under surface of the brain as a distinct hole if, when removing the brain from the skull, the infundibulum and the hypophysis (through the attached stalk) is torn, as easily happens. This hole is then seen to lead directly upwards into the third ventricle, and it is not unlikely that Galen came to his conclusion regarding the nature of the communication between infundibular recess and the underlying gland from such an observation. Moreover, for Galen to visualise the hypophysis, a further dissection has to be performed, which involves cutting part of the dura mater at the base of the skull:

> In the bone above the palate (sella turcica of the basi-sphenoid) upon which the spongy glandular flesh rests and supports itself, you find

---

πεπλατυσμένῃ σφαίρᾳ παραπλήσιον, αἰσθητὴν ἔχοντα καὶ αὐτὸν κοιλίαν. ἐκδέχεται δὲ τοῦτον ἠθμῷ τινι προσεοικὸς ὀστοῦν εἰς τὴν ὑπερῴαν περαῖνον. αὕτη μὲν ἡ τῶν παχέων περιττωμάτων ὁδός. II, pp. 7,22–8,10 Helmreich; III. 693–694 K.

[137] Cf. Sisson and Grossman, 1975, 955–959, 955.

[138] King, 1987, 280. Cf. Sisson, 1930, 836.

[139] ἡ μὲν ὑποδεχομένη τοὺς πόρους κοιλότης, ἣν οἱ μὲν ἀπὸ τοῦ σχήματος πύελον, οἱ δ' ἀπὸ τῆς χρείας χοάνην ὀνομάζουσιν, ἐκ μὲν τῶν ἄνω μερῶν οἷον δεξαμενῆς τινος ἐπέχει χρείαν· ἐκ δὲ τῶν κάτω, καθάπερ αὐτὸ τοὔνομα ἐνδείκνυται, χοάνην μεμίμηται· διατέτρηται γὰρ εἰς τὸ κάταντες αἰσθητῷ πόρῳ μέχρι τῆς κατὰ τὸν ἀδένα κοιλίας. *UP* II, p. 8,12–18 Helmreich; III. 694 K.

narrow perforations.[140] All this lies in this place.[141] Its structure (sc. the hypophysis) resembles that of a sieve or a strainer with which one strains and clarifies turbid beverages—by which I mean everything to which the name 'sieve' or 'strainer' is applied.[142] The first part of it (sc. the infundibulum) that you see is the duct of which I said that it descends from the brain (the infundibular recess), and the pia mater envelops it here. After that, in the place where this duct vanishes, sinking into the depth of the dura mater which is perforated in exactly the same manner as the pia mater,[143] you see the lupine-bean-shaped spongy flesh.[144] And at this flesh [*gland*] ends the effluent of the pool or cistern. When you have reached this place, then cut away the portion of dura mater which lies here below the site[145] of the brain and above the pituitary.[146] The substance of this gland is flabby and, round about

---

[140] May, 1968, 429 n. 7, states that: "The sella turcica of the sphenoid bone is pierced by foramina for blood vessels, and Galen may have thought that the residues descended through these into the body of the sphenoid." The sella turcica is covered by the "diaphragma sellae . . . a horizontal sheet of dura mater which bridges the sella turcica . . . It has an opening for the infundibular part of the hypophysis." Sisson and Grossman, 1975, 224, Cf. Owen, 1868, III, 206.

[141] I have taken the liberty to transpose the next two sentences of Duckworth's translation. As originally placed, the impression given is that the infundibular recess is the subject of Galen's disquisition about perforations. But these perforations must refer to the foramina in the sella turcica of the sphenoid and not to any perforations of the infundibular recess. Simon's edition indicates no hiatus in Hunayn's text (nor does Garofalo's translation). However, a lacunae, if it had already existed in the Greek text available to Hunayn, could have been filled in any number of ways by the translator (cf. Bergsträsser, 1925, 16). As Gotthard Strohmaier, points out, in the case of a lacuna in the Greek, "the translator was liable to make any sense out of a senseless statement." (personal communication). Accordingly, I have felt justified to make this change on the grounds of anatomical clarity and the consistency of Galen's exposition which he has already demonstrated in the extant Greek text of *De anatomicis administrationibus*.

[142] Although Galen states (*UP* II, pp. 8,21–9,1 Helmreich; III, 694–695 K) that "the body of the infundibulum" (τὸ σῶμα τῆς πυέλου) is made from the surrounding pia mater, this is not the same as saying the infundibulum possesses multiple perforations. It is better taken as a reference to the sella turcica of the sphenoid.

[143] As noted above, this perforation in the dura probably refers to the aperture noted when the brain is removed from the skull and the infundibulum is torn where it joins the hypophysis.

[144] An easily understood comparison in Galen's time. The bean-like seeds of the *Lupinus* species were an easily cultivated crop for cattle in winter, as well as a standby in cases of famine. Cf. *AA* IX.8; p. 7 Duckworth. See also Simon, 1906, II, 6 and n. 30; Garnsey, 1988, 52.

[145] Garofalo, 1991, II, 833, glosses the word 'site' ("regione") with a question mark. But there is nothing uncertain in this choice of word. It merely conveys that the place of interest is under the area where the relevant part of the brain lies.

[146] At first glance a curious sentence which seems unnecessary. After all, to get to this point in the dissection, Galen has just carried out the very procedure. However, it is possible that Galen is asking for the dissection to be widened in order to fully visualise the complex network of the retiform plexus. In duplicating

> it, is the so-called reticulate tissue (the retiform plexus), in which innumerable arteries travel entwined about one another in manifold fashion.[147]

Galen has progressively shifted the focus of the dissection from the interior of the brain to a consideration of a structure lying outside the brain substance in the base of the skull but outside the barrier of the dura mater, and which is directly linked to the brain by the third ventricle. This connection is made possible by Galen's "perceptible cavity" (the infundibular recess) to the underlying gland. According to Galen, the lower part of the infundibulum projects below the brain through a perforation in the dura and is attached to the gland lying below. To emphasise the delicate nature of this connection, Galen states that the lower part of the infundibulum is "membrane-like" (ὑμενοειδής) since it is *fastened* (σύνδεσμος) to the brain above and the gland below.[148] The surrounding pia mater provides both the material for the attachment and the construction of the lower part of the infundibulum itself.[149] The filtration of these residues is left to the hypophysis.[150] Its position suggests its role to Galen, who notes its topography with care: it lies *outside* the dura, and the space between the dura and the bone of the palate precisely equals the depth (i.e. the length) of this gland.[151] And it is the hypophysis' demonstrated connection with the third ventricle which enables Galen to state the usefulness of this gland. Indeed, Galen goes so far as to claim that its role in removing waste products from

---

this dissection, it was necessary to cut a space in the dura on either side of the sella turcica in order to expose the retiform plexus in the cavernous sinus. In *UP* II, p. 22,9–12 Helmreich; III. 713 K, it is stated that what lies beneath the dura in this region are the pituitary, retiform plexus and the passage which leads to the palate: οἷς δ' ἐπιβέβηκεν αὐτή τε καὶ ὁ κατ' αὐτὴν ἐγκέφαλος, ὅ τ' ἀδήν ἐστι καὶ τὸ δικτυοειδὲς πλέγμα καὶ ἡ εἰς τὴν ὑπερῴαν διάτρησις.

[147] *AA* IX.8; p. 7 Duckworth. My glosses in round brackets; square brackets Duckworth.

[148] *UP* II, p. 8,20–24 Helmreich; III. 694 K. Cf. II, p. 47,7–15 Helmreich; III. 747–748 K, where Galen says that the thick meninx gives off "membranous outgrowths" (ὑμενώδεις ἀποφύσεις) linking the brain to the perforations at the nostrils (i.e. the ethmoid), and the mouth (via the sphenoid).

[149] ἐπεὶ δ' ἄνωθεν μὲν αὐτὴν ἐχρῆν αὐτῷ τῷ ἐγκεφάλῳ συνῆφθαι, κάτωθεν δ' εἰς τὸν ἀδένα καταφύεσθαι, δεόντως ὑμενώδης ἐγένετο. περιέχοντος δὲ καὶ αὐτὸν τὸν ἐγκέφαλον ὑμένος λεπτοῦ, τῆς χοριοειδοῦς μήνιγγος, οὐκ ἦν εὔλογον ἄλλον τινὰ σύνδεσμον αὐτῇ ζητεῖν πρὸς τὸν ἐγκέφαλον. δεόντως οὖν μοῖρά τις ἀπὸ τῆς μήνιγγος ταύτης ἀποταθεῖσα τὸ σῶμα τῆς πυέλου συνεστήσατο. *UP* II, pp. 8,18–9,1 Helmreich; III. 694–695 K.

[150] Cf. Daremberg, 1841, 33–34.

[151] ὅτι δ' ἐκτὸς ἤδη τῆς παχείας μήνιγγός ἐστιν ὁ ἀδὴν οὗτος . . . καὶ ὡς τοσοῦτόν γε τὸ διάστημά ἐστι τοῦ κατὰ τὴν ὑπερῴαν ὀστοῦ πρὸς τὴν μήνγγα, ὅσονπερ καὶ τὸ τοῦ ἀδένος βάθος. *UP* II, p. 9,21–25 Helmreich; III. 696 K.

the brain is *obvious* (πρόδηλος), something which other anatomists have failed to recognise.[152] To be sure, given Galen's teleological standpoint, every structure of the body has a prescribed function.[153] In this light, the infundibulum is ideally situated for the discharge of these residues. Indeed, its could not have been better placed.[154]

The termination of the base of the third ventricle into the hypophysis is the second part of a dual system which, Galen maintains, is responsible for the "clearance of the waste products of nutrition" (ἐκκαθαίρειν τὰ περιττώματα τῆς τροφῆς) from the brain.[155] The first part consists of "fine openings",[156] the sutures of the skull, discussed in chapter three. The second part of this system, according to Galen, is designed to remove the heavy, thick residues, and comprises the passages from the anterior ventricles to the nostrils, as well as the set of ducts opening into the base of the third ventricle which pass these wastes to the hypophysis and thence to the palate. The only other gland associated with the third ventricle, the *pineal*, is for Galen a supporting structure for the veins in this region (and *inter alia* for the third ventricle itself). It is also an important guide to the anatomy of the canal connecting the third ventricle with the fourth.

Like the hypophysis, the pineal lies outside the third ventricle. The chief difference between them is that the pineal is within the meningeal covering of the brain; the hypophysis lies outside the dura mater covering the base of the skull. The pineal "lies upon the beginning of the canal connecting the middle ventricle with the posterior encephalon."[157] In substance it is a gland, and, according to Galen,

---

152 περὶ δὲ τῆς κατὰ τὸν ἐκδεχόμενον αὐτὴν ἀδένα χρείας, ὅτι μὲν διηθεῖ τὰ περιττώματα, πρόδηλον καὶ οὐδὲν τοῦτο μέγα τοῖς ἀνατομικοῖς ἔγνωσται. *UP* II, p. 9,1–3 Helmreich; III. 695 K.

153 The closing words of Book VIII of *De usu partium* reiterate this message; the parts of the brain are so arranged that any other would only result in improper function. Nature has arranged all the parts in a perfectly skilful way: δῆλον δ' ἤδη τοῖς γε προσεσχηκόσι τῷ λόγῳ τὸν νοῦν, ὡς, εἰ καὶ τῶν ἄλλων τι τῶν εἰρημένων μετακοσμηθείη, πολλαχῇ μὲν βλαβῆναι μόνον, ἔστιν ὅπῃ δὲ καὶ τελέως ἀπολέσθαι συμβήσεται τὴν ἐνέργειαν. ὥστ' ἐγὼ μὲν οὐκ ἔχω συμβαλεῖν, ὅτῳ τις ἂν ἐπιχειρήσειε τρόπῳ δεικνύναι ταῦτα μὴ οὐ τῆς ἀκριβεστάτης τέχνης ὑπάρχειν ἔργα. I, p. 496,2–9 Helmreich; III. 683 K.

154 ἡ γάρ τοι πύελος, . . . ἡ τὸν ἐκκαθαίροντα τὸν ἐγκέφαλον ἐν ἑαυτῇ περιέχουσα πόρον, ἐν τούτῳ τέτακται, καὶ ἦν ἀδύνατον ἑτέρωθι κάλλιον αὐτὴν μετατεθῆναι, μέλλουσάν γε τὸ περίττωμα πᾶν εἰς ὑπερῴαν ἐξερεύγεσθαι. *UP* II, p. 106,8–12 Helmreich; III. 831–832 K.

155 *UP* II, p. 1,11 Hemreich; III. 684–685 K.

156 *UP* II, p. 3,3 Helmreich; III. 686 K. Cf. May, 1968, 425 n. 2.

157 τὸ κατὰ τὴν ἀρχὴν τοῦ συνάπτοντος (sc. ἡ μέση κοιλία) πόρου πρὸς τὸν ὀπίσθιον ἐγκέφαλον, *UP* I, p. 489,11–13 Helmreich; III. 674 K.

takes its name from its shape.[158] In fact, it is called τὸ κωνάριον ("the small pine cone"), says Galen, "by those who know anatomy" (πρὸς τῶν περὶ τὰς ἀνατομὰς ἐχόντων).[159] Its function, as with other glands, is supportive; it sustains the "large vein" (μεγάλη φλέψ, sc. the great cerebral vein of Galen) at its point of bifurcation. This vein runs forward under the corpus callosum to enter the interventricular foramina, supplying the greater part of the veins which contribute to the choroid plexuses. When these veins enter the brain, the supporting action of the pineal no longer applies, and the veins (together with the pineal) are then supported by the brain itself.[160] The pineal also marks the beginning of the duct which links the third ventricle with the fourth (the aqueduct), and is a crucial landmark for understanding the topography of this channel. For Galen, the position of the pineal allows the dissector to take stock of the complex anatomy of the duct system of the third ventricle:

> Lying below in the base of the third ventricle there is a large duct which, as I said a little earlier, receives, via the holes, the residues from the anterior ventricles as well as from the overlying bodies. And the residues flow together into there, at the site where the veins from the pineal enter (sc. the interventricular foramina). But those who are completely ignorant of this (third) ventricle are, it follows, also ignorant of the duct extending backwards,[161] from where the pineal is affixed. And if it is stripped of the veins which surround it and broken off along the base, a certain opening is there observed high up.

---

[158] ἔστι δὲ τοῦτο τὴν μὲν οὐσίαν ἀδήν, τὸ δὲ σχῆμα κώνῳ μάλιστα παραπλήσιον, ὅθεν αὐτῷ καὶ τοὔνομα. *UP* I, p. 489,15–17 Helmreich; III. 675 K.

[159] *UP* I, p. 489,13–14 Helmreich; III. 674 K.

In the ox, the pineal is "elongated into a cylinder", Solly, 1847, 120. It is "long and fusiform", Sisson, 1930, 835; cf. Woollam, 1958, 15. In the sheep it is shorter and round to ovoid; cf. Sisson and Grossman, 1975, 958. The pineal in the human is ovoid. On the comparative anatomy of the pineal see Tilney and Warren, 1919.

[160] τὸν δ' ἀδένα τοῦτον τὸν τῷ κώνῳ παραπλήσιον, ἀναπληροῦντα τῆς μεγάλης φλεβὸς τὴν σχίσιν, ἀφ' ἧς ἅπαντα σχεδὸν τὰ κατὰ τὰς προσθίους κοιλίας χοριοειδῆ πλέγματα συνίσταται, τῆς αὐτῆς χρείας ἕνεκα τοῖς ἄλλοις ἀδέσι τοῖς στηρίζουσι τὰς σχιζομένας φλέβας ἡγοῦμαι γεγονέναι. καὶ γὰρ οὖν καὶ ἡ θέσις ἡ αὐτὴ κατὰ πάντ' ἐστὶν αὐτῷ τοῖς τοιούτοις ἀδέσι, τὴν μὲν κορυφὴν ἐκείνοις ἐνηδρακότι τοῖς μέρεσι τῆς φλεβός, ἵνα πρῶτον σχίζεται, τὸ δ' ἄλλο πᾶν ἀνάλογον τῇ διαστάσει τῶν ἐκ τῆς σχίσεως γεννωμένων ἀγγείων ἐπαυξανομένῳ καὶ μέχρι γε τοσούτου προϊόντι, μέχρι περ ἂν κἀκεῖνα μετέωρα φέρηται. ἡνίκα δ' ἂν πρῶτον ἐπιβῶσιν αἱ φλέβες αὗται τοῦ κατὰ τὸν ἐγκέφαλον αὐτὸν σώματος, ἀπολείπει μὲν αὐτὰς ἤδη τὸ κωνάριον, ἕδρα δ' αὐτῷ τε τούτῳ καὶ ταῖς φλεψὶν ἅμα γίγνεται τὸ ταύτῃ σῶμα τοῦ ἐγκεφάλου. *UP* I, p. 490,2–17 Helmreich; III. 675–676 K. Cf. Sisson, 1930, 689.

[161] Singer, 1956, 235, glosses this as the aqueduct; Daremberg, 1841, 34 n. 1, is uncertain, but believes it too refers to the aqueduct. Strictly speaking, for Galen it marks the *beginning* of the aqueduct. See section 4.6, below.

> This, to speak plainly, is just like a smoke hole, although the brain itself possess no vapour to breathe through this high lumen.[162]

According to Galen, the veins from the pineal enter the third ventricle through the interventricular foramina. Denuded of its surrounding veins, it can now be dissected from its position on the roof of the third ventricle ("the top of the duct"), and the nature of its internal opening demonstrated. The aperture in the stalk of the pineal is the pineal recess of the third ventricle. The surrounding veins and adnexa support the pineal; it cannot maintain its upright position without them. Denied this vascular support, the pineal falls backward on to a set of structures which are parts of the brain lying immediately in front of the cerebellum.[163] These will be discussed shortly. It is likely that the reason Galen has been so meticulous in delineating the anatomical relationships of the pineal is in order to assure himself that there is no further passage from it, as well as to ensure that anyone who performs a similar dissection is not misled into thinking that the blind passage in the pineal is a true passage—whether for vapour, air, pneuma or waste products. In any case, as Galen knows, lying over the pineal is the brain substance and the tentorium. It may be thought unnecessary for Galen to stress this anatomical barrier, since there is no hole at the tip of the pineal to allow any substance to pass through. But Galen's argument is more teleological than anatomical: a passage at this point would mean that Nature is acting without purpose.[164] The pineal acts as a marker for the interventricular foramina—and thus for the anterior part of the

---

162 καὶ μὲν δὴ καὶ πόρος τις ὑπόκειται μέγιστος ἐν τῇ βάσει τῆς τρίτης ταύτης κοιλίας, ἔκ τε τῶν προσθίων δεχόμενος τὰ περιττὰ κατὰ τὰς εἰρημένας ὀπάς, καὶ προσέτι τῶν ὑπερκειμένων σωμάτων, ὡς ὀλίγον ἔμπροσθεν ἐρρέθη. συρρεῖ δ' εἰς αὐτὴν ἐκεῖνα, καθ' ὃ μέρος αἱ ἀπὸ τοῦ κωναρίου φλέβες ἐμβάλλουσιν. ἀλλ' οἱ μὴ γνόντες ὅλως τὴν κοιλίαν ταύτην εἰκότως ἠγνόησαν ἅμ' αὐτῇ καὶ τὸν εἰς τὸ ὀπίσω πόρον ἐκτεταμένον, ἐφ' οὗ τὸ κωνάριον ἐστήρικται. καὶ εἰ γυμνόν τις αὐτὸ τῶν περικειμένων φλεβῶν ἀπορρήξει κατὰ τὴν βάσιν, ὀπή τις ἐνταῦθα φαίνεται μετέωρος, ἀτεχνῶς φάναι, καθάπερ τις καπνοδόχη, καίτοι μηδένα καπνὸν ἔχοντος τοῦ ἐγκεφάλου καθ' ἑαυτόν, ὃν ἀναπνεῦσαι δεήσεται διὰ μετεώρου τρήματος. *AA* II. 727 K.

163 ἀλλὰ τό γε κωνάριον, ἐπειδὰν γυμνώσῃς μὲν αὐτὸ τῶν περικειμένων, φυλάξῃς δ' ἐπικείμενον τῷ πόρῳ, καταπίπτειν εἴωθεν, οὐκ ὀρθὸν ἑστάναι, καθάπερ ὅτ' ἠμφίεστο τοῖς ὑμέσι μετὰ τῶν ἀγγείων· καὶ καταπίπτει γε τοὐπίπαν ὀπίσω ῥέπον. *AA* II. 728–729 K; p. 577 Garofalo, 2000.

164 οὐ μὴν οὐδ' εἰς τὸν πέριξ ἀέρα διικνεῖται τὸ στόμιον αὐτῆς, ἀλλ' ἐπίκειται πλεῖστον ἐγκεφάλου μέρος ἄνωθεν, εἶθ' ἡ σκληρὰ μῆνιγξ διπλῆ, καὶ μετὰ ταύτην τὸ τῆς κεφαλῆς ὀστοῦν, ὥστε μάτην ἂν ἡ φύσις εἰργάσατο τὴν ὀπὴν τήνδε, καίτοι γε οὐδὲν εἰκῇ ποιοῦσα. *AA* II. 727 K.

third ventricle—since by following the veins from the pineal one is lead to their point of entry into these foramina. By its position at the roof of the third ventricle, the pineal is also a guide to the cerebellum (via the aqueduct) and the infundibular region of the third ventricle.[165] When the pineal is stripped of its veins, it falls backward, says Galen, onto two pairs of rounded structures that are distinct parts of the brain and of a similar substance to it. The first pair are called "the gloutia" (τὰ γλουτία),[166] the second, "the testes" (οἱ ὄρχεις), or, with some modesty, "the twins" (τὰ διδύμια),[167] after their supposed resemblance to these parts of the body.[168] These structures, which will be discussed in the following section, help determine Galen's location of the aqueduct.

Before leaving the discussion of the pineal, Galen further stresses its supportive nature by stating what it does *not* perform. This provides some information on other ways of anatomical thinking which Galen addresses if only to refute. It is, says Galen, the opinion of *some* (τινες) unnamed doctors that the pineal functions similarly to that of the pylorus of the stomach, which acts, according to Galen, as a valve to prevent the contents of the stomach from prematurely entering the duodenum.[169] In the analogy used by these anonymous doctors, the pineal regulates the amount of psychic pneuma transmitted from the third ventricle to the fourth.[170] As will be discussed

---

[165] τὸν γάρ τοι πόρον, ᾧ τὸ κωνάριον ἐπιβέβηκεν, ὄψει διήκοντα πρὸς τὴν μέσην κοιλίαν, ὡς φαίνεσθαι κατὰ τήνδ' ὀπὰς οὐ μικρὰς δύο, μίαν μὲν εἰς τοὐπίσω φερομένην πρὸς τὴν ἐπικρανίδα . . . ἑτέραν δὲ τὴν ἐν τῷ πυθμένι τῆς μέσης κοιλίας κάτω φερομένην. *AA* II. 728 K.

[166] Cf. Greenhill, 1870, 547–548; Garofalo, 1991, 823 n. 27.

[167] "The diminutive form, found probably only in the plural διδύμια, is applied almost always to two of the *corpora quadrigemina* in the brain, which still bear the name of testes." Greenhill, 1871, 556–567. Hyrtl, 1880, §302, notes that, strictly, the term *corpora quadrigemina* should be *corpus quadrigeminum.*

[168] Ἐκδέχεται δ' αὐτοῦ τὴν κατάπτωσιν ἀτρέμα περιφερῆ σώματα, περιγραφὰς μὲν ἰδίας πως ἔχοντα, μόρια δὲ ὄντα τοῦ παντὸς ἐγκεφάλου, καὶ τὴν αὐτὴν οὐσίαν ἔχοντα αὐτῷ. καὶ καλοῦσιν ἀπὸ τοῦ σχήματος αὐτὰ τινὲς μὲν γλουτία, τινὲς δὲ διδύμια, ὅτι καὶ τοὺς ὄρχεις ὀνομάζουσι διδύμους, ὡς σεμνότερον ὂν οὕτως καλεῖν. *AA* II. 729 K. Reading γλουτία: Garofalo, 2000, 577, for γλουτά: Kühn. In contemporary anatomy they are known as the superior and inferior colliculi and together make up the *corpora quadrigemina* or *tectum of the mesencephalon.*

[169] χρείαν δ' αὐτοῦ τὴν αὐτὴν εἶναι νομίζουσί τινες, ἥνπερ καὶ τοῦ κατὰ τὴν γαστέρα πυλωροῦ. καὶ γὰρ οὖν κἀκεῖνον ἀδένα φασὶν ὑπάρχειν καὶ κωλύειν ἐκ τῆς γαστρὸς εἰς τὸ λεπτὸν ἔντερον μεταλαμβάνεσθαι πρὶν πεφθῆναι τὴν τροφήν. *UP* I, p. 489,17–21 Helmreich; III. 675 K.

[170] τὸ κωνάριον, ἐπὶ τῆς ἀρχῆς ἑστῶτα τοῦ πόρου τοῦ διαπέμποντος ἐκ τῆς μέσης κοιλίας εἰς τὴν ἐν τῇ παρεγκεφαλίδι τὸ πνεῦμα, φύλακά τινα καὶ οἷον ταμίαν ὑπάρχειν τοῦ ποσοῦ τῆς ἐπιπέμψεως. *UP* I, p. 489,22–26 Helmreich; III. 675 K.

in the next section, Galen does acknowledge that there is a need to control the flow of pneuma, but this one, according to him, is based on a mistaken idea of the pineal's anatomical location.

Although the function of the pineal given by these unnamed colleagues is different from that of Galen, the function they propose is a solution to an identical physiological problem: namely, the regulation of the flow of psychic pneuma in the ventricles of the brain. The same solution to an identical problem, albeit using a different part of the brain, is proposed. These doctors, if not sharing Galen's particular brand of encephalocentrism, nevertheless might have regarded parts of the structure of the brain in a not dissimilar way. To Galen, as will be discussed in the next section, the part of the brain that regulates the passage of psychic pneuma from the third ventricle to the fourth is the "worm-like outgrowth" (σκωληκοειδὴς ἐπίφυσις), the *cerebellar vermis*. It is true that the pineal lies at the beginning of the aqueduct; however, it is also above the third ventricle, not within it. If it were inside, it might be possible to state that it could act in a fashion similar to that ascribed it by Galen's unnamed colleagues. But for Galen the pineal is not a part of the brain. This also means that it is unaffected by the brain's own "contractions and dilatations" (συστολαὶ καὶ διαστολαί), and it also "lacks self-motion" (i.e. is not αὐτοκίνητος). Therefore the pineal cannot act on the third ventricle to control the flow of psychic pneuma.[171] In making the analogy with the pylorus of the stomach, Galen's colleagues ascribe too much to this gland and are ignorant of the activity of the cerebellar vermis.[172] The vermis, being a part of the brain, is able to move, and its posi-

---

[171] εἰ μὲν γὰρ αὐτοῦ γε τοῦ ἐγκεφάλου μόριον ἦν, ὥσπερ ὁ πυλωρὸς τῆς γαστρός, ἐνεδέχετ' ἂν αὐτὸν ταῖς ἐκείνου συστολαῖς καὶ διαστολαῖς συμμεταφερόμενον ἐκ τῆς ἐπικαίρου θέσεως ἀνοιγνύναι τε καὶ κλείειν ἐν μέρει τὸν πόρον. ἐπεὶ δ' οὔθ' ὅλως οὐδέν ἐστιν ἐγκεφάλου μόριον ὁ ἀδὴν οὗτος οὔτ' ἐκ τῶν ἔνδον μερῶν τῆς κοιλίας, ἀλλ' ἔξωθεν αὐτῇ προσήρτηται, πῶς ἂν δύναιτο τηλικαῦτα περὶ τὸν πόρον διαπράττεσθαι μὴ αὐτοκίνητός γε ὤν; *UP* I, pp. 490,21–491,2 Helmreich; III. 676–677 K.

[172] τὸ δ' οἴεσθαι τῆς διόδου τοῦ πνεύματος ἐπιτροπεύειν αὐτό, καὶ τὴν τῆς σκωληκοειδοῦς ἐπιφύσεως ἐνέργειαν ἀγνοούντων ἐστὶ καὶ πλέον ἢ κατὰ τὴν ἀξίαν ἀδένι χαριζομένων. *UP* I, p. 490,17–20 Helmreich; III. 676 K. In the following passage Galen's criticism is stronger: those colleagues, whose opinions are ignorant as well as unlearned, are not even aware of the *existence* of the cerebellar vermis: ταῦτα δ' ὅσης ἀγνοίας τε καὶ ἀμαθίας ἔχεται, τι δεῖ καὶ λέγειν; ὃ γὰρ ὀνειρώττουσι μέν, ὡς εἶναι δέον αὐτοῦ που περὶ τὸν πόρον ἐγκεφάλου μόριον τοιοῦτον, οἷον ἐπιτροπεύειν τε καὶ ἄρχειν τῆς διόδου τοῦ πνεύματος, εὑρεῖν δ' ἀδυνατοῦσιν, οὐ τὸ κωνάριόν ἐστιν, ἀλλ' ἡ κατὰ παντὸς ἐκτεταμένη τοῦ πόρου σκώληκι παραπλήσιος ἐπίφυσις. I, p. 491,8–14 Helmreich; III. 677 K.

tion, directly over the aqueduct, means it can control the movement of psychic pneuma into the fourth ventricle. The pineal, like the hypophysis, is not a part of the brain, and is outside its physiological sphere of influence. But Galen's anonymous colleagues, in ascribing the above function to the pineal, interpreted the anatomical evidence differently. This does not mean that they were incorrect, for Galen's description of the function of the cerebellar vermis is surely as valid as theirs of the pineal *qua* pneumatic regulator valve. The mention of these unknown doctors offers a brief glimpse of what may have been a lively debate regarding the structures of the brain.

The pineal completes the discussion of Galen's anatomy of the third ventricle. This may now be summarised. The anterior ventricle communicates with the third through the interventricular foramina. At the meeting point of these foramina, it is possible to probe downwards and thereby enter the base of the third ventricle. This is the passage which "goes straight downwards". This joins the "oblique duct" which begins at the proximal part of the aqueduct, and, conjoined with the first duct under the massa intermedia, empties via the "perceptible cavity" (the infundibular recess) and thence into the hypophysis. The anterior aspect of the roof of the third ventricle consists of the pineal, which acts as a supporting structure for veins, as well as indicating the true beginning of the channel joining the third to the fourth ventricle (the aqueduct). Any examination of the fourth ventricle must begin with an analysis of this canal.

4.6 *The "duct of Erasistratus" and the vermiform epiphysis*

The anatomical evidence Galen assembles to validate his claims that there is a channel between the anterior ventricles and the third ventricle is persuasive. For Galen's physiological requirements, it is imperative that there also exist a communication between third and fourth ventricles in order that psychic pneuma may have access to the nerves. However, before discussing the nature of this communication, criticism first articulated in the nineteenth century, that Galen did not demonstrate the true nature of such a communication, but instead created and explored a false passage, must be addressed.

The communication between the third and fourth ventricle has been known since the seventeenth century as the *Aqueduct of Sylvius* (hereafter, the aqueduct).[173] It is claimed by several commentators, beginning with Daremberg, that Galen missed this channel, or else that his description represented an artefact of dissection.[174] To Simon, Galen's aqueduct is "artifiziell".[175] Wiberg states that Galen, "can *not*—in any case as far as its anterior part is concerned—have been thinking of the *ductus sylvii* . . .".[176] Woollam voices a similar concern, stating that "whilst it is clear that Galen knew of the inter-ventricular foramen, he does not seem to have been aware of the real nature of the communication between the third and fourth ventricles . . . it seems more probable that he was simply describing the sub-arachnoid space on the dorsal surface of the mid-brain."[177] May describes Galen's failure "to see the aqueduct of Sylvius and the substituting for it a queer connection between the third and fourth ventricles by way of the upper surface of the brain stem . . . it is evident that the wish has been father to the thought and has persuaded the sight."[178] Those who maintain Galen accurately depicts the aqueduct in its entire course are Singer,[179] Duckworth,[180] and Clarke and O'Malley, who conclude: "Some have claimed that Galen was not describing the aqueduct of Sylvius, but there seems to be sufficient evidence . . . to indicate that he was."[181] Meyer states that the identification of the Aqueduct of Sylvius with Galen's duct, "from a broad functional view at least, appears to be justified."[182]

Three different views can be discerned from the above. The first holds that Galen constructed a false passage. The second maintains Galen did not create a false passage but was misled into thinking that a passage higher than the true aqueduct (the subarachnoid space of the midbrain) was the passage to the fourth ventricle. The third

---

[173] Cf. Clarke and O'Malley, 1968, 719; Meyer, 1971, 7 n. 3; Millen and Woollam, 1962, 16.

[174] 1854, I, 566 n. 1. Daremberg's thesis is silent on this point, stating only that: "Le *conduit* aqueduc de Sylvius est creusé sous les tubercules quadrijumeaux, dans un corps dur; son orifice est bouché par la piemère, le corps vermiforme repose sur lui." 1841, 30.

[175] 1906, II, 243 n. 6. Cf. Meyer-Steineg, 1912, 203; Garofalo, 1991, 821 n. 26.

[176] 1914, 99. Italics in original.

[177] 1957, 102. Cf. May, 1968, 420 n. 76; Millen and Woollam, 1962, 17; Wiberg, 1914, 98–101.

[178] 1968, 41–42; cf. 417 n. 70; See also idem, 1970, 171.

[179] 1956, 236.

[180] 1962, 1. Cf. Garofalo, 1981, II, 827 n. 33; Simon, 1906, II, 242–243, nn. 1, 6.

[181] 1968, 711.

[182] 1971, 7 n. 3.

states that Galen is correct according to contemporary anatomical teaching. What matters, of course, is the consistency of Galen's exposition in relation to his methodology. Before embarking on an examination of Galen's anatomical record on this question, it is necessary to understand Galen's necessity for such a communication between the anterior and posterior parts of his ventricular system.[183] According to Galen, there is only one large passage or canal connecting the third and fourth ventricles, that is, between the cerebrum and cerebellum. Galen did not discover the passage. It seems to have been first noticed by Erasistratus, as the following verbatim excerpt reproduced in *De placitis Hippocratis et Platonis*, attests:

> And we also viewed the nature of the cerebrum, and it was bipartite in shape, as in other animals, and ventricles were lying there, elongated in form. They were united by a perforation at the point of contact of the parts. From there a passage led to what is called the cerebellum, and where there is another small ventricle.[184]

As discussed in chapter one, Galen states that Erasistratus "wrote accurately" (ἀκριβῶς ἔγραψεν) about the four ventricles of the brain.[185] The earlier citation from *De usu partium*, where Galen states that the cerebellum is separated from the cerebrum by a double fold of dura mater (the *tentorium cerebelli*), augments Erasistratus' observations of the nature of these dural folds.[186] The only connection between cerebrum and cerebellum is the passage beneath the crown of the head, together with the bodies which encompass it.[187] These bodies, already noted in the last section, are the *corpora quadrigemina* of contemporary anatomy, and surround the aqueduct.[188] Galen's description of the path of the aqueduct is given in *De anatomicis administrationibus* as follows:

> Then the duct which comes through from the middle to the posterior ventricle, between the gloutia, is covered by its own tunic, this being of

---

[183] See the citations from *De usu partium* on pp. 136–137 and 138.

[184] ἐθεωροῦμεν δὲ καὶ τὴν φύσιν τοῦ ἐγκεφάλου καὶ ἦν ὁ μὲν ἐγκέφαλος διμερής, καθάπερ καὶ τῶν λοιπῶν ζῴων, καὶ κοιλίαι παραμήκεις τῷ εἴδει κείμεναι· συντέτρηντο δ' αὗται εἰς μίαν κατὰ τὴν συναφὴν τῶν μερῶν· ἐκ δὲ ταύτης ἔφερεν εἰς τὴν ἐπεγκρανίδα καλουμένην καὶ ἐκεῖ ἑτέρα ἦν μικρὰ κοιλία. *PHP* p. 440,27–31 De Lacy; V. 602–603 K. Cf. Solmsen, 1961, 189 n. 20.

[185] *PHP* p. 442,11–12 De Lacy; V. 604 K.

[186] Cf. *PHP* p. 440,31–34 De Lacy; V. 603 K.

[187] καὶ γὰρ οὖν καὶ διείργεσθον ἐνδιπλουμένῃ τῇ σκληρᾷ μήνιγγι καὶ κατὰ μόνον τὸν ὑπὸ τῇ κορυφῇ τῆς κεφαλῆς κείμενον πόρον καὶ τὰ τοῦτον περιέχοντα σώματα συνάπτεσθον. *UP* I, p. 461,17–21 Helmreich; III. 637 K.

[188] See n. 168, above, and n. 191, below.

> the same substance as the meninx which unites together all the vessels of the brain. Therefore pay attention as you try to denude it from what lies above, knowing that, if negligent, it will be torn apart. Lying on it is a circumscribed part of the brain, having an outward appearance like the worm which grows in wood.[189] And thence the custom by anatomists to call the body covering all the duct the vermiform outgrowth.[190]

Here Galen states quite clearly that the passage between the third and fourth ventricle lies *between* (μεταξύ) the gloutia, the two bodies making up half of the corpora quadrigemina.[191] There is no reference to any other channel linking the third ventricle with the fourth. Galen articulates the care which must be taken in such a dissection, implying an awareness on his part of the possibility of creating a false passage. Galen then states that this duct is covered by the same meninx which invaginates the vessels of the brain. It is likely that Galen regarded this membrane as consisting of the pia together with a contribution from the anterior medullary velum, which, as will be seen below, is part of the vermiform epiphysis and forms the anterior part of the roof of the fourth ventricle.[192]

Lying above the membrane that immediately covers the gloutia, according to Galen, is the σκωληκοειδὴς ἐπίφυσις, the "worm-like outgrowth", the vermiform epiphysis of the cerebellum.[193] Although

---

[189] Cf. Greenhill, 1866, 273.

[190] ὁ δὲ εἰρημένος πόρος, ἐκ τῆς μέσης κοιλίας εἰς τὴν ὀπίσω διήκει, μεταξὺ τῶν γλουτίων τούτων [ἐστὶ] σκεπόμενος ἰδίῳ χιτῶνι, τοιαύτην ἔχοντι οὐσίαν, οἵα περ ἡ συνάπτουσα πάντα τὰ κατὰ τὸν ἐγκέφαλον ἀγγεῖα μήνιγξ. διὸ προσέχων τὸν νοῦν ἐπιχείρει γυμνοῦν τῶν ἐπικειμένων αὐτόν, εἰδὼς διασπασθησόμενον, ἢν ἀμελήσῃς. ἐπίκειται αὐτῷ τι μόριον ἐγκεφάλου, περιγραφὴν ἔχον ἐν σχήματι παραπλησίως τῷ κατὰ τὰ ξύλα γεννωμένῳ <σ>κώληκι. ἐντεῦθεν καὶ τοὔνομα, καλεσάντων τῶν ἀνατομικῶν ἐπίφυσιν σκωληκοειδῆ τοῦτο τὸ σκέπον ἅπαντα τὸν πόρον σῶμα. II. 729 K; p. 577, Garofalo, 2000. Reading γλουτίων: Greenhill, Garofalo, 2000, for γλουτῶν: Kühn. Vesalius mirrors Galen's account: "The other passage, which is farther back, is an important part of the third ventricle itself. It descends between the testes and nates, above the origin of the dorsal marrow, and reaches back to the fourth ventricle." (tr. Singer, 1952, 36).

[191] These four bodies are the most conspicuous part of the midbrain. In ruminants, the rostral pair (superior or anterior colliculi) are much larger and higher than the caudal pair (inferior or posterior colliculi), and are almost hemispherical. A median longitudinal groove runs between the anterior colliculi; the pineal lies at the beginning (rostral end) of this groove. At its posterior end, a white band, the *frenulum veli*, connects with the anterior medullary velum and thence to the vermiform epiphysis. Together, the frenulum and the velum are for Galen the covering membrane which is prolonged onto the cerebellar vermis.

[192] Cf. *AA* II. 591–596 K.

[193] In common with the two cerebellar hemispheres, its surface consists of numerous convolutions (*folia*), separated by narrow sulci, running chiefly transversely (cf. Sisson and Grossman, 1975, 210, 1072–1073). Galen interprets these sulci as the

Galen states that the duct lies between the gloutia, he does not say how deep within them the passage lies. Galen provides a more extended description of this area in *De usu partium*. Here, the gloutia of the corpora quadrigemina are noted as being "on either side" (ἑκατέρωθεν) of the duct.[194] The duct is not said to be situated above the gloutia. The other two bodies making up the corpora quadrigemina, the *testes*, are said by Galen to be associated with the pineal, while the gloutia are next after these bodies.[195] It is the gloutia which:

> . . . form the left and right parts of the duct, and this is covered above with a thin, yet by no means weak, membrane, attached to the gloutia on both sides.[196] The membrane that reaches as far as the posterior ventricle below is the extremity of the vermiform epiphysis, which has nothing in common (structurally) with the testes and gloutia. For on the one hand the epiphysis is articulated of many parts, while the testes as well as the gloutia are entirely similar and not wholly composite. For truly the epiphysis is articulated of many parts; further, these parts are all similar though not entirely cohered together. But besides the many parts of which it is formed, the vermiform epiphysis seems composed of many parts, yet united by thin membranes, and this is its noteworthy feature. At the posterior ventricle its extremity is curved and thin, where, as I have said, it ends in the membrane lying on it. But from there it gradually increases in size as well, with its dorsum having nearly the same width as the interval between the gloutia, and by means of its stretching out along the length of the duct, it completely stops up the entire passage.[197]

---

parts which make up the vermis, and connected by the same membrane which constitutes the medullary velum.

[194] ἑκατέρωθεν τοῦ πόρου λεπταὶ καὶ προμήκεις εἰσὶν ἐξοχαὶ τοῦ ἐγκεφάλου, γλουτία καλούμεναι. I, p. 491,18–20 Helmreich; III. 677–678 K.

[195] τινὲς δὲ τὰ μὲν ὁμιλοῦντα τῷ κωναρίῳ σώματα διδύμια καλοῦσι. τὰ δ' ἐφεξῆς αὐτῶν γλουτία. I, pp. 491,23–492,2 Helmreich; III. 678 K. Cf. *AA* II. 729 K. It is clear that Galen distinguished the four bodies which make up the corpora quadrigemina (*pace* Wiberg, 1914, 99, who believes Galen only knew the διδύμια).

[196] Here the description indicates that the covering membrane of the duct *is* part of the velum. Cf. n. 191 above.

[197] τὰ μὲν οὖν ἀριστερὰ καὶ δεξιὰ μέρη τοῦ πόρου τὰ τούτων αὐτῶν ἐστι σώματα, τὰ δ' ἄνωθεν ὑπὸ λεπτοῦ μέν τινος ὑμένος, οὐ μὴν ἀρρώστου γε σκέπεται, συναπτομένου τοῖς γλουτίοις ἑκατέρωθεν, ὅστις ὑμὴν ἄχρι τῆς ὀπίσω κοιλίας ἐκτεταμένος τὸ κάτω πέρας ἐστὶ τῆς σκωληκοειδοῦς ἐπιφύσεως, οὐδὲν ὅμοιον ἐχούσης τοῖς διδύμοις τε καὶ γλουτίοις. ἡ μὲν γὰρ πολυειδῶς διήρθρωται, τὰ δ' ἐστὶν ὅμοιά τε πάντη καὶ οὐ πάνυ τι σύνθετα. πρὸς δὲ τῷ πολυειδῶς διηρθρῶσθαι καὶ δοκεῖν ἐκ παμπόλλων συγκεῖσθαι μορίων ὑμέσι λεπτοῖς συναπτομένων ἔτι καὶ τοῦθ' ἡ σκωληκοειδὴς ἐπίφυσις ἐξαίρετον ἔχει. κυρτὸν μὲν καὶ λεπτόν ἐστιν αὐτῆς τὸ κατὰ τὴν ὀπίσω κοιλίαν πέρας, ἵνα περ εἰς τὸν ἐπικείμενον ὑμένα τελευτᾶν ἐλέγετο· τὸ δ' ἀπὸ τοῦδε κατὰ βραχὺ προσαυξανομένη τε καὶ πλατυνομένη σχεδὸν ἴσον ἴσχει τὸν νῶτον τῇ τῶν γλουτίων διαστάσει, καὶ διὰ τοῦτο μακρὰ μὲν ἐκταθεῖσα κατὰ τοῦ πόρου τελέως αὐτὸν ὅλον ἐπιφράττει, *UP* I, p. 492,2–19 Helmreich; III. 678–679 K.

The nature of the vermiform epiphysis is the key to understanding Galen's interpretation of the anatomy of this area.[198] Galen seeks to make the comparison that while the testes and gloutia are homogeneous structures, the epiphysis is made of discrete parts joined together (like a worm) but not so as to form a totally homogeneous entity. It is indisputably a part of the brain.[199] Its physiological role is to regulate the flow of psychic pneuma to the fourth ventricle. The above citation should also dispel the notion that the duct between cerebrum and cerebellum in any way runs *above* the gloutia.[200] The membrane which covers the duct above is in intimate association with—and forms part of—the vermiform epiphysis.[201] The Greek part of *De anatomicis administrationibus* discusses the anatomical relationships of the vermiform epiphysis in detail:

> [The vermiform epiphysis], as can be seen, indeed has two extremities. In order, one part is, as I have said, located anteriorly near the pineal, the other is posterior, but is not as yet observable. This is because it is covered by the whole upper part of the substance of the rear encephalon. Now hold the extremity which is situated near the origin of the spinal marrow, attempting to carry it forward, rolling it, for instance, until there appears another peculiar body of this kind, like a worm. And therefore, when this is found, strip off little by little most of the overlying bodies, so that only those on the duct are left, the end being double on each side, resembling in form the worms mentioned previously.[202] And further, you see the thin bodies binding the anterior[203] outgrowth of the vermiform epiphysis on each side to

---

198 Cf. Meyerhof, 1928, 18.26–33.

199 The epiphysis is said to be firm to resist injury, but is not so hard as to disqualify it from being part of the brain: καὶ μὲν δὴ καὶ σκληρὰν αὐτὴν ἀπειργάσατο δυσπαθείας ἕνεκα, οὐ μὴν εἰς τοσοῦτόν γε σκληράν, ὡς μηκέτ' ἐγκεφάλου μέρος ὑπάρχειν, ἀλλὰ κἀνταῦθ' ἀκριβῶς πάνυ μετρήσασα τὴν χρείαν, εἰς ὅσον οἷόν τ' ἦν ἐγκέφαλον ἔτι μένουσαν αὐτὴν γενέσθαι σκληράν, εἰς τοσοῦτον προήγαγεν. *UP* I, p. 495,13–18 Helmreich; III. 682–683 K.

200 "[I]t is certain that he (sc. Galen) missed the real channel deep within the brainstem." May, 1968, 420 n. 76.

201 This membrane cannot be the pia mater, which is both thin and weak. May, 1968, 421, glosses this membrane as "the arachnoid". Galen did not identify this layer, which is a very fine avascular membrane between the dura and pia mater. But his descriptions of the pia are strictly speaking, depictions of the pia-arachnoid in which run the brain's blood vessels. Cf. Millen and Woollam, 1962, 103–122.

202 What is left is each end of the vermiform epiphysis, lying on the aqueduct, when the rest of the cerebellum has been removed.

203 Garofalo, 1991, II, 825, reads: "posteriore", which does not make topographic sense, nor is it a correct translation of πρόσθεν. He also glosses this part of the vermiform epiphysis as *crura cerebelli*. The crura, however, are located on the ven-

> the parts of the encephalon which lie on either side of the gloutia. These are called "tendons" by some anatomists.[204]

The vermiform epiphysis covers the duct, whilst the membrane of the anterior medullary velum constitutes the anterior termination of the epiphysis, and, given its close apposition to the gloutia, Galen may well have concluded that the velum is adherent to the pia between the gloutia. In a fresh ox brain, the cerebellum completely covers the dorsal surface of the midbrain, including the corpora quadrigemina. The vermiform epiphysis of the cerebellum thus overlies the aqueduct, and it is likely Galen observed this relationship and inferred its function. The instructions Galen gives for the removal of the rest of the cerebellum, which will be discussed shortly, tend to reinforce the importance of the epiphysis to the aqueduct. The cerebellum is attached to the sides of the brain by three pairs of stalks (or peduncles); anterior (rostral), middle, and posterior (caudal).[205] By successively cutting through each of these the cerebellum may be removed completely. But Galen would also have been required to have dissected the cerebellum in such a way so as to remove the cerebellar hemispheres but leave the midline epiphysis *in situ*. The *tendons* (τένοντες) that Galen describes above may be identified as the *brachia conjunctiva*—the portion of the anterior cerebellar peduncle on either side of the midline which is connected to the anterior medullary velum, and runs down to just below the inferior colliculi. From Galen's perspective, the close relationship of the tendons with the medullary velum of the epiphysis would strongly suggest the former's role as a stabiliser of the epiphysis. And Galen states that the function of these

---

tral aspect of the brain. The attention of the dissector is here focussed on key structures of the dorsal aspect of the brain. These "tendons" are known in contemporary anatomy as the superior cerebellar peduncles (brachia conjunctiva). See below.

[204] καὶ διττόν γε τὸ πέρας αὐτοῦ θεάσῃ, τὸ μὲν ἕτερον ἐν τοῖς πρόσω μέρεσιν ἐφεξῆς, ὡς εἴρηται, τῷ κωναρίῳ τεταγμένον, ἕτερον δὲ ὀπίσω, μηδέπω φαινόμενον· ἐπίκειται γὰρ αὐτῷ τὸ ὑψηλὸν μέρος ἅπαν τῆς κατὰ τὸν ὀπίσθιον ἐγκέφαλον οὐσίας, ἧς τοῦ πέρατος λαβόμενος, ὃ τέτακται πλησίον τῆς ἀρχῆς τοῦ νωτιαίου μυελοῦ, πειρῶ πρὸς τὸ πρόσω φέρειν, ὥσπερ ἀνακυλὶνδων, ἄχρις ἂν ἴδῃς ἕτερον σῶμα τοιοῦτον, οἷον σκώληξ ἐστί. καὶ τοίνυν, ὅταν εὕρῃς τοῦτο, περιαίρει κατὰ βραχὺ τὰ πλείω τῶν ἐπικειμένων σωμάτων, ὥστε μόνα λείπεσθαι τὰ κατὰ τοῦ πόρου, διττὸν ἔχοντα πέρας ἑκατέρωθεν, τοῖς εἰρημένοις σκώληξι τὸ σχῆμα παραπλήσιον. ὄψει δὲ ἐν τῷδε καὶ τὰ συνδοῦντα σώματα λεπτὰ τὴν προσθίαν ἐπίφυσιν σκωληκοειδῆ πρὸς τὰ παρακείμενα τοῖς γλουτίοις ἑκατέρωθεν ἐγκεφάλου μόρια· τένοντας δὲ αὐτὰ καλοῦσιν ἔνιοι τῶν ἀνατομικῶν. II. 729–30 K; p. 579 Garofalo, 2000. Reading γλουτίοις: Garofalo, 2000, for γλουτοῖς: Kühn.

[205] Cf. Sisson and Grossman, 1975, 641.

tendons (or *bands*, σύνδεσμοι) is to prevent the vermiform epiphysis from sliding off the gloutia and losing its place over the canal.[206] The vermiform epiphysis sets the limits of the extent of the aqueduct. Galen states that the aqueduct has this length in order that the epiphysis may have as full a range of movement as possible.[207] Such an interpretation is easily converted by Galen into further evidence for the "most precise skill" of Nature.[208]

The function of the vermiform epiphysis is further described in the opening chapter of the Arabic continuation of Book IX of *De anatomicis administrationibus*:

> If you now repeat the movement of that body [*the vermis*] which covers the passage (the aqueduct) and, simultaneously with this movement, you raise it upwards somewhat, then you can see the end of the passage which opens out into the posterior ventricle. This is closed up and covered in by the vermiform process [*vermis inferior*] which, from behind, forms a sort of lid for it [*the ventricle*] as soon as you draw back the whole of the body which covers the passage towards the rear, and it (the end of the aqueduct)[209] opens itself as soon as this body goes forwards. This is the moment to introduce from in front into the posterior ventricle whatever smooth cylindrical instrument you have available,[210] the calibre of which corresponds to the opening of the passage, . . . The Greeks are accustomed to name all instruments of such use with the generic term sounds.[211] For my part, I have sometimes in the past, when I had no other thing at hand, inserted into this passage from the middle ventricle the writing reed,[212] though I did

[206] ἐπεὶ δ᾽ ἐκ τούτων ἁπάντων εὐκίνητός τε καὶ πολυκίνητος ἔμελλεν ἔσεσθαι, καὶ κίνδυνος ἦν κυρτοῖς ἐποχουμένην αὐτὴν τοῖς τῶν γλουτίων νώτοις ἀποκυλισθήσεσθαί ποτ᾽ ἐκτὸς αὐτῶν καὶ καταλείψειν τὸν πόρον, ἐτεχνήσατό τινας αὐτῇ πρὸς τὰ γλουτία συνδέσμους ἡ φύσις, οὓς οἱ περὶ τὰς ἀνατομὰς δεινοὶ τένοντας ὀνομάζουσιν, ὑφ᾽ ὧν ἑκατέρωθεν ἐσφιγμένη τε καὶ κατεχομένη πλανᾶσθαι κωλύεται. *UP* I, p. 495,5–12 Helmreich; III. 682 K.

[207] ὅ τε πόρος ὅλος ἐπιμήκης ἐγένετο δι᾽ οὐδὲν ἀλλ᾽ <ἢ> ἵν᾽ ἡ κίνησις αὐτῆς πολλὴν ἔχῃ τὴν ἐν τῷ ποσῷ διαφοράν, *UP* I, p. 494,24–26 Helmreich; III. 682 K.

[208] δῆλον δ᾽ ἤδη τοῖς γε προσεσχηκόσι τῷ λόγῳ τὸν νοῦν, ὡς, εἰ καὶ τῶν ἄλλων τι τῶν εἰρημένων μετακοσμηθείη, πολλαχῇ μὲν βλαβῆναι μόνον, ἔστιν ὅπῃ δὲ καὶ τελέως ἀπολέσθαι συμβήσεται τὴν ἐνέργειαν. ὥστ᾽ ἐγὼ μὲν οὐκ ἔχω συμβαλεῖν, ὅτῳ τις ἂν ἐπιχειρήσειε τρόπῳ δεικνύναι ταῦτα μὴ οὐ τῆς ἀκριβεστάτης τέχνης ὑπάρχειν ἔργα. *UP* I, p. 496,2–9 Helmreich; III. 683 K.

[209] Duckworth's gloss: "the ventricle". Garofalo, 1991, II, 827 n. 33, states that this should be read as indicating: "L'estremità del canale, non il ventricolo."

[210] Galen then states that the sound may be made of wood, copper, iron, silver, or gold. If wood, boxwood is to be preferred, as it is both strong and smooth. Cf. *AA* II. 711 K; Milne, 1907, 16, 55, and Pl. X, fig. 2.

[211] This seems an odd sentence for Galen, who would probably use an expression such as "the anatomists." It may represent a gloss by Hunayn.

[212] Cf. Milne, 1907, 114–5; Simon, 1906, II, 2, n. 2.

> not introduce the tapered end with which one writes, for this is sharp, but the other rounded-off end, and I pressed it steadily and gradually onwards, until I brought it into the posterior ventricle without boring through any part of the surroundings.[213]

Galen stresses how he has passed the sound through from the middle to the posterior ventricle "without boring through any part of the surroundings", but this of course does not exclude the creation of a false passage, and Galen's instructions highlight this hazard. Galen now takes the dissection a stage further, in an attempt to reduce the chances of making just such a mistake:

> When you have seen all this, now remove the whole body with which two vermiform processes are connected, after you have previously taken care for the undamaged preservation of the membrane (sc. the anterior medullary velum) covering the passage in order that you may have a clear view of it.[214] You will then be in a better position than on the first occasion if you wish to introduce an instrument to penetrate through the passage, and you can also dilate it[215] by means of a cannula.[216] When you have done that, now is the time to cut into the membrane or to open it up, so that the passage may be laid open to view.[217]

This account should at least help lay to rest the idea of Galen having made an *entirely* artificial communication between third and fourth ventricle. The bulk of the cerebellum has been removed, leaving the vermiform epiphysis, together with the attached anterior medullary velum, in place *over* the duct. Galen then introduces a sound from the third ventricle and through the aqueduct, dilating the passage if necessary (and if the sound were passing through the delicate subarachnoid space over the gloutia, there would be no need for any such dilatation). Galen's manoeuvre puts the probe *between* the gloutia but

---

[213] IX.6; p. 1 Duckworth. My glosses in round brackets; square brackets Duckworth.

[214] In removing the epiphysis to expose the aqueduct, the anterior medullary velum is torn from the roof of the aqueduct and goes with the epiphysis. By this action, Galen's description of the posterior portion of the aqueduct is correct. It is not that he has created a false passage; rather, his regarding the epiphysis as a discrete structure with a membranous termination means that both anterior medullary velum and vermiform epiphysis are considered by him part of the one body, and Galen's removal of the one necessarily results in the removal of the other.

[215] Reading "gonfiarlo" (*tanfuḫa*): Garofalo, 1991, 828, n. 35, for "aprirlo" (*taftaḫa*): Hunayn; Simon, 1906, II, 2; Duckworth: "open it up".

[216] Is it certain that Galen meant a cannula? Examples of cannulae suggest their diameter may have been too large for this particular task. It is more likely Galen used a fine sound, or perhaps a hog's bristle.

[217] *AA* IX.6; p. 2 Duckworth. My gloss in round brackets.

also *under* the anterior medullary velum of the epiphysis. Only after this has been accomplished, does Galen state that this is the moment to incise or open the membranous velum so that the aqueduct is completely exposed to view. Such a procedure would greatly facilitate the dissector's ability in unequivocally locating the aqueduct and also tend to minimise the possibility of creating a false passage. This exposure of the passage is reminiscent of instructions Galen has given in regard to the floor of the third ventricle: when the probe is placed in the base of the third ventricle and lifted upwards, it tears through the interthalamic adhesion and exposes the floor of the ventricle. Here, once the exposure of this part of the roof of the aqueduct has been carried out, the entire extent of the canal is apparent.

If the above interpretation of Galen's dissection of the region of the aqueduct is correct, there is no inconsistency in his account of its course nor of its topographical relationship to the gloutia and vermiform epiphysis. To be sure, Galen never states that the aqueduct runs *under* or *below* (ὑποκάτω) the gloutia. But if Galen had created a false passage from the base of the pineal between the gloutia, then the sound would already be lying immediately beneath the velum and Galen's instructions to "open it up by means of a cannula", or to "also dilate it by means of a cannula", would be superfluous. Galen is correct in his description of the beginning and end of the aqueduct and probably based his demonstrations on the work of Erasistratus. In any case, Galen, by lifting the distal portion of the vermiform epiphysis exposes the roof of the distal part of the aqueduct and the upper part of the roof of the fourth ventricle, delineating a canal which for him is the only anatomically verifiable route between the third and fourth ventricle, and which, on the balance of probabilities, is not entirely an artefact of dissection.

4.7 *The "cavity of Herophilus": the fourth ventricle*

The fourth ventricle is associated with the name of Herophilus, but, according to Galen, there have been mistakes regarding its true location. In *De usu partium*, some anatomists maintain the cavity under the fornix is a fourth ventricle.[218] As has been discussed above, for

---

[218] καὶ οἷς γε τετάρτη τις αὕτη κοιλία νενόμισται, κυριωτάτην εἶναί φασιν αὐτὴν

Galen this cavity corresponds to the anterior ventricles and the first part of the third. Galen describes the anatomy of the fourth ventricle in a way which clearly indicates that the notable feature of its structure was delineated by the great Alexandrian anatomist:

> Next pay attention to how, when it (sc. the vermiform epiphysis) is bent forward, the result is that the posterior ventricle, the fourth, is exposed, and, when it is moved backward, the greater part of the ventricle is covered and that part only is visible which Herophilus compared to the carved-out groove of a pen he used for writing (sc. *calamus scriptorius*). For indeed it is really like a pen, as it has in the centre a hollow resembling an incision,[219] and on either side of this each of the two lateral parts[220] extends up to as great a height as they rise in pens from the line in the centre. Especially in Alexandria do they carve out writing pens in this manner, and since Herophilus lived there, it is probable that, when he was dissecting, he applied this name, being induced to do so by the resemblance of the image.[221]

This is the *locus classicus* of the *calamus scriptorius* (κάλαμος), that part of the fourth ventricle which Herophilus, using a familiar, everyday object, compared to a writing instrument. It is a term which Galen does not feel any need to improve upon.[222] The fourth ventricle is a diamond-shaped duct running the entire length of the under part of the cerebellum.[223] In the ox brain, it is a relatively long and narrow channel.[224] The roof of the fourth ventricle is covered by the anterior and posterior medullary vela, which for Galen are part of

---

ἁπασῶν τῶν καθ' ὅλον τὸν ἐγκέφαλον. 'Ηρόφιλος μὴν οὐ ταύτην, ἀλλὰ τὴν ἐν τῇ παρεγκεφαλίδι κυριωτέραν ἔοικεν ὑπολαμβάνειν. I, p. 484,11–15 Helmreich; III. 667 K.

[219] This is the median sulcus, which divides the floor of the fourth ventricle (also known as the *rhomboid fossa*) into two symmetrical lateral halves.

[220] At about the midpoint of the fourth ventricle, these lateral halves become prominent and are known as the *facial colliculi*.

[221] εἶτα πρόσεχε τὸν νοῦν, ὅπως ἀνακλωμένου μὲν αὐτοῦ πρόσω γυμνοῦσθαι συμβαίνει τὴν ὀπίσω κοιλίαν τὴν τετάρτην, ἔμπαλιν δὲ κινουμένου κατακαλύπτεσθαι μὲν αὐτῆς τὸ πλεῖστον μέρος, ἐκεῖνο δὲ φαίνεσθαι μόνον, ὅπερ 'Ηρόφιλος εἴκαζεν ἀναγλυφῇ καλάμου, δι' οὗ γράφομεν. ὄντως γάρ ἐστι τοιοῦτον, ἐν μὲν τῷ μέσῳ κοιλότητά τινα κεκτημένον οἷον τομήν· ἑκατέρωθεν δ' αὐτῆς τοσοῦτον εἰς ὕψος ἀνατεινόμενον ἑκάτερον τῶν πλαγίων μερῶν, ὅσον ἐν τοῖς καλάμοις ἀπὸ τῆς μέσης ὑψοῦται γραμμῆς. καὶ μάλιστά γε κατὰ τὴν 'Αλεξάνδρειαν οὕτω γλύφουσι τοὺς καλάμους οἷς γράφομεν, ἔνθα διατρίβοντα τὸν 'Ηρόφιλον ἡνίκ' ἀνέτεμνεν εἰκὸς δήπου τῇ τῆς εἰκόνος ὁμοιότητι προσαχθέντα τοὔνομα θέσθαι. II. 731 K; p. 579 Garofalo, 2000. Cf. T 79 *Herophilus*. This citation concludes the extant Greek text of *De anatomicis administrationibus*.

[222] Von Staden, 1989, 158, describes the calamus scriptorius as "a cavity in the floor of the fourth ventricle." Strictly, it is a groove, which Galen has noted.

[223] The diagram in Singer, 1956, 277, Fig. 26, shows this well.

[224] Cf. Sisson and Grossman, 1975, 1073.

the vermiform epiphysis. In making this exposure of the calamus scriptorius, Galen would have removed most of the cerebellum, as he must already have done in order to chart the course of the aqueduct. Such a manoeuvre inevitably tears the delicate posterior medullary velum, carrying it away together with the postero-medial part of the cerebellum. In so doing, the choroid plexus of the fourth ventricle, which lies on the underside of the posterior part of the cerebellum and is covered by the posterior medullary velum, is also removed. This may explain Galen's silence on the question of the choroid plexus in the fourth ventricle.[225]

For Galen, the fourth ventricle has a crucial relationship with the cerebellum and the spinal marrow. To take the cerebellum first. As noted above, the caudal medullary velum of the cerebellar vermis forms the roof of the fourth ventricle. The cerebrum is composed of large convolutions each demarcated by the pia mater. The cerebellum is different, being formed of many small bodies, which give it, according to Erasistratus, a more intricate structure.[226] Although structured differently from the rest of the brain, the cerebellum is nonetheless part of it, for Galen reiterates that the entire *encephalon*

---

[225] May, 1968, 417 n. 70, states that "it is evident that Galen failed to see or chose to ignore the choroid plexus of the fourth ventricle". If so, he was not the last. Vesalius, in the *Epitome*, noted that the "gaze of the dissector encounters no peculiar body in this fourth ventricle as may be found in the three former ventricles." (tr. Lind, 1969, 68). Although Galen mentions the choroid plexus in the anterior ventricles, and its presence may be inferred in the middle ventricle, he does not explicitly refer to the presence of a choroid plexus in the fourth ventricle. Since the choroid plexus of the fourth ventricle lies *above* the posterior medullary velum, Galen may have thought, as pointed out above, that the plexus was in apparent continuity with the velum and was thus reckoned by him to be part of the cerebellum. The choroid plexus of the fourth ventricle appears intimately bound to the cerebellar cortex, and in this respect is "hidden" (cf. Ariëns Kappers, Huber and Crosby, I, 1936, 49 n. 14). Galen may have simply missed it as a separate structure. Further, the choroid plexus is more laterally situated, and Galen's focus, from the beginning of the aqueduct to the fourth ventricle, is with midline structures. Although the anterior part of the medullary velum is relatively firm, its caudal part is thin and partially replaced by the choroid, and if Galen were following the velum throughout its length, he may have thought that the choroid was part of the velum and thus was not a special vascular feature of the fourth ventricle. Galen of course made mistakes in anatomy, but in this instance the "mistake" probably relates more to how Galen classified structures than to any error in his method of dissection.

[226] οὐδὲ γὰρ ἐξ ἑλίκων μεγάλων τῇ λεπτῇ μήνιγγι διειλημμένων, ὥσπερ ὁ ἐγκέφαλος, ἀλλ' ἐκ πολλῶν μὲν καὶ αὐτῶν πάνυ σμικρῶν σωμάτων, οὐ μὴν τὸν αὐτὸν ἐκείνῳ τρόπον σύγκειται . . . Ἐρασίστρατος δέ, ὅτι μὲν ἐγκεφάλου σύγκειται ποικιλώτερον ἡ ἐπεγκρανίς, οὕτω γὰρ αὐτὴν ὀνομάζει, καλῶς ἀποφαίνεται· *UP* I, p. 488,3–16 Helmreich; III. 672–673 K.

(i.e. cerebrum and cerebellum) is the "source of the origin of nerves for the entire body."[227] In addition, the cerebellum is said to contain a "very great amount" (πάμπολυς) of psychic pneuma and the alleged connections between its parts serve as "paths for pneuma" (ὁδοὶ τοῦ πνεύματος).[228] It would be reasonable for Galen to infer that part of the psychic pneuma in the fourth ventricle permeates the cerebellum through these connections (which are also similar in nature to the medullary vela). There is a precedent. Galen holds that the pia mater (and, to a lesser extent, the dura), possess fine perforations. The permeable nature of the membranes binding the discrete bodies which make up the cerebellum could therefore afford passage of psychic pneuma both into and through this part of the brain.

The fourth ventricle is also linked to the spinal cord since its distal portion ends at the beginning of the *spinal marrow* (νωτιαῖος μυελός).[229] In *De anatomicis administrationibus*, Galen examines this part of the fourth ventricle:

> Do you now turn your attention to and consider particularly what I am about to describe for you. It is that the posterior ventricle is bare in the region where it borders upon the first origin of the spinal marrow, and has no sort of body which overlaps it, as is the case with the three other ventricles and those portions of this fourth ventricle which apply themselves to the passage between it and the middle ventricle. As for what relates to that portion of this [fourth] ventricle which lies at the upper end of the spinal marrow, it is completely free from any structure of which the substance resembles this [*spinal marrow*], except that it is covered by the dura mater only.[230]

Galen has followed the canal of the fourth ventricle until it is seen to terminate, or as he implies, to merge with the beginning of the

---

[227] ὅλον τὸ σῶμα νεύρων ἀρχὴν γενήσεσθαι. *UP* I, p. 488,11 Helmreich; III. 673 K. The cranial nerves are excluded since they take their origin from the cerebrum.

[228] ἐπειδὴ γὰρ δι' ὅλου τοῦ κατὰ τὸν ἐγκέφαλον σώματος οὐκ ἐν ταῖς κοιλίαις αὐτοῦ μόναις, . . . πάμπολυ περιέχεται πνεῦμα ψυχικόν, ἡγεῖσθαι χρὴ καὶ κατὰ τὴν παρεγκεφαλίδα, μέλλουσάν γε τῶν καθ' ὅλον τὸ σῶμα νεύρων ἀρχὴν γενήσεσθαι, πλεῖστον τοῦτο περιέχεσθαι τὸ πνεῦμα, καὶ τὰς μεταξὺ δὴ χώρας ἐκείνας, τὰς συναπτούσας αὐτῆς τὰ μόρια, τὰς ὁδοὺς εἶναι τοῦ πνεύματος. *UP* I, p. 488,6–14 Helmreich; III. 673 K.

[229] The term is not a Galenic neologism (cf. Onians, 1951, 118, 149). As Galen puts it: "It makes no difference to the exposition of anatomy which we have in mind whether . . . we name the spinal cord according to the customary usage of the Greek language 'vertebral marrow', or simply 'marrow'". *AA* XV.1; p. 223 Duckworth. Cf. Simon, 1906, II, 203 nn. 642–43.

[230] IX.6; pp. 1–2 Duckworth; square brackets Duckworth.

spinal marrow.[231] Here Galen is also implicitly arguing against the thesis of the followers of Praxagoras and Philotimus who maintain that the brain is nothing more than a *growth* (βλάστημα) of the spinal cord. Galen refutes this on two related grounds: first, the posterior part of the brain, continuous with the spinal medulla, has little in the way of the convoluted structure of the rest of the brain. Second, the spinal marrow is continuous only with the base of the brain, the only part of the encephalon which "has not been twisted" (οὐκ ἐλήλικται), i.e. is without convolutions.[232] In *De anatomicis administrationibus*, Galen describes where other, unnamed, anatomists have placed the end of the fourth ventricle:

> .... and further, you see the posterior ventricle, and you find that the end of it resembles a reed which one trims down in order to write with it. But although some have asserted that it extends downwards as far as the third vertebra, yet nevertheless no one is in a position to demonstrate that this ventricle descends visibly as far as the second vertebra (the axis), let alone as far as the third vertebra. But in many animals it descends as far as the site of the sockets of the skull (the condyles of the occipital bone),[233] in many as far only as the interval between the first vertebra (the atlas) and the skull (the atlantooccipital articulation), that is the place at which the persons called the 'slaughterers of oxen'[234] insert the small knife.[235] Only in very few animals does one find this ventricle reaching as far as the first vertebra, but it does not go beyond this vertebra and reach as far as the second vertebra.[236]

In fixing the end of the fourth ventricle at the point where animals are slaughtered (the region of the upper vertebral bodies in the neck), Galen reinforces both its critical importance and the legitimacy of

---

[231] Cf. *PHP* p. 446,20 De Lacy; V. 609 K: καί τις ἄλλος εἷς ἀζυγὴς ἐμβάλλει τῇ πρώτῃ γενέσει τοῦ νωτιαίου. Cf. Simon, 1906, II, 2, n. 3; Ariëns Kappers, Huber and Crosby, I, 1936, 335; Sisson and Grossman, 1975, 1073. The medulla oblongata is not recognised by Galen as a separate structure of the brain; for him it is part of the fourth ventricle and the first part of the spinal marrow.

[232] *UP* I, p. 487,4–17 Helmreich; III. 671–672 K.

[233] These are the occipital condyles (cf. *AA* II. 460–463 K). Duckworth's gloss: "condylar fossa on first cervical vertebra", would be correct if the text read "sockets *for* the skull." (cf. Simon, 1906, II, 9 n. 45; Garofalo, 1991, II, 841: *condili*).

[234] Cf. Garofalo, 1991, II, 841 n. 59.

[235] The knife would pass through the dorsal atlanto-occipital membrane and penetrate the spinal cord. It is well known to butchers that: "the medulla may be cut by passing a knife between the skull and the atlas, less easily, between the atlas and axis." Macgregor, 1980, 99. That is, between the skull and the first cervical vertebra in the first case, and between the first and second cervical vertebra in the second.

[236] IX.10; p. 14 Duckworth. My glosses in round brackets.

his interpretation. The reason the butcher's wound is fatal here, according to Galen, is that the fourth ventricle has been opened, not that the dura has been incised (see further below). Galen also fixes the end of the fourth ventricle to that part of the surface of the brain where Galen's seventh (and last) cranial nerve pair arise (the *hypoglossal*). Since the termination of the ventricular system marks the end of the brain, the outgrowth of this cranial nerve establishes the brain's distal limits.[237]

The spinal marrow arises from the brain, and Galen—reversing the order of the followers of Praxagoras and Philotimus—refers to it as an "offshoot from the substance of the brain."[238] "Its own nature is nearly like the brain."[239] It is important enough to be considered "like a second encephalon" for those parts below the head.[240] Yet the spinal cord is also different from the brain:

> The texture of the spinal cord is just the same as that of the brain from which it starts out, but it is more indurated than the brain tissue. The degree of increase in this induration corresponds to the degree in which the nerve arising from the spinal cord is more indurated than the cord itself.[241]

The spinal marrow is the source of all the hard nerves of the body, the nerves of motion.[242] Cranial nerves, some of which have sensory as well as motor tasks, are derived from the base of the brain, which for Galen is harder than the corresponding part of the brain above. Psychic pneuma in the ventricles must be able to pass into the spinal marrow and hence the nerves. It would be no more than reasonable then, to expect a passage of some description at the beginning of the spinal marrow. Although Galen does not explicitly state in *De*

---

[237] Περὶ δὲ τῆς ἑβδόμης συζυγίας τῶν ἀπ' ἐγκεφάλου νεύρων ἤδη ῥητέον . . . ὁρμᾶται δὴ ταῦτα τὰ νεῦρα, καθ' ἃ παύεται μὲν ὁ ἐγκέφαλος, ἄρχεται δ' ὁ νωτιαῖος. *UP* II, p. 36,1–9 Helmreich; III. 731–732 K. Cf. Savage-Smith, 1971, 188–90. Galen's seventh nerve is the twelfth cranial nerve of contemporary anatomy.

[238] Dissection provides this evidence clearly, but even butchers know it: ὃς ὅτι μὲν ἀποβλάστημα τῆς κατὰ τὸν ἐγκέγαλον οὐσίας ἐστίν, ἐναργῶς φαίνεται διὰ τῶν ἀνατομῶν, ὡς μηδὲ τοὺς μαγείρους ἀγνοεῖν. *PHP* p. 476,15–17 De Lacy; V. 645 K. The notion of the spinal cord as the outgrowth of the brain is also picked up in *UP* II, p. 189,18–19 Helmreich; IV. 11 K, where the spinal cord is "like the trunk of a large tree" (οἶον πρέμνου δένδρον μέγα). It flows from the fount of the brain "like a river" (οἶον ποταμόν, II, p. 215,21–22 Helmreich; IV. 47 K).

[239] ἡ φύσις αὐτοῦ παραπλήσιος ἐγκεφάλῳ. *UP* II, p. 218,9 Helmreich; IV. 51 K.

[240] *UP* II, p. 216,9 Helmreich; IV. 48 K.

[241] *AA* XV.1; p. 223 Duckworth.

[242] *UP* II, pp. 30,24–31,15 Helmreich; III. 724–725 K.

usu partium that there is a demonstrable channel between the fourth ventricle and the spinal marrow, in *De placitis Hippocratis et Platonis* he compares the opening of the first part of the spinal medulla with the alleged foramen in the optic nerves. Each has a "perceptible pore" (πὸρος αἰσθητικός).[243] In the Arabic part of *De anatomicis administrationibus*, the channel in the beginning of the spinal marrow is also compared to the passage in each olfactory tract:

> Previously also, if you will, inspect thoroughly the discharging effluents on the two sides [R. and L.] of the anterior ventricles, and remove completely all the parts lying around them, so as to get a fair view of the duct which comes out from the end of each of the two ventricles, noticing how for a wide stretch it descends in the same way as the commencement of the spinal marrow [*central canal*]. However, the top of that duct does not resemble the point of the calamus scriptorius. For it has in this spot no sort of vaulting; on the contrary each of the two ventricles steadily narrows and diminishes, so that the duct comes into being thereout.[244]

The final piece of anatomical information that can be gleaned from the fourth ventricle, alluded to earlier in this section, is Galen's observation that the distal part of fourth ventricle, where it borders on the first part of the spinal marrow, is not covered by any other part of the brain except the dura.[245] If, however, the dura is incised or damaged in any way, then this is interpreted as an injury to the fourth ventricle and its contents, and not to any other part of the brain or its covering membranes. In this way, Galen refutes the Erasistratean notion that it is damage to the dura alone which is responsible for the lack of sensation and motion in an animal when an incision is made in this area.[246] The physiological implications of this will be discussed in the next chapter.

---

[243] *PHP* p. 452,18 De Lacy; V. 617 K.

[244] IX.7; p. 3 Duckworth; square brackets Duckworth.

[245] καθ' ὃ μέρος μάλιστα τῆς παχείας μήνιγγος τρωθείσης ὁ πόρος ὅλος γίγνεται γυμνὸς ἅμα τῷ πέρατι τῆς ὄπισθεν ἐγκεφάλου κοιλίας. *PHP* p. 446,21–22 De Lacy; V. 609 K. Galen here also accurately describes the posterior third of what is recognised as the medulla, which is only covered by the meninges and blends imperceptibly with the spinal cord.

[246] See the discussion in chapter 1.5.

4.8 *Conclusion*

Galen's account of the cerebral ventricles is an impressive anatomical *tour de force*. It is also a tightly controlled narrative. Although Galen presents us with the views of other anatomists, he also claims originality in parts of his exposition. Other sources for the anatomy of the ventricles are, inevitably, filtered through him. Yet whilst the bulk of such source material is subject to Galen's polemic, he sometimes informs us of other opinions on the anatomy of the ventricles, even if a complete discourse on these differences by selective verbatim quotation from other sources is not entirely forthcoming. In the case of Herophilus and Erasistratus, Galen does, it is true, sometimes provides a word for word account. It also serves Galen's agenda since such citations highlight his own achievements. But how *original* is Galen as far as ventricular anatomy is concerned? It is true that he augments the anatomical account of the ventricles first laid down by Herophilus and Erasistratus. Galen's originality lies rather in the skill with which he crafts a detailed and consistent anatomical account of the ventricles. Having created an empirically verifiable—and elegant—anatomy of the ventricles, Galen proceeds to experiment upon them. These experiments serve as the prototype for all Galen's investigations of the brain and nerves.

PART THREE

A PHYSIOLOGY OF THE BRAIN

CHAPTER FIVE

A METHODOLOGY OF VENTRICULAR FUNCTION

I make my start with those parts where symptomatic changes follow when one divides them, bruises them, ligatures them with a cord, presses them with the fingers or twists them or stretches them unduly far. For when they are exposed to such conditions, the [normal] activity of such organs will be either completely or partially suppressed.

De anatomicis administrationibus.[1]

5.1 *Introduction*

Galen's understanding of the function of the ventricles is grounded in experimentation and observation underpinned by a thorough knowledge of anatomy. The last chapter has examined the nature of Galen's knowledge of the ventricles. His experiments on them reflect the methodology that characterises his entire approach to the physiology of the brain and the nerves. His experimental reasoning is also coloured by his teleological imperative: each structural component of the body is purposefully designed by Nature. As Crombie states: "The logic of experiment entailed at once the confirmation of true theories and the falsification of their incompatible rivals, but there was a subtle relation between this and Galen's antecedent choice of possibly true theories on grounds not simply of observation but also of reason: that is of an acceptable view of nature as the product of intelligent design as opposed to blind chance."[2] Although to a certain extent Galen must be taken on trust in the accounts he provides of his ventricular experiments and observations, it is possible to determine his experimental method and the validity of the conclusions which he draws from them. Galen's examination

[1] X,1; p. 27 Duckworth; square brackets Duckworth.
[2] 1994, 214.

of the brain encapsulates "a standard, persistent program of physiological inquiry."[3] Furley and Wilkie conclude that: "It is not unreasonable to ask whether Galen actually performed the experiments he describes, but the evidence seems strongly to favour the assumption that he did. He certainly vivisected animals. His experiments on the nervous system, in particular, carry total conviction as accounts of what he had himself actually done and seen."[4] The experiments Galen performs on the ventricles of the brain are crucial for understanding his perception of the brain as the organ of the rational soul. There are three places in the Galenic corpus where explicit reference is made to ventricular experimentation.[5] The first, in *De placitis Hippocratis et Platonis*, Book I, gives an overview of Galen's experimental results. The second, more extensive, is in Book VII, and occurs in the context of a discussion on the physical location of the soul, and whether it is correct to regard psychic pneuma as either the soul itself or as its "first instrument" (πρῶτον ὄργανον). The third occurs in the Arabic part of Book IX, *De anatomicis administrationibus*, and describes how such experiments were performed. This chapter will attempt to show that the results of Galen's brain experiments are consistent, given his interpretation of the anatomical topography of the ventricular system. They are also consistent with his choice of physiological agent, psychic pneuma.

5.2 *Galen's method of ventricular experimentation*

As noted in chapter 1, Herophilus' physiological interests seemed to have been focused primarily on the fourth ventricle. It is not possible to state whether the methodical investigation of the ventricles is Galen's specific initiative, but Galen's work can be viewed as a continuation of the research programme promulgated by Herophilus and Erasistratus. What is beyond question is that, in order to determine ventricular function, Galen must have exploited, as Herophilus had

[3] Hall, 1969, 163.

[4] 1984, 48. "If we deny this we have to assume a degree of duplicity on the part of Galen that seems totally incredible.", idem, 48 n. 2. Cf. Prendergast, 1930, 1136–1137.

[5] There are also important references to pathological effects on the ventricular system which are to be found in *De locis affectis*, and these will be examined in the course of this chapter.

done in his differentiation between sensory and motor nerves, "the use of vivisection for heuristic purposes."[6] Galen's anatomical works are not simply textbooks but interactive manuals of dissection in which he maintains and demands an impressively high standard. The descriptions of his vivisectional experiments are designed to reflect and reinforce those same criteria. Dissection makes possible an informed vivisection. Only dissection can enable the anatomist to move beyond mere *beliefs* concerning the structure and function of the living body. This forms the basis of Galen's attacks on certain of his peers, who, whilst being referred to as *anatomists*, nevertheless perform no dissections. If they regard dissection as unimportant, then it follows they will remain ignorant of the possibilities of determining function in the living animal by vivisection. Why then, concludes Galen, should they bother "to cut or ligate" (τέμνειν ἢ βρόχους περιβαλεῖν) the living animal?[7] Galen's accounts of his vivisectional experiments do not make for pleasant reading, yet these experiments show him to be, apart from anything else, well-trained in a set of surgical techniques which makes vivisection not only possible but consistently repeatable.[8] Galen's experience in surgery as physician to the gladiators at Pergamum would have given him ample opportunity to develop those skills.[9]

---

[6] Von Staden, 1989, 153. On Galen as an experimenter see Debru, 1994, 1718–1756; Grmek, 1996, 101–122.

[7] πολλὰ τοιαῦτα καθ' ὅλον τὸ ζῷον ὠλιγωρημένα τοῖς ἀνατομικοῖς ἐστιν εὑρημένα, ὀκνήσασι μὲν ἀκριβῶς ἀνατέμνειν, ἀποφηναμένοις δὲ τῶν δοξάντων ἑαυτοῖς τὸ πιθανώτατον. ὥστε οὐδὲ θαυμάζειν χρὴ τὸ πλῆθος τῶν ἀγνοηθέντων αὐτοῖς, ἔτι περιόντος τοῦ ζῴου. ὅπου γάρ, ἃ θεάσασθαι χρὴ μόνον ἐπιμελῶς ἀνατεμόντα, παρέλιπον ὀλιγωρήσαντες, ἦπού γ' ἐπετηδεύσαντ' ἄν ποτ' αὐτοὶ τέμνειν ἢ βρόχους περιβαλεῖν μορίοις ἔτι ζῶντος τοὺ ζῴου χάριν τοῦ μαθεῖν, ἥτις ἐνέργεια βλάπτεται; *AA* II. 232 K.

[8] Cf. Mani, 1991, 26–49. Toledo-Pereyra, 1973, 361–5, offers a summary of Galen's surgical skills, although the paper draws many of its examples from the pseudo-Galenic *Introductio sive medicus*. The older, more experienced Galen castigates the surgeons of the legions in Germany during the Macromannic Wars for not dissecting German corpses (*Comp. Med. Gen.* XIII. 604 K). But as Davies, 1970c, 88, notes, "the Roman doctors may well have been too busy looking after the casualties and may not all have shared Galen's enthusiasm for public demonstrations of dissection." (cf. idem, 1970b, 101–6 and 1970a; Nutton, 1969; Scarborough, 1968). The overriding factor against carrying out dissection was practical: an army at war is no place for research. Galen's scorn comes from a position of tranquillity as well as privilege.

[9] Although Galen already had the nucleus of such technical expertise, for by his own account he had obtained the position as physician to the gladiators by performing a public demonstration of his anatomical prowess with a living ape whose

In order to expose the ventricles in a living animal, and to keep it alive, it is obvious that Galen must perform delicate and sophisticated surgery, and what he records of such procedures represent nothing less than major operations on the brain.[10] In *De anatomicis administrationibus*, the method of exposing the brain in the living is similar to that employed in a dissection:

> But as for what concerns the vivisection itself, it should proceed on both animals in all details after the same fashion. That is to say, every cut that you impose should travel in a straight line, just as it travels in the dead animal.[11]

Exposing the brain of a living animal requires, at the very least, the same techniques refined upon the dead. The crucial difference between dissection and vivisection is that in a vivisection, the skilled operator must also be prepared to deal with blood loss and its consequences.[12] That the animals survived the post-operative period for sufficient time for Galen to record any meaningful results is testimony to the high level of his operative skill and is indicative of Galen having developed a set of techniques that enabled him to minimise the inevitable trauma and blood loss. With the top of the skull removed and haemostasis achieved in the living animal, Galen instructs the experimenter to expose the brain itself:

> Insert a hook into the dura mater and draw it upwards. Then first cut through the piece of it that has been raised, so that it may not

---

abdomen was opened, the intestines exposed and returned to the abdominal cavity, and the wound sutured with a minimum of blood loss. Cf. *Opt. Med. Cogn.* p. 105,4–14 Iskandar.

[10] There is no evidence that Galen undertook his later experiments on the ventricular system in order to "verify the conclusions which he had already drawn from surgical experience." Siegel, 1973, 43. Siegel also states that Galen "reported that pressure, created by a traumatic depression of the skull, especially in gladiators, seemed to result in a displacement of the cerebral ventricles and, in turn, slowed or completely suppressed the respiratory movements." As evidence, Siegel quotes *PHP* p. 442,30–35 De Lacy; V. 605 K. However, there are no references to gladiators in this citation, but to those who have undergone trepanation. Granted, Galen probably trephined gladiators who had sustained skull fractures, but these would not have formed his sole source of information.

[11] IX.11; p. 15 Duckworth.

[12] "Nothing so disturbs any surgical procedure as haemorrhage" (οὐδὲν δ' οὕτως χειρουργίαν ἅπασαν ἐν ζῴου σώματι ταράττειν εἴωθ' ὡς αἱμορραγία. *AA* II. 628 K). See also *AA* IX.11; pp. 15–16 Duckworth, where Galen describes the minimisation of blood loss when the pericranium is incised and retracted. On Galen's haemostatic methods, see Majno, 1975, 403–404; Salazar, 2000, 43–44.

> make contact with the part of the brain lying beneath it. Next insert two hooks, one on each side of the first hook, from the ends of the margins of the incision; also with these two hooks draw upon the dura mater found above, and cut through the whole raised portion of it, without touching any of the underlying parts of the brain. If you do this well, you can also introduce your fingers upwards beneath the part of the dura mater which you have incised, and you can split it until you have uncovered the whole of the brain lying beneath it.[13]

Galen next reminds the dissector to follow the dural foldings that divide the cerebrum and separate it from the cerebellum. After these have been dissected away, the brain is now fully exposed, and the animal is examined for any adverse effects from the procedure so far performed:

> Now you can cut away these three parts of the meninx [*dura mater*] and thus expose the underlying portions of the brain. Two of these three parts lie in the region beneath the parietal bone, the third is the portion overlying the hind brain. When you have done that, then make an inspection and ascertain for yourself whether the animal is being deprived of respiration, voice, movement or sensation, or whether none of these defects is showing itself in it, either at the time when the incision was made upon it or else soon afterwards. The latter may quite well be the case, when it happens that the air is warm.[14]

Galen is seeking to establish that, with the brain substance untouched, the animal's behaviour is unchanged. In his last sentence, Galen also implies that deleterious effects will not necessarily occur immediately, but "soon afterwards" if the air is warm. The corollary is that cold air either increases the severity of distress shown by the animal, or else retards its recovery. Galen clarifies this by noting his observations on the effects of external temperature upon the exposed living brain; again, Galen attempts to minimise the number of outside influences on the brain before experimentation begins:

> But if the air is cold, then in a degree corresponding to the amount of the cold air streaming in upon the brain, each single one of these functions of the brain that we have mentioned weakens; the animal remains for a certain length of time unconscious, and then expires. Therefore it is best that you should take in hand the detachment of the dura mater from the skull in the summer time, or, if you perform

[13] IX.12; pp. 17–18 Duckworth.
[14] IX.12; p. 18 Duckworth; square brackets Duckworth.

> it at another season, no matter which that season may be, you should heat the room in which you intend to dissect the animal, and warm the air.[15]

This methodological preparation of the right conditions for experimentation cannot be too highly stressed. That some animals will die from such a procedure is blatantly obvious. Galen attempts to minimise this by control of blood loss, as noted above. A further factor which Galen considers is the temperature of the surrounding environment, and cold air is noted by him to reduce significantly the usefulness of the animal for experimentation.[16] When these factors have been optimised, Galen continues with the next phase, experimentation upon the brain itself:

> Should the dissection be thus performed, then after you have laid open the brain, and divested it of the dura mater, you can first of all press down upon the brain on each of its four ventricles, and observe what derangements have afflicted the animal.[17]

The methods employed in this type of experimental procedure are cogently summarised in *De placitis Hippocratis et Platonis*:

> Following the excision of the bone of the head of the animal, with it still being alive, and after the thick meninx has been completely exposed, and when you have with hooks retracted it on either side of the midline where it has a double fold and dips into the brain, then if you simply cut it or else remove it completely, the animal does not become senseless or motionless; nor does it lose them if you merely cut or if you remove that portion of it which covers all the rear part of the brain. And even if you cut away the brain itself in any manner, even then the animal does not become motionless or senseless until the incision reaches one of the ventricles.[18]

---

[15] Ibid.

[16] Cf. *Loc. Aff.* VIII. 161, 162 K, where cold affecting the brain produces torpor and somnolence.

[17] *AA* IX.12; p. 18 Duckworth.

[18] ἐκκοπέντος τοῦ τῆς κεφαλῆς ὀστοῦ ζῶντος ἔτι τοῦ ζῴου καὶ γυμνῆς της παχείας μήνιγγος γενομένης ἐὰν ἑκατέρωθεν τῆς μέσης εὐθείας, καθ' ἣν ἐγκαταβαίνει τῷ ἐγκεφάλῳ διπλουμένη, δι' ἀγκίστρων ἀνατείνας αὐτὴν ἢ τέμῃς μόνον ἢ ἐκτέμῃς ὅλην, οὔτ' ἀναίσθητον οὔτ' ἀκίνητον γίγνεται τὸ ζῷον, ὥσπερ οὐδ' εἰ τὸ σκέπον αὐτῆς μέρος ὅλον τὸν ὄπισθεν ἐγκέφαλον ἢ τέμοις μόνον ἢ ἐκτέμοις. οὐ μὴν οὐδ' εἰ τὸν ἐγκέφαλον αὐτὸν ὁπωσοῦν ἐκτέμνοις, οὐδ' οὕτως τὸ ζῷον ἀκίνητον ἢ ἀναίσθητον γίγνεται, πρὶν ἐπί τινα τῶν κοιλιῶν αὐτοῦ τὴν τομὴν ἐξικέσθαι. p. 442,22–30; De Lacy; V. 605 K.

Galen's observation that the animal does not become "senseless or motionless" (ἀναίσθητον οὔτ' ἀκίνητον) when the dura is either cut or removed may also be a reference to the earlier Erasistratean thesis that apparently fixed the *hegemonikon* in the meninges.[19] Galen next summarises the effects of incision on each ventricle, linking every observed symptom to the relative age of the animal:

> (Incising) the posterior (ventricle) harms the animal most, and next after that the middle (ventricle). (Incising) each of the anterior (ventricles) causes a less serious injury, but of a greater degree in older animals, a lesser degree in the young.[20]

To these observations, Galen adds a rider that picks up the theme cited earlier in *De anatomicis administrationibus* IX.12, that *pressure* (θλῖψις) on the ventricles will produce similar effects:

> Pressure on the ventricles has the same effect as incising them, and we see this happen sometimes, not deliberately, but with every attempt to guard against it, in men who are being trepanned, when the bones of the head have been fractured.[21]

Galen's intention is to include as many contributions from various sources as possible to reproduce (or to be seen to have reproduced) repeatable effects on various parts of the ventricular system. The physiological agent affected by these various injuries—pressure, incision, as well as trauma (although trauma has a different, more generic status than the first two)—is psychic pneuma. Galen makes this explicit in what immediately follows from the above citation:

> From these phenomena you might assume either of two things about the pneuma in the ventricles of the brain: if the soul is without body, the pneuma is, as it were, its first home; or if the soul is embodied, then pneuma is the soul. But when after a short time, following the closure of the ventricle,[22] the animal regains sensation and motion, it

---

[19] A view Erasistratus apparently recanted. See chapter 1.5. However, the fact that Galen stresses that the dura plays no part in these events could indicate that the Erasistratean meningeal thesis might still have enjoyed support in some quarters.

[20] μάλιστα μὲν οὖν ἡ ὀπίσω βλάπτει τὸ ζῷον, ἐφεξῆς δ' ἡ μέση· τῶν προσθίων δ' ἑκατέρα βραχυτέραν ἐργάζεται τὴν βλάβην καὶ μᾶλλον μὲν ἐπὶ τῶν πρεσβυτέρων ζῴων, ἧττον δ' ἐπὶ τῶν νέων. *PHP* p. 442,30–32 De Lacy: V. 605 K.

[21] τὰ δ' αὐτὰ ταῖς εἰς τὰς κοιλίας τομαῖς αἱ κατ' αὐτῶν ἐργάζονται θλίψεις, ἃς οὐδ' ἐπιτηδευόντων, ἀλλὰ καὶ πάνυ φυλαττομένων ὁρῶμεν ἐνίοτε γιγνομένας ἐπὶ τῶν ἀνατιτραμένων ἀνθρώπων, ὅταν τὰ τῆς κεφαλῆς ὀστᾶ καταγῶσιν. *PHP* p. 442, 32–35 De Lacy; V. 605 K.

[22] If the incision was made by a fine scalpel, as it would most likely have been,

> is no longer possible to accept either alternative concerning pneuma. It is better, then, to accept that the soul resides in the actual body of the brain, whatever its substance may be—for the examination has not yet arrived at this question—, and that the soul's first instrument for all the sensations of the animal and for its voluntary motions is this pneuma; and therefore, when the pneuma has escaped, and until it collects again, whilst it does not deprive the animal of life, it makes it incapable of sensation and motion. For indeed, if pneuma itself were the substance of the soul, the animal would instantly die along with the escape of the pneuma.[23]

This is a good argument for the existence of a physiological agency. The question of the substance of the soul is a matter for supposition and is a physiological irrelevance. Galen limits himself to demonstrating not the actions of the rational or hegemonic soul, but only of its governing agency which he fixes within the ventricular system of the brain.

The above citations summarise Galen's vivisection techniques. They inform us that the functional importance of the brain is delineated by the topography of the ventricles. And Galen attempts to show that the two principal activities of the rational soul, sensation and voluntary motion, may be successfully manipulated by experiment. Galen does not formally quantify his observations, beyond noting the presence or absence of the key parameters of motion and sensation. An ambiguous estimation of the time taken to return to a normal state is as far as Galen goes in attempting to quantify his results.[24] However, what Galen stresses is the relative differences in

---

then such a wound could have sealed itself over time, or, more likely, the surrounding brain would have acted as a wound tamponade.

[23] Ἐκ τούτων οὖν τῶν φαινομένων ἴσως ἄν τις ὑπονοήσειε τὸ κατὰ τὰς κοιλίας τοῦ ἐγκεφάλου πνεῦμα δυοῖν θάτερον, εἰ μὲν ἀσώματός ἐστιν ἡ ψυχή, τὸ πρῶτον αὐτῆς ὑπάρχειν, ὡς ἂν εἴποι τις, οἰκητήριον, εἰ δὲ σῶμα, τοῦτ᾽ αὐτὸ [πνεῦμα] τὴν ψυχὴν εἶναι. ἀλλ᾽ ὅταν γε συναχθεισῶν τῶν κοιλιῶν ὀλίγον ὕστερον αὖθις αἰσθάνηται καὶ κινῆται τὸ ζῷον, οὐκέτ᾽ <οὐδέτερον> οἷόν τε φάναι τῶν εἰρημένων ὑπάρχειν τουτὶ τὸ πνεῦμα. βέλτιον οὖν ὑπολαβεῖν ἐν αὐτῷ μὲν τῷ σώματι τοῦ ἐγκεφάλου τὴν ψυχὴν οἰκεῖν, ἥτις ποτ᾽ ἂν ᾖ κατὰ τὴν οὐσίαν—οὔπω γὰρ περὶ τούτου σκέψις ἥκει—, τὸ πρῶτον δ᾽ αὐτῆς ὄργανον εἴς τε τὰς αἰσθήσεις ἁπάσας τοῦ ζῴου καὶ προσέτι τὰς καθ᾽ ὁρμὴν κινήσεις τοῦτ᾽ εἶναι τὸ πνεῦμα, διὸ καὶ κενωθέν, ἄχρις ἂν αὖθις ἀθροισθῇ, τὴν μὲν ζωὴν οὐκ ἀφαιρεῖσθαι τὸ ζῷον, ἀναίσθητον δὲ καὶ ἀκίνητον ἐργάζεσθαι. καίτοι γε, εἴπερ ἦν αὐτὸ ἡ τῆς ψυχῆς οὐσία, συνδιεφθείρετ᾽ ἂν αὐτῷ κενουμένῳ παραχρῆμα τὸ ζῷον. *PHP* pp. 442,36–444,11; De Lacy; V. 605–606 K. Cf. p. 446,11–17 De Lacy; V. 609 K.

[24] The results of Galen's experiments on carotid ligation are dealt with in similar fashion (see chapter 6). Temkin, 1961, 470, in an important study, notes that: "Instances of a quantitative approach to physiological problems are not very numer-

the time taken for the animal to return to "normal", based on which ventricle is affected and the nature of the injury. This information, together with correlating the age of the experimental animal concerned and incorporating evidence (direct and anecdotal) from clinical cases, is gathered by Galen to support his hypothesis of ventricular function. That function is expressed in pneumatic terms. And although Galen never articulates a doctrine of ventricular or brain localisation, his observations that different physical actions may be elicited depending on which ventricle is pressed or incised laid the groundwork for later concepts of localisation.[25]

In the Arabic part of *De anatomicis administrationibus*, Galen describes the experiments performed on the exposed brain. Quoting this passage in its entirety gives a fundamental insight into Galen's experimental methodology:

> Should the dissection be thus performed, then after you have laid open the brain, and divested it of the dura mater, you can first of all press down upon the brain on each of its four ventricles, and observe what derangements have afflicted the animal. I will describe to you what is always to be seen when you make this dissection, and also before it, where the skull has been perforated, as soon as one presses upon the brain with the instrument which the ancients call 'the protector of the dura mater'.[26] Should the brain be compressed on both the two anterior ventricles, then the degree of stupor which overcomes the animal is slight. Should it be compressed on the middle ventricle, then the stupor of the animal is heavier. And when one presses down upon that ventricle which is found in the part of the brain lying at the nape of the neck (the fourth or posterior ventricle), then the animal falls into a very heavy and pronounced stupor. This is what happens also when you cut into the cerebral ventricles, except that if you cut into these ventricles, the animal does not revert to its natural condition as it does when you press upon them. Nevertheless it does sometimes do this if the incision should become united.[27] This return to the normal

---

ous in Greek physiology." See also Bylebyl, 1977 (but cf. Gourevitch, 1987, 276 and n. 24); Grmek, 1990, 17–43, esp. 40ff.; idem, 1996; Shryock, 1961. Given the stress in parts of Greek physiology on the importance of qualitative change, then the scope for using or expanding a quantitative methodological approach was limited. This is not to say that a quantitative method could not be employed to assist in solving a physiological problem involving qualitative change.

[25] See Appendix One.

[26] See section 3 below.

[27] Galen is deducing wound closure due to tamponade, where the brain substance overlying the ventricles seals the wound. Cf. n. 22, above.

> condition follows more easily and more quickly, should the incision be made upon the two anterior ventricles. But if the incision encounters the middle ventricle, then the return to the normal comes to pass less easily and speedily. And if the incision should have been imposed upon the fourth, that is, the posterior ventricle, then the animal seldom returns to its natural condition; although nevertheless if the incision should be made into this fourth ventricle, provided that you do not make the cut very extensive, that you proceed quickly, and that in the compression of the wound in some way or other you employ a certain amount of haste,[28] the animal will revert to its normal state, since the pressure upon the wound is then temporary only—and indeed especially in those regions where no portion of the brain overlies this ventricle, but where the meninx only is found. You then see how the animal blinks with its eyes, especially when you bring some object near to the eyes, even when you have exposed to view the posterior ventricle. Should you go towards the animal while it is in this condition, and should you press upon some one part of the two anterior ventricles, no matter which part it may be, in the place where as I stated the root of the two optic nerves lies, thereupon the animal ceases to blink with its two eyes, even when you bring some object near to the pupils, and the whole appearance of the eye on the side on which lies the ventricle of the brain upon which you are pressing becomes like the eyes of blind men.[29]

This passage is without question one of the most impressive accounts of physiological experimentation extant in Western Antiquity. The claims made are far reaching and will be explored in the following two sections. But whether Galen deals with the effects of pressure or incision, these results are remarkable. Galen presents his readers and students with a formidable—even overwhelming—combination of factors that few could manipulate successfully, let alone concurrently. His techniques are able to account for the function of each ventricle. This account should be read as a composite of Galen's investigations of the ventricles, since clearly they cannot have been the results of any single experiment. And underlying this experimental *tour de force* is the message that failure to observe what Galen has expounded means only that the procedure has been improperly carried out; not that the methodology or the results can be called into question. As noted in earlier chapters, these manipulative and

---

[28] This stresses the importance of the fourth ventricle as much as it does Galen's dexterity.

[29] IX.12; pp. 18–19 Duckworth. My gloss in round brackets.

polemic techniques form the mainstay of Galen's experimental methodology of the brain.[30] Galen next deliberately excludes the rest of the brain from having a bearing on these experiments:

> . . . the animal, when one pierces or incises the thin meninx, sustains no derangement as a result, just as none such befalls it if the brain should be incised without the incision reaching as far as to one of its ventricles.[31]

In this way, Galen reinforces the notion that the ventricles are of first importance—not the covering meninx nor the brain substance.[32] Galen also serves notice that his ventricular experiments may further be manipulated to produce a satisfactory outcome—a small incision, performed quickly and with prompt compression applied to the wound may result in the animal returning to its normal state. Experiments of this nature obviously required a large supply of animals; especially pigs and goats.[33] With these preliminary observations from Galen's experiments in mind, his interpretations of the effects of pressure on the ventricles may next be examined.

5.3 *The effects of pressure: chance and the youth from Smyrna*

In *PHP* p. 442,22–35 De Lacy, above, Galen noted the effects on the ventricles of *trepanation* (ἀνάτρησις).[34] A *trephine* (χοινικίς) is a toothed drilling instrument that, with modifications in its calibre, was also employed to drain body cavities.[35] The earliest Western use of trepanation is found in the Hippocratic Corpus, where trephining the

[30] Apart from incision and pressure, the other experimental technique Galen employed was that of *ligation* (βρόχος), used principally to determine nerve function.

[31] IX.12; p. 20 Duckworth.

[32] Cf. *PHP* pp. 440,20–442,18 De Lacy; V. 602–604 K.

[33] Cf. *AA* IX.11; p. 15 Duckworth.

[34] On the history of trepanation see Arnott, Finger, Smith (eds.), 2002; Nunn, 1996, 168–169; Roberts and Manchester, 1995, 79–85, 91–94. On Galen's use of trepanation see Rocca, 2002a.

[35] The Pseudo-Galenic *Introductio sive medicus* XIV. 783 K, states that the χοινικίς is a tool associated with "the ancients." The Hippocratic text *Internal Affections* (*Int.*) VII. 226 L, advises that fluid accumulating in the lung as a result of *dropsy* (ὕδερος) be drained by inserting a "straight-pointed trephine" (περητήριον) at the level of the third lowest rib (cf. Milne 1907, 131). Galen defines this instrument in his Hippocratic *Glossary* XIX. 129 K, as a variant of the trephine. Galen also uses the term ὀρθοπρίων (*Gloss.* XIX. 126 K), for which there is no reference in any other

skull was usually performed in cases of *fracture* (ῥωγμή or κάταγμα) and its sequelae.[36] By Galen's time trepanning was an established procedure for dealing with skull trauma and the consequences of a *depressed fracture* (ἐμπίεσμα), where there was danger of bone fragments pressing directly onto the outer meninx of the brain.[37] Galen limits the use of trephining to the relief of pressure and its consequences; that is, to skull trauma and in cases such as the draining of phleg-

---

medical author or text. Both terms were likely to have been more sophisticated variants of the standard χοινικίς, and possibly, to judge by their rare citations, infrequently used.

[36] In *Places in Man* (*Loc.*), one of the earliest works in the Corpus, trephining is recommended for what seems to be a depressed fracture, with the aim of preventing the accumulation of purulent *fluid* (ἰχώρ) within the wound (*Loc.* 32.1; p. 70,20–27 Craik; VI. 324 L). The procedure itself is not described. Three cases are described in *Epidemics V* (pp. 10–11, 16–18 Jouanna; V. 216, 226–228 L). In the third case, the presence of exposed suture lines in the wound is an indication for prompt trepanation. A fourth case is given in *Epidemics VII* (pp. 73–74 Jouanna; V. 404 L). The trepans in the Hippocratic Corpus are varying sizes of a *crown trephine* or *modiolus* (πρίων, πρίων χαρακτός, σμικρὸν τρύπανον), all of which lacked a guard. *On Wounds to the Head* (*VC*) describes the conditions for their use (pp. 90,7–92,9 Hanson; III. 256–260 L). The text cautions against the too enthusiastic use of the trephine, even in cases of depressed and comminuted fractures, since there is a significant risk of damage to the dura with the use of the instrument (p. 86,4–10 Hanson; III. 248 L). A fracture that is not comminuted is an indication for trephining (pp. 82,5–84,1 Hanson; III. 238–242 L). Cf. Adams, 1849, 465 n. 2; 442–3 and Plates I & II; Grmek, 1983, 287–8; Marganne, 1998, xi–xiii, 74–83, 156, 160; Milne, 1907, 131–2. The skull was also trephined by some physicians in cases of epilepsy and paralysis (cf. Temkin, 1971, 76). The term ῥωγμή is used by Galen to refer to an uncomplicated fracture; κάταγμα implies a compound fracture. On ἐμπίεσμα see n. 37 below.

[37] ἐμπίεσμα is a relatively late development of a technical expression in anatomical pathology (its only instance in the Hippocratic Corpus is the verb ἐμπιέξεται, where the brain is said to exert pressure on the channels of the ears (*De glandulis* p. 120 Joly; VIII. 568 L). In the pseudo-Galenic *Introductio sive medicus* XIV. 782 K, a sophisticated classification of five types of skull fracture is presented (and a sixth admitted by some, based on the Hippocratic five-fold classification in *VC*). This classification is increased to eight in the pseudo-Galenic *Definitiones Medicae* XIX. 431 K. However, ἐμπίεσμα is not part of the terminology employed in this system. Nor does it appear in Rufus or Aretaeus. Celsus has no specific name for it, although Soranus, *De signis fracturarum* I.1 p. 157,16–26 Ilberg, uses it to define a depressed skull fracture, a depressed fracture to the sternum, and a displaced fracture of the spine. The fullest definition before Galen is found in Heliodorus, a surgeon probably active in Alexandria in the first century AD (*apud* Oribasius, *Collectiones medicae* XLIV. pp. 118, 120, 164, 166, 168, 176 Raeder). Heliodorus outlines his concern for the integrity of the dura whilst trephining in cases of ἐμπίεσμα: skull fracture with depression of bone fragments. Cf. Marganne, 1986 and 1988; Sigerist, 1920, 1–9. Galen's use of ἐμπίεσμα is singular, in the context of trepanning the skull and, like Heliodorus, he employs it in order to stress the need to protect the dura from the effects of the depressed bone fragment(s). And see n. 47 below.

matous lesions on the head.[38] In these instances, whilst different trephines were sometimes employed, a special tool was introduced to safeguard the outer meninx. Two types of trepanation instruments were employed for the purposes of removing bone and relieving pressure.[39] The first, noted above, was the *crown trephine*, χοινικίς, or *modiolus*, used for bone excision and relief of pressure in small wounds, and which lacked a guard.[40] Galen comments that the unguarded trephine is used only by those who are "either cowardly . . . or careful."[41] This is not simply polemic, but indicative of the high risk involved in using such an instrument. The second type of instrument was the *drill trepan* (τρύπανον) for larger wounds, in which the goal was to perforate the affected bone at circumferential points to facilitate its subsequent removal.[42] A variant of this instrument was employed to minimise possible damage to the dura and was known as the *abaptiston* or *terebrum non profundans* (ἀβάπτιστον, ἀβάπτιστον τρύπανον).[43] It could not have been an easy procedure.[44] The employment of either instrument invariably meant that a larger portion of bone was excised than if the trephine alone was employed. Therefore, to protect the dura when the bone fragments were levered out, the use of the "protector of the meninx" (μηνιγγοφύλαξ, *meningophylax, membranae custos*) was recommended.[45] Regardless of the instrument

---

[38] Cf. *Hipp. Off. Med.* XVIIIB. 808 K.

[39] Cf. Celsus, *Med.* VIII, pp. 363,3–364,6; 374,23–382,16 Marx.

[40] Durling, 1993, 339, defines this as "a kind of trepan." It is most likely the πρίων χαρακτός of *VC* (cf. Adams, 1849, 465 n. 3). See also Brothwell, 1974, 209–211; Como, 1925, 160, fig. 6, 1–5; Jackson, 1990, 18, fig. 5.8; Künzl, 1996, Pl. IV, figs. 1–3, 2587; Majno, 1975, 166–169.

[41] εἴτε δειλοτέρους . . . εἴτ' ἀσφαλεστέρους. *Meth. Med.* X. 447–448 K.

[42] Cf. *Comp. Med. Sec.* XII. 821 K. See also Milne, 1907, 126–9. In the pseudo-Galenic *Intr.* XIV. 783 K, the χοινικίς is contrasted with the τρύπανον by the use of the word κεφαλοτρύπανον, a term not found in any other ancient medical source. The κεφαλοτρύπανον may have represented something of an advance over the χοινικίς and was regarded as an instrument specifically for the head, hence the prefix.

[43] Cf. *Meth. Med.* X. 446–447 K. This instrument, as Greenhill, 1864, 556, notes, is "a sort of trepan, with a ring or knob a little above the extremity, in order to prevent its penetrating the cranium too suddenly, and so injuring the brain. Hence the name, διὰ τὸ μὴ βαπτίζεσθαι, says Galen, because it could not be suddenly *plunged* or *immersed* into the brain." Italics in text. See also Horne, 1894, 14–18, 24–27; Milne, 1907, 129–130, Plate XLII, figs. 3, 4, 5.

[44] Cf. Horne, 1894, 17. See Caton, 1914, for a description of the drill-bow which drove the trephine.

[45] Cf. Milne, 1907, 126. The use of the *meningophylax* was not restricted to the skull. Galen describes its use as a protector of the underlying pleura during excision of a rib (*AA* II. 686 K). Cf. Milne, 1907, 59–60.

employed, trepanation carried obvious attendant risks, and the skill of the operator was paramount to prevent perforation of the dura and damage to the brain.[46] Galen's accounts of trepanation of the skull are not solely based on second-hand reports but speak of personal experience with a set of procedures with which he was highly skilled.[47]

The use of any trepanation instrument, either by himself, or witnessed in the hands of others, would have afforded Galen ample opportunity for observing symptoms which, he maintained, ensued from the consequent ventricular disturbance.[48] These are bluntly stated in *De locis affectis*:

> If one presses down too hard while treating the fractured bones of the skull by trepanation, the [patients] immediately lose all sensation and becomes motionless.[49]

Only the general, unqualified effects of loss of sensation and motion are given. They are not localised by Galen to any specific ventricle. Similarly, the description at the end of *De instrumento odoratus* tells of a *stuporous-like* state (καρώδης) in cases of skull fracture when excess pressure is applied to the ventricles of the brain by trepanation.[50] In both of these texts, Galen's intention is to stress that when unwarranted pressure is applied to the brain, the two key losses are those which encompass the principal activities of the *hegemonikon*, namely, sensation and motion. Galen's codicil to the above citation from *De locis affectis* also reinforces the site of the *hegemonikon* by noting that,

---

[46] It is perhaps worth noting the effects of the trephine in the hands of another skilled operator before the era of antiseptic surgery. One of the foremost English surgeons of the eighteenth century, Percivall Pott, 1768, 145–157, described some eighteen cases of fractures to the skull. Of these, five were not trephined (and only one survived). Of the thirteen trephined, six died. Admittedly, these numbers are not statistically significant, but they are indicative of a high risk procedure, and Pott was well aware of the hazards of trephining, regardless of the instrument used.

[47] See *Hipp. Epid. III.* p. 25,14–21 Wenkebach; XVIIA. 521–522 K, where the need to protect the dura from the trephine is emphasised and the effects of the depressed skull fracture—from lack of movement and sensation to apoplexy—are noted. Cf. *Hipp. Off. Med.* XVIIIB. 670 K.

[48] On Galen as a diagnostician, see García-Ballester, 1994; Nutton, 1993b.

[49] ἐπί τε γὰρ τῶν ἀνατιτραμένων τὸ κατεαγὸς ὀστοῦν τῆς κεφαλῆς, εἴ τις βιαιότερον ἐπιθλίψειεν, ἀναίσθητοί τε παραχρῆμα καὶ ἀκίνητοι γίνονται, VIII. 128 K.

[50] ἀνατιτράμενοι γοῦν πολλοὶ παραχρῆμα καρώδεις γίνονται τῶν ὑπηρετουμένων τοῖς χειρουργοῦσιν ἀμετρότερον ἐνίοτε θλιψάντων τὰς κατὰ τὸν ἐγκέφαλον κοιλίας. p. 64,1–3 Kollesch; II. 886 K. It is not clear whether the pressure here is due to the drill itself or to the subsequent use of the *meningophylax*.

as a result of such pressure, the *mind* (διάνοια) is permanently damaged.[51] A more detailed account of the effects of a head injury and its interpretation in ventricular terms is given by Galen in *De usu partium*:

> At Smyrna in Ionia I once witnessed an unexpected sight, a youth who had suffered a wound in one of the anterior ventricles and yet survived, apparently by divine will; however if both (ventricles) had been simultaneously wounded, he could not have survived for even a moment. Similarly, if some trouble other than a wound affects one of the ventricles while the other remains sound, the animal will be in less danger of its life than if both are affected simultaneously.[52]

Galen gives no details as to the nature of the *wound* (τραῦμα) and what steps were taken to treat it, although it is likely the injury consisted of a fracture to the skull and that trephining would have formed part of the treatment process. For the young man to have survived such a wound was indeed an "unexpected sight" (παράδοξον θέαμα). Since Galen mentions that the wound was sustained "in one of the anterior ventricles", this also means that the dura mater had been perforated. Any "injury to the dura" (μηνιγγότρωτον) was a matter for concern. In one of his earlier works, Galen remarks how:

> ... as often as it were possible for me, when I inspected a meningeal wound, to know clearly whether [the person] will die in every circumstance, or for the most part, or rarely, or doubtfully.[53]

In *De locis affectis*, the effect of such an injury is interpreted by Galen not strictly in terms of the injury to the dura alone but to the ventricle. The text does not permit the conclusion that Galen saw this wound immediately following its infliction. However, that it was localised to one side of the anterior part of the head seems clear. And Galen does not hesitate to correlate this observation to his own

---

[51] ἡ τῆς διανοίας οὖν βλάβη γίνεται διὰ παντός, VIII. 128 K.

[52] ἡμεῖς ποτε τὸ παράδοξον ἐκεῖνο θέαμα τὸ ἐπὶ τῆς Ἰωνίας ἐν Σμύρνῃ γενόμενον ἐθεασάμεθα, τὸν τρωθέντα νεανίσκον εἰς τὴν ἑτέραν τῶν προσθίων κοιλιῶν, εἶτα περιγενόμενον, ὡς ἐδόκει, βουλήσει τοῦ θεοῦ· ἀλλ' οὐκ ἂν οὐδ' ἐπ' ὀλίγον ἐξήρκεσεν ἀμφοτέρων ὁμοῦ τρωθεισῶν. οὕτως οὖν, εἰ καὶ χωρὶς τοῦ τρωθῆναι πάθημά τι περὶ τὴν ἑτέραν συμπέσοι, τῆς λοιπῆς μενούσης ὑγιοῦς, ἧττον εἰς τὴν ζωὴν βεβλάψεται τὸ ζῷον, ἢ εἰ ἀμφότεραι πάθοιεν ἅμα. I, pp. 481,22–482,5 Helmreich; III. 664 K.

[53] ὁποσάκις ἂν εἴη μοι θεασαμένῳ τὸν μηνιγγότρωτον ἀκριβῶς εἰδέναι, πότερον διὰ παντὸς ἤ ὡς <ἐπὶ> τὸ πολὺ ἢ σπανίως ἢ ἀμφιδόξως τεθνήξεται, *On Medical Experience* Ch. VII, p. 95,18–20 Walzer; p. 104,9–11 Deichgräber. The context is an attack on Empiricist doctors and their attitude to clinical observation.

experimental observations, expressed in the change from human patient to *animal* (ζῷον) subject. In mentioning what he has witnessed in Smyrna, Galen moves quickly from discussing the clinical condition to a similar picture that he produced in an animal. The injury of the youth from Smyrna allows Galen to emphasise the critical importance of the integrity of both anterior ventricles. This case illustrates the importance Galen attributes to paired structures in living things, reiterating the value he places in symmetrical organisation, albeit in a weak sense. This is outlined in a passage that immediately precedes Galen's description of the young man from Smyrna:

> It is better that there exist two [ventricles] not one, as the inferior opening (sc. interventricular foramen) was created double, as well as all sense organs and even the brain itself has a double form . . . But yet the first and most common use of the double organs is that if one is affected, the remaining one will come into service.[54]

The loss of function of one ventricle may partially be compensated by the other. Galen makes such compensation explicit:

> And if two (ventricles) have both become at the same time equally affected (πάθημα),[55] then truly this is as if from the outset there was only one (ventricle) which had become affected. Therefore there is surety in a double organ, where possible, rather than a single one. But it is not possible in every instance.[56]

The last sentence may be considered Galen's escape clause, since not only does it qualify the support a paired structure can provide, but it sets a further limit on how far Galen can take the analogy of paired symmetry in his account of ventricular function. A severe enough wound inflicted on one side of the skull would render void any form of contralateral compensation. But, equally clearly, the third and fourth ventricles, being unpaired, possess no symmetrical advantage, and therefore Galen cannot appeal directly to the func-

---

[54] δύο δ' ἦν ἄμεινον αὐτάς, οὐ μίαν ὑπάρχειν, ὡς ἂν καὶ τῆς κάτω συντρήσεως διττῆς γεγενημένης καὶ τῶν αἰσθητηρίων ἁπάντων διδύμων καὶ αὐτοῦ τοῦ ἐγκεφάλου διφυοῦς . . . ἀλλ' ἥ γε πρώτη τε καὶ κοινοτάτη πάντων τῶν διφυῶν ὀργάνων ἥδ' ἐστίν, ἵν', εἰ καὶ θάτερον αὐτῶν πάθοι, τὸ λοιπὸν ὑπηρετοῖ. *UP* I, p. 481,14–22 Helmreich; III. 663–664 K.

[55] Galen's use of πάθημα is of course context dependent, but conveys the varied nature of what may befall a ventricle, unless a disease process is specifically mentioned.

[56] καὶ μὴν ἴσον ἐστίν, ἢ δυοῖν οὐσῶν ἀμφοτέρας ἅμα παθεῖν ἢ μιᾶς ἐξ ἀρχῆς γενομένης μίαν εἶναι τὴν πάσχουσαν. ὥστ' ἀσφαλέστερον ἐν οἷς ἐγχωρεῖ τὸ διφυὲς τοῦ μονοφυοῦς. ἐγχωρεῖ δ' οὐ πάντη. *UP* I, p. 482,5–8 Helmreich; III. 664 K.

tional security of bilateral symmetry. He seeks therefore to qualify the importance of symmetrical structure by an appeal to obvious structural constraints. For example, there cannot be two spinal columns (and hence two spinal cords), for such a thing is impossible. That there is only one spinal cord means that there can only be one cavity in the cerebellum, that is, the fourth ventricle, as the spinal cord arises from this cavity.[57] This weak argument depends for its validity on Galen's conception of the symmetrical wisdom of Nature, who would not provide an animal with two spinal columns in the first place. This allows Galen not only to bypass the restrictions imposed by his argument from symmetry when it suits him, but also enables him to stress the critical importance of the unpaired fourth ventricle—depending on the extent of trauma sustained to the head in any particular circumstance.

5.4 *Pressure as an experimental technique*

The untoward effects of trepanation and the observations made from those who have sustained head injuries provide Galen with information that he compares to that obtained from his vivisection experiments. As noted in the citation from *De anatomicis administrationibus* IX.12; p. 18 Duckworth, Galen points out that pressure may be used as a specific experimental technique to determine the functional status of each ventricle.[58] Galen's intention is to try, as far as possible, to limit the disturbance sustained so that some form of recovery can take place and observations made. The following discussion begins with the anterior ventricles.

Pressure applied to both anterior ventricles results in *stupor* (κάρος) which Galen describes as "slight". As will be seen, Galen employs stupor as an index of ventricular damage, albeit not in a quantitative sense.[59] In *De locis affectis*, Galen gives the following definition of stupor:

---

[57] δύο γοῦν γενέσθαι ῥάχεις ἑνὸς ζῴου παντάπασιν ἦν ἀδύνατον· εἰ δὲ τοῦτο, καὶ δύο μυελοὺς νωτιαίους· εἰ δὲ ταῦτα, καὶ τὴν τῆς παρεγκεφαλίδος κοιλίαν οὐχ οἷόν τ' ἦν γενέσθαι διττήν, ἐξ αὐτῆς γε τοῦ νωτιαίου μυελοῦ βλαστάνοντος. *UP* I, p. 482, 9–13 Helmreich; III. 664–665 K.

[58] See n. 17 above. Cf. Neuburger 1897, 67–70; Clarke and O'Malley, 1968, 492–497.

[59] For instance, what is meant by a "slight stupor" and how is stupor to be defined so that its meaning will be universally understood?

> When the entire anterior part of the brain is affected, the foremost ventricle is necessarily affected by sympathy and the activities of the mind are harmed in similar fashion. And (the patient) lies in this way without sensation and motion, but respiration is unharmed. This condition is called stupor.[60]

Clearly, any adverse effect can elicit *sympathy* (συμπάθεια); this general term is used by Galen to interpret an injury from a chosen theoretical viewpoint. Immediately prior to this citation, Galen notes that movement of the tongue is retained in conditions affecting the anterior part of the brain, and that it is still present when sensation and motion of the facial muscles is absent. To Galen the reason is anatomical: the nerves which supply the tongue musculature are derived from the posterior parts of the brain, whereas the nerves to the face are derived from the affected anterior part of the brain.[61] It is not explicitly stated, but part of Galen's clinical picture of stupor must include speechlessness. This particular presentation was noted in a set of clinical experiments made by the French physiologist Auburtin in 1861:

> . . . During the interrogation [of the patient] the blade of a large spatula was placed on the anterior lobes; by means of light pressure speech was suddenly stopped; a word that had been commenced was cut in two. The faculty of speech reappeared as soon as the compression ceased. It has been claimed that this observation proved nothing because pressure could be transmitted to the other parts of the brain; but this pressure was directed in such a way that only the anterior lobes were affected, and, besides, it produced neither paralysis or loss of consciousness.[62]

There is a significant difference from Galen's account, for Auburtin elicited only loss of speech; consciousness as well as voluntary motion was preserved. In Galen's case, sensation and motion are noted to be absent; any loss of phonation is not recorded here.[63] Auburtin's work emphasises that Galen's experiments exerted pressure on larger

[60] ἐὰν γὰρ ὅλον ποτὲ πάθῃ τὸ πρόσθιον ἐγκεφάλου, συμπάσχειν μὲν ἀναγκαῖόν ἐστι καὶ τὰ περὶ τὴν ὑψηλοτάτην αὐτοῦ κοιλίαν, βλάπτεσθαι δὲ καὶ τὰς διανοητικὰς αὐτῶν ἐνεργείας. καὶ κεῖται ἀναίσθητος μὲν καὶ ἀκίνητος ὁ οὕτως παθὼν, οὐδὲν δ' εἰς τὴν ἀναπνοὴν βλάπτεται, καὶ καλεῖται τὸ πάθος τοῦτο κάρος. VIII. 231 K. Cf. Soury, 1899, 322–323; *ODB* 58–59.

[61] VIII. 230–231 K.

[62] Cited in Clarke and O'Malley, 1968, 492–493.

[63] But see the citation from *PHP* on pp. 195–196.

parts of the anterior brain than Auburtin's, since Galen considered the anterior ventricle to be larger in extent and hence covered by a far more extensive part of the brain. Auburtin, by contrast, localised his experiments to a more circumscribed area in the anterior lobes of the brain substance. Yet his results showed that definite and repeatable clinical symptoms could be elicited from pressure applied to the anterior part of the brain. Galen's case in *De locis affectis* shows that an unspecified pathological condition of the anterior part of the brain affects the anterior ventricles, and that this clinical state should be regarded as a close approximation of the experimental result of pressure applied to the anterior ventricles in *De anatomicis administrationibus* (IX.12; p. 19 Duckworth).[64] In this latter text, Galen also exploits what he maintains is the anatomically verifiable link between the anterior ventricle and the eye to affirm the importance of the anterior ventricles in motor and sensory activities. With the brain exposed, Galen notes that the animal continues to blink its eyes, "even when you have exposed to view the posterior ventricle." Galen cites the posterior ventricle in this way to emphasise that it has no effect on the physiology of the eye. But when pressure is placed upon the anterior ventricles then, as one approaches the animal, it ceases to blink, and Galen further observes that the "whole appearance of the eye on the side on which lies the ventricle of the brain upon which you are pressing becomes like the eyes of blind men."[65] The pupil ceases to react. If one then turns to *De placitis Hippocratis et Platonis*, it is inferred that the reason for this lack of reaction is that the flow of psychic pneuma from the optic nerves to the eyes is impeded.[66]

---

[64] The discussion in *Loc.Aff.* VIII. 232 K, concerns the development of the group of symptoms associated with epilepsy. However, Galen also speaks more generally of "diseases of the anterior part of the brain." In apoplexy, Galen notes that the body of the brain is chiefly affected, whereas stupor and epilepsy are primarily an affection of the ventricles. Stupor is more associated with the anterior part of the brain. See further below.

[65] According to Galen, it does not seem to matter where on the anterior ventricles pressure is applied.

[66] The "luminous pneuma" (αὐγοειδὲς πνεῦμα) in the optic nerves mixes with the *vitreous humour* (ὑαλοειδὲς ὑγρόν) of the eye (p. 458,19–22 De Lacy; V. 624 K). At pp. 450,10–452.7; V. 614–616 K, Galen argues that this pneuma is responsible for changes in the diameter of the pupil, citing in part the cases of those who are blind (but not due to cataract). In these, when one eye is closed, the other pupil remains unchanged, as if the "pneumatic substance" is prevented from reaching the eye. The reason is that the *pores* (πόροι) in the optic nerves are impacted or *blocked* (ἔμπηξις).

Galen states that vision is dependent on a large amount of "pneuma-like substance" (πνευματικὴ οὐσία) that flows from the brain to the eye.[67] In *De anatomicis administrationibus*, the blockage is due to pressure on the anterior ventricle in response to pressure exerted on the part of the brain that overlies it. However, Galen's observation of what happens to each pupil, correlated both with his physiological theory and clinical experience, serves primarily to demonstrate the importance of the anterior ventricles, not the nature of their contents.

In *De anatomicis administrationibus* IX.12, Galen has also noted that pressure on the middle ventricle results in a stupor that is "heavier" in degree. This may be augmented by the following from *De locis affectis*, where Galen describes the effects of trepanation in the context of both the middle ventricle and his pneumatic physiology:

> When the middle cerebral ventricle is compressed by trepanning the bone, the person will go into a stupor without convulsions or difficulty in breathing . . . Just as stupor can be caused by lack of attention during trephining, if one lowers the blade guarding the meninges deeper than is permissible, and in the same manner when a fractured bone presses hard on the cerebral ventricle, especially the middle one, stupor follows. And such an affliction is attended by violent pain, when the tension of the psychic pneuma in the ventricles has fallen.[68]

That the middle ventricle is singled out in this way indicates that Galen regarded it as a more extensive structure than is recognised today. To him, the middle ventricle encompassed part of what is now classified as the anterior ventricle. Here, *pressure* (θλῖψις) on the middle ventricle results in stupor, but the patient continues to breathe and does not convulse (the patient is unconscious). The description admits of such an interpretation, for stupor involves loss of sensation and of voluntary motion, two indicators construed by Galen as part of an unconscious state.[69] Convulsion and difficulty in breathing are themselves indices of posterior ventricular damage, as will be discussed below.

---

[67] *PHP* p. 458,3 De Lacy; V. 623 K.

[68] ὅτ' ἂν δὲ ἀνατιτραμένου τινὸς ὀστοῦ ἡ μέση κοιλία θλιφθῇ, κάρος καταλαμβάνει τὸν ἄνθρωπον ἄνευ τοῦ σπᾶσθαί τε καὶ δυσχερῶς ἀναπνεῖν . . . ὥσπερ δὲ ἐν ταῖς ἀνατρήσεσιν, ὅτ' ἂν ἀμελῶς τις πιέζων τῷ μηνιγγοφύλακι τὴν μήνιγγα θλίψῃ περαιτέρω τοῦ προσήκοντος, ὁ κάρος γίγνεται, κατὰ τὸν αὐτὸν τρόπον, ὅτ' ἂν ὀστοῦν συντριβὲν σφοδρῶς τοῦ κρανίου θλίβῃ τὰς κοιλίας αὐτοῦ, καὶ μάλιστα τὴν μέσην, ὁ κάρος συμπίπτει· ἕπεται δὲ καὶ σφοδραῖς ὀδύναις τὸ πάθος τοῦτο, καταπίπτοντος ἐν αὐταῖς τοῦ κατὰ τὸ ψυχικὸν πνεῦμα τόνου· VIII. 232–233 K.

[69] Cf. Clarke and O'Malley, 1968, 493.

The above citation is also important as Galen introduces psychic pneuma to account for these changes in clinical status. As a result of trepanation, the *tension* (τόνος) of the pneuma in the ventricle is decreased, which accounts for the pain experienced, notwithstanding that caused by the procedure itself. The concept of τόνος was by the second century AD a familiar one in the medical and philosophic lexicon and there does not appear to be any significant reworking of its earlier antecedents by Galen, who adapts a general concept for his needs.[70] Galen is careful to say that the ventricle is only compressed; there is no indication that ventricular integrity is compromised. Therefore, an alteration to the balance of pneumatic tension by pressure alone seems sufficient to create a pathological condition. Galen provides a speculative concept of a balance in pneumatic tension within the ventricles, which is easily upset in cases of surgical intervention. The resulting imbalance is used by him to interpret the experimental effects of pressure on the ventricles.

In *De anatomicis administrationibus* IX.12, Galen states that pressure over the posterior ventricle produces "a very heavy and pronounced stupor." In *apoplexy* (ἀποπληξία), which can affect the posterior part of the brain, respiration in particular is said to be affected. In *De locis affectis*, the effects of apoplexy are the absence of sensation and motion, and slow breathing with effort. Unlike stupor and epilepsy, which primarily affect the ventricles, Galen maintains that the substance of the brain (both anterior and posterior parts) is more affected in apoplexy.[71] Convulsions or respiratory distress may also occur with the posterior ventricle affected. These convulsions are not necessarily epileptic in nature; Galen earlier in *De locis affectis* states that exact knowledge of a convulsion is not possible. What is important to know is that the brain itself is effected.[72] Not all clinical conditions which

---

[70] Stoic τόνος is used to explain bodily cohesion as well as character states. Cf. *PHP* p. 270,25–33 De Lacy; V. 404 K. See also Inwood, 1985, 31–32, 40, 162–164, 301 n. 119. Galen describes muscles as possessing a (pneumatic) tension necessary to carry out movement (*Mot. Musc.* IV. 402–403 K). In *Trem. Palp. Conv. Rig.* VII. 589–596 K, Galen differentiates τόνος from *palpitation* (παλμός). The former is associated with voluntary motion; the latter with involuntary motion and disease states.

[71] VIII. 231–232 K.

[72] εἴς γε μὴν τὴν τῆς θεραπείας εὕρεσιν οὐδὲν ὀνίνησιν ἡ περὶ τούτων ἀκριβὴς γνῶσις· ἱκανὸν γὰρ εἰς τὸ καλῶς θεραπεύειν ἐπίστασθαι τὸν μὲν πεπονθότα τόπον ἐγκέφαλον εἶναι, χυμὸν δὲ τὸν γλίσχρον, ἢ παχὺν ἐν ταῖς κοιλίαις ἀθροιζόμενον αὐτοῦ. VIII. 175 K. This does not exclude the possibility that if the *entire* ventricular system is affected, epileptic seizures will *not* occur.

Galen states involve the posterior part of the brain result in respiration being affected. In *catalepsy* (κατάληψις, κατοχή) respiration remains unimpaired.[73] However, Galen does state that *apnoea* (ἄπνοια) may also be seen in some cases of stupor, as well as in catalepsy.[74] Clearly, each condition has a range of clinical severity and should not be used as an absolute yardstick in determining localisation of damage to the ventricles. For example, epilepsy is sometimes caused by "a thick, melancholic humour", which, like "a thick, phlegmatous humour", obstructs the flow of pneuma in either the middle or posterior ventricle.[75] Nevertheless, from the standpoint of the severity of symptoms, it is the posterior ventricle which is the most important. In *De locis affectis*, Galen underlines this by stating that psychic pneuma is the "first instrument" (πρῶτον ὄργανον) of the soul, especially the pneuma in the posterior ventricles. From this perspective, the other ventricles are not as important.[76] Interruption to psychic pneuma within the posterior ventricle causes the most severe effects. For Galen, as for Herophilus, the posterior ventricle is essential for life.

5.5 *The effects of ventricular incision*

In the citation quoted earlier from *De anatomicis administrationibus* (IX.12; p. 19 Duckworth), Galen states that effects similar to that of pressure are noted when each ventricle is incised. The techniques required to employ the experimental knife are more precise and exacting than that of pressure alone. It might be argued that the effects of pressure are too general and do not focus exactly on the ventricle under consideration. The technique of incision, underwritten by Galen's wide-ranging knowledge of ventricular topographic

---

[73] VIII. 232 K. Cf. 156, 270, 300.

[74] VIII. 282 K. Cf. 201.

[75] VIII. 177 K. At 173–175, Galen states that it is reasonable to assume that epilepsy is caused by a thick humour obstructing the outlets of pneuma in the ventricles. The symptoms of epilepsy are due to the irritation of the nerve roots in the brain by this humour. *Loc. Aff.* VIII. 173–174 K. Cf. VIII. 180–181 K.

[76] τὸ πρῶτον δ' αὐτῆς ὄργανον εἰς ἁπάσας τὰς αἰσθητικάς τε καὶ προαιρετικὰς ἐνεργείας εἶναι, τὸ κατὰ τὰς κοιλίας αὐτοῦ πνεῦμα, καὶ μᾶλλόν γε κατὰ τὴν ὄπισθεν· οὐ μὴν οὐδὲ περὶ τῆς μέσης κοιλίας ἀπογινώσκειν προσῆκεν ὡς οὐ κυριωτάτης· πολλὰ γὰρ εὔλογα καὶ πρὸς ταύτην ἡμᾶς ἄγει, καθάπερ γε τῶν ἐμπροσθίων δυοῖν ἀπάγει, VIII. 175 K.

anatomy, offers a way of gainsaying that argument inasmuch as it further highlights Galen's dexterity.

If incising a ventricle can create the same effect as that of the application of pressure, it is not necessarily reversible. In *De anatomicis administrationibus* IX.12, Galen remarks that the experiment may be manipulated to produce a satisfactory outcome: a small incision, rapidly performed and with prompt compression to the wound may result in the animal returning to its normal state. In *De placitis Hippocratis et Platonis,* Galen notes that an incision into the posterior ventricle harms the animal most, next affected is the middle ventricle, whilst each anterior ventricle is least harmed by an incision.[77] To this apparent scale of morbidity, Galen, in *De anatomicis administrationibus*, emphasises the consequences of incising the region of the posterior ventricle:

> ... should the spinal marrow that lies between the skull and the first vertebra be severed, or the meninx which protects the end of the posterior ventricle of the brain be cut through, then at once the whole body of the animal becomes deprived of movement. It is just here that you will see, in the temples of the gods, the oxen receive the stab when the so-called sacrificers of oxen cut into them.[78]

Apart from experimental evidence, observations at animal sacrifices yield important clues for Galen. The instantaneous death of oxen that occurs when the sacrificial cut is made at the level of the first cervical vertebra, is interpreted by Galen in terms of the topography of the posterior ventricle and its pneumatic contents, as the region incised is precisely where Galen maintains the posterior ventricle ends and the spinal cord begins.[79] Erasistratus also observed a similarly fatal effect in a sacrificial animal, but interpreted this in terms of damage to the covering meninx.[80] Like Galen, Erasistratus noted that damage to the brain in the area of the first cervical vertebra usually resulted in death. As that part so incised is covered by

---

[77] See page 177 and n. 20 above.
[78] IX.14; p. 25 Duckworth.
[79] See chapter 4.7.
[80] ὅπερ οὐχ ἥκιστα τὸν Ἐρασίστρατον ἠπάτησεν, ὡς οἰηθῆναι διὰ τὴν τῆς μήνιγγος τρῶσιν ἀκίνητον αὐτίκα γίγνεσθαι τὸ ζῷον· ἑώρα γὰρ ἐπὶ τῶν κατὰ τὸν πρῶτον σπόνδυλον τιτρωσκομένων βοῶν ἅμα τῷ διαιρεθῆναι τὴν μήνιγγα ἀκίνητον αὐτίκα τὸ ζῷον γινόμενον. ἀλλ' οὐ τῷ πάθει τῆς μήνιγγος, ἀλλὰ τῷ γυμνοῦσθαι τὴν ὀπίσω κοιλίαν γίγνεται τοῦτο. *PHP* p. 446,22–27 De Lacy; V. 609–610 K.

dura mater, Erasistratus reasoned that death was due to damage to the dural meninx and made the valid empirical deduction that the dura was thereby responsible for nervous transmission in some way.[81] To Galen, Erasistratus is almost correct; but the importance of damage to the meninx for Galen is that it exposes the posterior ventricle and its contents, as the posterior ventricle (according to him) empties via a single channel into the beginning of the spinal medulla.[82] Galen validates his interpretation of these observations, and dismisses the Erasistratean thesis, by pointing out that damage to the meninx anywhere else does not result in the animal becoming motionless.[83] In *De anatomicis administrationibus*, Galen states that:

> ... the animal, when one pierces or incises the thin meninx (sic), sustains no derangement as a result, just as none such befalls it if the brain should be incised without the incision reaching as far as to one of its ventricles.[84]

If the effects of incising the ventricles are as broadly similar to those elicited by pressure, why does Galen perform them? Although they further highlight Galen's skill as an experimenter, the answer seems to lie in the contents of the ventricles. For Galen, ventricular incision is the first of a two-stage experimental process, the second of which creates the conditions for resealing the ventricle, allowing the animal to recuperate, and then observing the results in each case. These results are interpreted in *pneumatic* terms. That the animal recovers is ascribed to the replenishment of pneuma following closure of the incised ventricle. As Galen puts it, no other substance is as capable of *emptying* (κενοῦσθαι) or of *collecting* (ἀθροίζεσθαι) again so easily. Pneuma is also capable of moving into the body *instantaneously* (ἐν ἀκαρεῖ χρόνῳ).[85] This can be used to explain why the ani-

---

[81] Fr. 42A Garofalo. Cf. Phillips, 1973, 148.

[82] καί τις ἄλλος εἷς ἄζυγὴς ἐμβάλλει τῇ πρώτῃ γενέσει τοῦ νωτιαίου, καθ' ὃ μέρος μάλιστα τῆς παχείας μήνιγγος τρωθείσης ὁ πόρος ὅλος γίγνεται γυμνὸς ἅμα τῷ πέρατι τῆς ὄπισθεν ἐγκεφάλου κοιλίας, *PHP* p. 446,20–22 De Lacy; V. 609 K. Cf. Soury, 1899, 303 n.1.

[83] An Erasistratean could object that this proves nothing; the meninx may well be at its most vulnerable in this area, and that any damage to it will inevitably have a fatal outcome.

[84] IX.12; p. 20 Duckworth. Cf. *PHP* p. 446,27–29 De Lacy; V. 610 K: δῆλον δ' ἐστὶν ἐκ τοῦ κατὰ πάντα τἄλλα μέρη τὴν μήνιγγα μηδὲν τοιοῦτον τιτρωσκομένην ἐργάζεσθαι Duckworth's "thin meninx" may be read as "thick meninx."

[85] τί γὰρ οὕτω ῥᾳδίως ἀθροίζεσθαί τε καὶ κενοῦσθαι δυνατὸν ἄλλο πλὴν πνεύματος; ... πάντῃ γὰρ ἐξικνεῖται τοῦ σώματος ἐν ἀκαρεῖ χρόνῳ ῥᾳδίως τε καὶ

mal recovers when the incision has been sealed. For Galen, only pneuma accounts for the results of ventricular incision. But does Galen interpret his experimental findings in exactly this way? In the following from *De placitis Hippocratis et Platonis*, which summarises his ventricular experiments as part of his polemic against Chrysippean cardiocentrism, Galen concludes that psychic pneuma is contained within the ventricles of the brain:

> And all who wish may learn from me by the procedure I have had cause to demonstrate many times to no few disbelievers forthwith by comparison with another animal or many times in one and the same animal, whatever is the case, the number as well as the type of misfortunes that seize the body when one of the ventricles of the brain is wounded. However why do I say 'wounded'? For if even before inflicting a wound, pressure is applied to any of the ventricles the animal will forthwith be without motion as well as sensation, breathless as well as voiceless. The same thing is shown to happen in persons who have themselves undergone trepanation. For when we chisel out the fragments of bone we are compelled for safety to put underneath the so-called protectors of the meninx, and if these are pressed a little too heavily on the brain, the effect is to render the person senseless as well as incapable of all voluntary motion, but this does not occur when pressure is applied to the exposed heart. For I know I once let someone hold a heart with a pair of smith's fire tongs, since it leapt from his fingers because of its violent palpitation;[86] and even then the animal did not suffer any loss in sensation or voluntary motion, but gave a great cry and breathed unhindered and all its limbs were in violent motion. For when the heart is separated off in this way only the motion of the arteries is hindered, but the animal is not otherwise affected, and as long as it lives it moves all its limbs and breathes. But when pressure is applied to the brain in the manner noted above, all the symptoms are reversed: the arteries pulsate according to Nature together with the heart, but no limb moves and the animal not even breathes nor cries out. Now once all this has been established, another more important point is revealed as a consequence, that neither the

ἀκωλύτως. ταχεῖαν οὖν κένωσιν, ἢ πλήρωσιν, ἢ διάστασιν, ἢ συνίζησιν, ἢ θέσιν, ἢ ἔπαρσιν, ἤ τινα ἄλλην κίνησιν οὐδὲν ἂν τῶν ἁπάντων ἐργάσαιτο πλὴν πνεύματος. *Trem. Palp. Conv. Rig.* VII. 596–597 K. The context is the role of pneuma as the cause of palpitation. Manuli, 1988, 196, suggests that the movement of the passionate part of the soul is reflected in the expansion and contraction of the pneuma in the blood, which is a somatic representation of what is occurring at the level of the psychic faculty. At *PHP* p. 416,5–7 De Lacy; V. 573 K, the heart is described as the origin of the "pneuma-like and boiling blood".

[86] Cf. *Diff. Puls.* VIII. 723 K. *Palpitation* (παλμός) is an involuntary motion and is also "contrary to nature" (παρὰ φύσιν). *Trem. Palp. Conv. Rig.* VII. 594 K.

> heart needs the brain for its proper movement nor the brain the heart.[87] But it was not for the sake of this that I mentioned the appearances from anatomy, but rather to indicate that the psychic pneuma is contained in the ventricles of the brain.[88]

Galen then states that Chrysippus mistakenly sites pneuma—rendered as "pure and unalloyed pneuma" (εἰλικρινές τι καὶ καθαρὸν πνεῦμα)—in the heart. Galen does not disagree with Chrysippus' use of pneuma in this way, only his placement of it. To Galen, the point of the above extensive disquisition is to show that the only correct location for psychic pneuma is the ventricles of the brain. Galen also infers that, as a result of his experiments on the ventricles, psychic pneuma is responsible for sensation and voluntary motion. Only psychic pneuma, Galen's effector agent of the soul, can be placed with empirical confidence within the ventricular system of the brain.[89]

5.6 *The soul and the ventricles*

For Galen, the rational soul and psychic pneuma are two discrete entities that are linked in that the former manifests its activities

---

[87] But for Galen the brain does need the heart for the elaboration of vital pneuma which is then sent to the brain for transformation into its psychic form.

[88] καὶ τοῦτ' ἔνεστι τῷ βουλομένῳ μαθεῖν παρ' ἡμῶν, ὥσπερ ἀμέλει πολλάκις ἐδείξαμεν οὐκ ὀλίγοις τῶν ἀπιστούντων εὐθὺς ἀντιπαραβάλλοντες αὐτοῖς ἐφ' ἑτέρου ζῴου, πολλάκις δὲ καὶ καθ' ἓν ὁτιοῦν <καὶ> ταὐτόν, ὁπόσα τε καὶ ὁποῖα καταλαμβάνει τὸ σύμπ[τ]αν σῶμα παθήματα κοιλίας ἐγκεφάλου τρωθείσης. καίτοι τί λέγω τρωθείσης; εἰ γὰρ καὶ πρὶν τρῶσαι θλίψεις ἡντινοῦν αὐτῶν ἀκίνητόν τε καὶ ἀναίσθητον, ἄπνουν τε καὶ ἄφωνον εὐθὺς ἔσται τὸ ζῷον. οὕτω δὲ κἀπὶ τῶν ἀνθρώπων αὐτῶν ἀνατιτραμένων φαίνεται συμπῖπτον. ἐν γὰρ τῷ τὰ κατεαγότα τῶν ὀστῶν ἐκκόπτειν ὑποβάλλειν ἀναγκαζομένων ἡμῶν ἀσφαλείας ἕνεκα τοὺς καλουμένους μηνιγγοφύλακας, εἰ βραχεῖ βιαιότερον ἐπιθλίψει τις αὐτοῖς τὸν ἐγκέφαλον, ἀναίσθητός τε καὶ ἀκίνητος ἁπασῶν τῶν καθ' ὁρμὴν κινήσεων ὁ ἄνθρωπος ἀποτελεῖται, οὐ μὴν τήν γε καρδίαν γυμνωθεῖσαν θλιβόντων. ἀλλ' ἔγωγε οἶδα καὶ πυράγρᾳ ποτὲ χαλκέως ἐπιτρέψας τινὶ περιλαβεῖν αὐτήν, ἐπειδὴ καὶ τῶν δακτύλων ἐξεπήδα βιαίως παλλομένη· ἀλλ' οὐδὲν οὐδὲ τότε τὸ ζῷον ἔπασχεν οὔτε εἰς αἴσθησιν οὔτε εἰς κίνησιν τὴν καθ' ὁρμήν, ἀλλ' ἐκεκράγει τε μεγάλα καὶ ἀκωλύτως ἀνέπνει καὶ πάντα ἐκίνει σφοδρῶς τὰ κῶλα. μόνη γὰρ ἡ τῶν ἀρτηριῶν βλάπτεται κίνησις οὕτω διαληφθείσης τῆς καρδίας, ἄλλο δὲ οὐδὲν πάσχει τὸ ζῷον, ἀλλ' ἄχρι περ ἂν ζῇ καὶ κινεῖ πάντα τὰ μέλη καὶ ἀναπνεῖ. ἐγκεφάλου δὲ οὕτω θλιφθέντος ἔμπαλιν ἅπαντα συμπίπτει· σφύζουσι μὲν αἱ ἀρτηρίαι κατὰ φύσιν ἅμα τῇ καρδίᾳ, κινεῖται δὲ οὐδὲν μέλος οὐδὲ ἀναπνεῖ τὸ ζῷον οὐδὲ ἐκφωνεῖ. ἤδη μὲν οὖν ἐκ τῶνδε καὶ ἄλλο τι μεῖζον ἐκφαίνεται, τὸ μήτε τὴν καρδίαν ἐγκεφάλου τι προσδεῖσθαι πρὸς τὴν οἰκείαν κίνησιν μήτε τὸν ἐγκέφαλον τῆς καρδίας. οὐ μὴν τούτου γε ἕνεκεν ἐμνημόνευσα τῶν ἐκ τῆς ἀνατομῆς φαινομένων, ἀλλ' ὅπως ἐπιδείξαιμι τὸ ψυχικὸν πνεῦμα κατὰ τὰς τοῦ ἐγκεφάλου κοιλίας περιεχόμενον. pp. 78,27–80,18 De Lacy; V. 185–187 K.

[89] Cf. *Loc. Aff.* VIII. 174–175 K.

through the latter. It is enough for him to localise the soul in very general terms as residing within the brain, yet at the same time exclude it from the ventricular system. To place the soul directly in the ventricles would represent for Galen the irrevocable step of equating soul with pneuma. Apart from anything else, this could expose Galen to the charge, perhaps first expressed *inter alia* by Aristotle, that the activities of the soul are promoted at the expense of the actions of pneuma and its effector organs.[90]

Galen also states that the soul lies in the body of the brain, where thought takes place and memory is stored.[91] Galen is careful not to say that he has proved the soul lies within the brain; rather, considering the empirical evidence, he assumes that the soul abides in some part of the brain that does *not* constitute the ventricular system.[92] Since the rational part of the soul is the *hegemonikon*, and that is located in the brain, and as the pneuma in the ventricles is not the soul, then the soul resides somewhere else in the brain. Galen's use of the ventricles as the focal points of the *hegemonikon* reveal his power of dissection and how it may be applied to validate an empirically coherent theory of brain physiology. For Galen, the ventricles are distinctive parts of the brain and as such possess particular properties—just as the heart and the liver also have discrete functions and abilities.[93] Galen's arguments for the ventricles as the functional unit of the brain are impressive, and—as no alternative models were propagated in late Antiquity—successful, if longevity of a concept is a measure of its success. The ventricles process psychic pneuma largely from its vital form. As noted in the previous section, psychic pneuma in the ventricles of the brain is said to account for sensation and voluntary motion, albeit on the soul's behalf. Psychic pneuma being the soul's "first instrument" (πρῶτον ὄργανον), means that for

---

[90] Cf. Solmsen, 1961, 170.

[91] τοῖς γὰρ ἐκ τῆς ἀνατομῆς φαινομένοις ἀπολουθοῦσιν ἡμῖν εὔλογον ἐφαίνετο, τὴν μὲν ψυχὴν αὐτὴν ἐν τῳ σώματι τοῦ ἐγκεφάλου κατῳκῆσθαι, καθ' ὃ καὶ τὸ λογίζεσθαι γίγνεται, καὶ ἡ τῶν αἰσθητικῶν φαντασιῶν ἀποκειται μνήμη· *Loc. Aff.* VIII. 174–175 K. The importance of psychic pneuma as the first instrument of the soul is stressed again at VIII. 233 K.

[92] Placing the soul within the brain but not in the ventricles also allows Galen to dispense with trying to clarify the nature of the soul (*PHP* p. 600,6–19 De Lacy; V. 794 K; see also *De propriis placitis* 3,1 pp. 58,22–60,6 Nutton). Cf. De Lacy, 1988; Manuli, 1986, idem, 1988; Vegetti, 1986, 233–235.

[93] For Galen, "the distinctive properties of the liver and heart serve as signs (σημεῖα or τεκμήρια) of the essential nature of these organs." Tieleman, 1996b, 272.

Galen the index of the actions of the soul can be measured in pneumatic terms.[94] The strength of Galen's argument of using psychic pneuma in this way allows pneuma to have ascribed to it functional *differentiae* which serves to vitiate the localisation of soul within the ventricular system. In this way, the *exactness* (ἀκρίβεια) of the intellect, an important attribute of the rational soul, may be given by Galen in pneumatic terms. Intelligence itself depends on what Galen calls the "correct temperament of the substance of the *intelligent body* (νοοῦν σῶμα), whatever that may be, and not on the intricacy of its structure."[95] This "body" should be taken as a reference to the substance of the brain. With the ventricles excluded by Galen, there is no other place for intelligence to reside.[96]

5.7 *Conclusion*

Thanks in large part to his interpretation of "the appearances from anatomy", Galen gives the ventricles a strong, almost noumenal presence. The experiments he performs on them are crucial for his conceptualisation of the brain as the organ of the rational soul. For Galen, the brain is the *hegemonikon* for voluntary motion and sense perception. The ventricular system is the locus of this controlling centre. The results of Galen's experiments show that damage to the brain—interpreted in terms of impairment affecting the ventricular system—produces a range of *affective states* (παθήματα) from reversible stupor to death. Galen thereby creates a scale of function based on relative and absolute incapacity as recorded by experiment. The ventricles therefore have a clearly defined status within Galen's physiological system in that their interruption at various points leads to empirically verifiable disturbances. Galen's painstaking ventricular experiments demonstrated to him the absolute importance of the ventricles insofar as their function accurately reflects the *hegemonikon*

---

[94] Cf. *PHP* p. 446,11–15 De Lacy; V. 609 K.

[95] ἄμεινον δ' ἦν ἄρα τῇ τῆς οὐσίας εὐκρασίᾳ τοῦ νοοῦντος σώματος, ὅ τι ποτ' ἂν ᾖ τοῦτο, τὴν σύνεσιν ἕπεσθαι νομίζειν, οὐ τῇ ποικιλίᾳ τῆς συνθέσεως. *UP* I, p. 488,24–26 Helmreich; III. 673–674 K.

[96] οὐδὲ γὰρ τῷ πλήθει τοῦ πνεύματος τοῦ ψυχικοῦ χρῆναι δοκεῖ μοι μᾶλλόν περ ἢ τῇ ποιότητι τὴν ἀκρίβειαν τῆς νοήσεως ἀναφέρειν. *UP* I, pp. 488,26–489,2 Helmreich; III. 674 K. On later concepts of ventricular localisation, see Appendix One.

of the rational soul. Within the context of these experiments, Galen's handling of the processing of psychic pneuma is more problematic. Although the value of psychic pneuma as the effector agent of the rational soul lies in its precise and qualitative distillation in the ventricles of the brain, the manner of its elaboration, as chapter 6 will attempt to show, is far from straightforward.

CHAPTER SIX

THE ELABORATION OF PSYCHIC PNEUMA

From the outside air, pneuma is drawn in by the rough arteries and receives its first elaboration in the flesh of the lungs, its second in the heart and the arteries, especially those of the retiform plexus, and then a final elaboration in the ventricles of the brain which completes its transformation into psychic pneuma.

De usu partium.[1]

6.1 *Introduction*

For Galen, the usefulness of the ventricles of the brain lies in their ability to qualitatively change a substance presented to it. This material substrate is vital pneuma, which, when transformed to its final, psychic form, becomes the effector agent of Galen's physiology of the brain and nerves. Within the brain are two vascular structures whose function is to elaborate it. These are the *retiform plexus* (δικτυοειδὲς πλέγμα), a network of fine arteries at the base of the brain, and the *choroid plexuses* (χοροειδῆ πλέγματα), appendages of small arteries and veins within the anterior and middle ventricles. Yet the mechanism of action of the retiform and choroid plexuses is not a settled issue for Galen. Moreover, the retiform plexus has been the subject of much post-Galenic controversy and confusion, and has been invested with what amounts to an almost totemic status.[2] This chapter will begin with an examination of Galen's account of its structure and function. This will be followed by a similar discussion of the choroid plexus. That there are puzzles in Galen's account of the function of choroid plexuses, especially their supposed role in the

---

[1] τὸ δ' ἐκ τῶν τραχειῶν ἀρτηριῶν πνεῦμα τὸ ἔξωθεν ἑλχθὲν ἐν μὲν τῇ σαρκὶ τοῦ πνεύμονος τὴν πρώτην ἐργασίαν λαμβάνει, μετὰ ταῦτα δ' ἐν τῇ καρδίᾳ τε καὶ ταῖς ἀρτηρίαις καὶ μάλιστα ταῖς κατὰ τὸ δικτυοειδὲς πλέγμα τὴν δευτέραν, ἔπειτα τὴν τελεωτάτην ἐν ταῖς τοῦ ἐγκεφάλου κοιλίαις, ἔνθα δὴ καὶ ψυχικὸν ἀκριβῶς γίγνεται [πρότερον]. I, pp. 393,23–394,6 Helmreich; III. 541–542 K.

[2] See Appendix Two for a survey of some of these later accounts.

elaboration of psychic pneuma from the outside air, will be the subject of the succeeding section, which examines Galen's experiments on carotid ligation. Galen's handling of these experiments enables a wider evaluation of the roles of both the retiform and choroid plexuses. These structures might be viewed as a vascular confluence acting in accord to ensure production of psychic pneuma; but however helpful this might be in constructing a concise account of Galen's physiology of the brain, it is not certain if such an interpretation was ever entirely Galen's intention.

6.2 *The anatomy of the retiform plexus*

Galen's depiction of the retiform plexus in *De usu partium*, while illustrating the familiar conception of a purposeful Nature, is also as good an example of descriptive anatomical writing as anything in the Galenic Corpus:

> The plexus known as retiform by anatomists is the most marvellous of the structures in this area. It surrounds as a circle the gland itself, and for the most part extends as far as the rear, since this plexus is immediately under all but a little of the base of the brain. It is not merely a net-like structure but looks as if you had taken several fisherman's nets and stretched one out over the other. But it is characteristic of Nature's net that the meshwork of one layer is always attached to the other, thereby making it impossible to remove any one net by itself; for, one after the other, all the rest follow the one you remove, since they are all attached to each other. Naturally, because of the fineness of the members composing this network, and their intimate conjunction, you could neither compare this plexus to any man-made net, nor ascribe its formation to chance. On the contrary, Nature appropriated as the material for this marvellous network the greatest part of the [internal carotid] arteries that ascend from the heart to the head.[3]

[3] Τὸ δὴ καλούμενον ὑπὸ τῶν ἀνατομικῶν δικτυοειδὲς πλέγμα μέγιστον θαῦμα τῶν ἐνταυθοῖ τέτακται, περιλαμβάνον μὲν ἐν κύκλῳ καὶ αὐτὸν τὸν ἀδένα, παρῆκον δὲ καὶ εἰς τοὐπίσω μέχρι πλείστου. πᾶσα γὰρ ὀλίγου δεῖν ἡ τοῦ ἐγκεφάλου βάσις ὑποτεταγμένον ἔχει τοῦτο τὸ πλέγμα. ἔστι δ' οὐχ ἁπλοῦν τὸ δίκτυον, ἀλλ' ὡς εἰ καὶ ταῦτα τὰ δίκτυα τῶν ἁλιέων πλείω λαβὼν ἐπ' ἀλλήλοις ἐκτείναις. πρόσεστι δὲ τῷ τῆς φύσεως ἐκείνῳ δικτύῳ τὸ τὰς ἐπιβολὰς ἀεὶ θατέρου συνῆφθαι θατέρῳ καὶ μὴ δύνασθαι μόνον ἓν ὁτιοῦν λαβεῖν ἐξ αὐτῶν· ἕπεται γὰρ καὶ τἆλλα τῷ ληφθέντι κατὰ στοῖχον ἁπάντων ἑξῆς ἀλλήλοις συνημμένων. οὐ μὴν οὐδὲ τῇ λεπτότητι

The analogy of fisherman's nets piled on top of the other is fitting, for the retiform plexus is remarkably similar to those twisted and coiled skeins seen in any fishing port in the Mediterranean. Galen provides a readily verifiable metaphorical label for the retiform plexus. Moreover, the plexus is a structure whose existence is already known to other anatomists, for it is they who describe it as *retiform* (δικτυοειδές). There is therefore nothing untoward or disproportionate about the retiform plexus as it is understood by Galen or other anatomists. To expose the retiform plexus, it is necessary, after removing the brain, to examine a small area within the base of the cranial cavity that, covered by the dural mater, is indistinguishable from the rest of the dural covering. To reach the plexus, therefore, it is first necessary to know where to look. In *De usu pulsuum*, Galen provides the following:

> For in this place the carotid arteries, on their way to the brain, before they penetrate the hard meninx, are divided by it into many branches, and twist themselves into many rows, as they would if you were to imagine several nets on top of each other, and they occupy a great space known as the base of the brain, which indeed is their destination from the very beginning.[4]

Galen prefaces this account by stating that the retiform plexus was first named either by Herophilus or by "those about Herophilus"

τῶν συντεθέντων οὐδὲ τῇ πυκνότητι τῆς συνθέσεως ἔχοις ἂν αὐτῷ παραβαλεῖν οὐδὲν τῶν ἀνθρωπείων τούτων δικτύων. οὐ μὴν οὐδ' ἐξ ὕλης τῆς ἐπιτυχούσης γέγονεν, ἀλλὰ τῶν ἀπὸ τῆς καρδίας ἐπὶ τὴν κεφαλὴν ἀναφερομένων ἀρτηριῶν τὴν μὲν μεγίστην μοίραν ὕλην ἡ φύσις ὑπεβάλετο τῷ θαυμαστῷ τούτῳ πλοκάνῳ. II, pp. 10,9–11,2 Helmreich; III. 696–697 K. Cf. the similar description of the formation of the retiform plexus in *PHP* p. 444,20–29 De Lacy; V. 607–608 K: ἐνταῦθα γὰρ οὐκ ὀλίγην χώραν ἡ φύσις οἷον θαλάμην τινὰ τῷ δικτυοειδεῖ τῷδε παρεσκεύασε πλέγματι περιεχομένην ὑπὸ τῆς παχείας μήνιγγος, εἰς ἣν χώραν ἀπὸ τῶν καρωτίδων ὀνομαζομένων ἀρτηριῶν οὐ σμικρά τις ἀφικνουμένη μοῖρα, καθ' ἑκάτερον μέρος ἓν ἀγγεῖον, εἶτα κατασχιζόμεναι πολυειδῶς οὐχ ἁπλοῦν ἐργάζονται καὶ τὸ δίκτυον, ἀλλ' ἐπ' ἀλλήλοις κείμενα πολλὰ μετὰ τοῦ συνῆφθαί τε καὶ συμπεφυκέναι πάντα. καὶ πάλιν γε κατὰ τοῦ πλέγματος τοῦδε τηλικοῦτον ζεῦγος ἀρτηριῶν ἐκφυόμενον ἡλίκον ἐξ ἀρχῆς ἦν τὸ παρὰ τῶν καρωτίδων ἧκον, εἰς τὸν ἐγκέφαλον ἀναφέρεται τά τ' ἄλλα μέρη διαπλέκον αὐτοῦ παμπόλ<λ>αις ἀπονεμήσεσι καὶ τὰ κατὰ τὰς κοιλίας ἐργαζόμενον πλέγματα.

[4] ἐκεῖ γὰρ αἱ ἐπὶ τὸν ἐγκέφαλον ἀνιοῦσαι καρωτίδες ἀρτηρίαι, πρὶν διελθεῖν τὴν σκληρὰν μήνιγγα, σχίζονται πολυειδῶς ὑπ' αὐτῆς, περιπλεκόμεναι κατὰ πολλοὺς στίχους, ὡς εἰ νοήσαις ἀλλήλοις ἐπικείμενα δίκτυα πλείω, καὶ χώραν παμπόλλην ἣν καλοῦσιν ἐγκεφάλου βάσιν καταλαμβάνουσιν, ἐνὸν αὐταῖς εὐθὺς μὲν διεκπεσεῖν τὰς μήνιγγας, ἐμφῦναι δὲ εἰς τὸν ἐγκέφαλον, οὗπερ ἐξ ἀρχῆς ἵενται. p. 200 Furley-Wilkie; V. 155 K. Cf. T 121 *Herophilus*.

(ἀμφὶ τὸν Ἡρόφιλον). Phrases such as "those about X", or "followers of X" can be employed periphrastically here to imply Herophilus.[5] And strictly, Galen's description of the retiform plexus refers only to the name of the structure, not to its discovery.[6] But for the purposes of this argument, let it be assumed that the discovery was made either by Herophilus or under Herophilean mediation. However, the retiform plexus does not exist in man, and the Herophilean provenance would appear to be unusual, given that Herophilus, unlike Galen, had access to human material. According to Von Staden: "This description (sc. of the *rete*) cannot readily be reconciled with human anatomy."[7] Harris writes that Galen: "No doubt confident that Herophilus had derived his knowledge of this structure from the dissection of the human brain . . . took it for granted that the human brain also contained such a network of arteries."[8] It is true that the description Herophilus gives of other vascular structures in the brain (the torcular Herophili and the choroid plexuses), entirely accord with human anatomy.[9] It is highly likely that Herophilus had only a limited number of human subjects available for brain dissections, and relied upon animal subjects to correlate and augment his human findings (the choroid plexus and torcular Herophili are also present in animals). Content with the knowledge that the retiform plexus is found in some animals he had dissected, Herophilus probably extrapolated this structure to the human brain.

A further possibility involves Herophilus' method of brain dissection. Only when the brain has been removed from the cranial cavity and the dural lining of the base of the skull opened can the retiform plexus be exposed. The accurate localisation and anatomical descriptions of the torcular Herophili and the choroid plexuses are not dependent on the brain being removed from the skull. If Herophilus performed brain dissections without removing the brain from the skull, this might also explain why he missed the *circulus arteriosus* (the Circle of Willis), a vascular anastomosis of two branches of the carotid arteries at the base of the brain, and which is present in many animals.[10] Since Herophilus makes no comment on it

---

[5] Lloyd, 1980, 187. Cf. Greenhill, 1872, 276.
[6] As pointed out by Vivian Nutton (personal communication).
[7] 1989, 179; cf. 140 n. 3.
[8] 1973, 357–358.
[9] Cf. Von Staden, 1989, 179.
[10] All mammals are supplied with a Circle: "The appearance of the circle and the pattern of its major branches are remarkably constant among mammals." Dyce,

whatsoever, it might indicate that structures at the base of the brain were not part of his regular dissection protocol of animals as well as humans.[11] This is admittedly speculative, but if Herophilus' brain examinations were entirely limited to dissecting the brain *in situ*, it would not have been possible for him to have discovered the retiform plexus. Perhaps certain of Herophilus' followers, unable to perform human dissections, may have continued with animal subjects and made this discovery, even if there is no clear evidence for an Herophilean anatomical tradition.[12] These disciples could have discovered the retiform plexus and transferred it to Herophilean provenance.[13] A subsequent transfer to human anatomy would then have been an easy step to take, either by themselves or through the agency of other, unknown, anatomists. But whatever heuristic process of anatomical investigation occurred in Alexandria as far as the retiform plexus is concerned, it is clear that by the end of the third century BC, the retiform plexus was considered a part of *human* anatomy. Galen then, cites an existing anatomical tradition, initiated by the authority of Herophilus.

The retiform plexus exists in mammals such as the ox, goat, pig and sheep.[14] In other words, it is present in those animals Galen routinely employed for brain dissection.[15] That the retiform plexus

Sack, Wensing, 1996, 304. Cf. Baldwin and Bell, 1963; King, 1987, 1; Rogers, 1947. Spillane, 1981, 25, states that the Circle is "absent in oxen." Siegel, 1968, 113 n. 36, makes the same mistake. The retiform plexus is not analogous to the Circle of Willis (stated by Garofalo, 1991, II, 833 n. 46). Neither is it accurate to state that the region occupied by the retiform plexus is taken by the Circle of Willis in man (claimed by Souques, 1933, 315; cf. May, 1968, 47 n. 203). In primates, the Circle, "betrays no feature not met within the range of variation for man." Hill, 1970, 465.

[11] Galen also never mentions an arterial circle at the base of the brain, even though it is present in the ox. But, as will be discussed in the following section, what is more important for Galen's pneumatic physiology is a *structurally complex* vascular network. The retiform plexus fits this criterion admirably. The Circle of Willis one the other hand, is a single-layered, relatively simple polygonal arrangement of arteries. The retiform plexus is also located in what for Galen is an advantageous position. It is protected within the cranial fossa and lies under a double fold of dura. By contrast, unless great care is taken in removal of the brain from the skull, the arterial circle at the base of the brain is inevitably torn, and its symmetrical arrangement lost.

[12] Cf. Von Staden, 1989, 445–446; 1992, 234–236.

[13] Or this could have been done at some stage in the doxographic tradition.

[14] It is absent in the rat, rabbit, dog and in primates. Cf. De Gutiérrez-Mahoney and Schechter, 1972, 146.

[15] Cf. Daniel, Dawes and Prichard, 1952–54, 173–208. See also Baldwin, 1964, 101–118, and Plates 1–3; Barnett, Harrison and Tomlinson, 1958, 467; King, 1987, 7; Popesko, 1977, 41–42; Sisson and Grossman, 1975, 956, 964, 965, 968, 1315.

is absent in apes is another indication that Galen's use of this animal was restricted to dissections of the body excluding the brain, as discussed in chapter 2. In whatever animal the retiform plexus is found, its most distinguishing feature is its complexity. This is well illustrated by Galen in the formal description of the retiform plexus and its relationship to the surrounding structures in the Arabic part of *De anatomicis administrationibus*:

> . . . you see that just as two arteries, one on each side, ascend from below and reach the brain and unite themselves and pass over the one into the other at the place called the 'pool' or 'cistern' [*infundibular recess*] . . . Adjoining the pool . . . and lying below it there is found a flabby spongy flesh, similar in formation and size to a lupine bean [*pituitary gland*] . . . Further, you find that the retiform plexus which surrounds this, being set together from arteries, extends backwards over a wide expanse, but occupies a restricted space towards the front and at the sides. In the bone above the palate (sella turcica of the basisphenoid) upon which the spongy glandular flesh rests and supports itself, you find narrow perforations. All this lies in this place.[16] Its structure resembles that of a sieve or a strainer with which one strains and clarifies turbid beverages—by which I mean everything to which the name 'sieve' or 'strainer' is applied. The first part of it (sc. the infundibulum) that you see is the duct of which I said that it descends from the brain (the infundibular recess), and the pia mater envelops it here. After that, in the place where this duct vanishes, sinking into the depth of the dura mater which is perforated in exactly the same manner as the pia mater, you see the lupine-bean-shaped spongy flesh. And at this flesh [*gland*] ends the effluent of the pool or cistern. When you have reached this place, then cut away the portion of dura mater which lies here below the site of the brain and above the pituitary. The substance of this gland is flabby and, round about it, is the so-called reticulate tissue, in which innumerable arteries travel entwined about one another in manifold fashion. Nor do they resemble a simple plain network, but rather numerous nets, laid the one upon the other, and moreover interwoven and plaited with one another. And the whole of that mass is produced from the two arteries of the neck which one names 'the vessels of stupor'[17] [*carotids*].[18]

[16] For the particular arrangement of this citation, see chapter 4, n. 141.

[17] In Hunayn's translation of Galen's *On Nerves, Veins and Arteries* XVI, 194, he states that 'carotid' in Greek means "the places of slaughter." Savage-Smith, 1969, 216, notes that this mistake is due to a lacuna in the text. Hunayn correctly identifies the external jugular veins as "the place where a beast is slaughtered." But Savage-Smith, 215, incorrectly states that the retiform plexus, "occupies the area where the Circle of Willis is in humans."

[18] IX.8; pp. 6–7 Duckworth. My glosses in round brackets; square brackets Duckworth.

It is clear from the above account that the exposure of the retiform plexus involves meticulous and painstaking dissection. The key lies in locating the infundibulum and its duct, removing the brain (or as much as is necessary) in order to expose the dura mater which covers the cranial fossa behind the crista galli and on either side of the hypophyseal fossa. When this is done, the dissector can, "cut away the portion of dura mater which lies here below the site of the brain and above the pituitary." This opens a cavity known to contemporary anatomy as the *cavernous sinus*, which is only a venous space in humans. In animals such as the ox, the densely packed meshwork of the retiform plexus is found within this sinus.[19] The relationships of the structures at the base of the skull to the retiform plexus are also stressed in the above citation, and later in the same text Galen once more takes some pains to describe as precisely as he can the vascular nature of the retiform plexus:

> This network encloses, in those regions which lie in front of it, the spongy flesh [*glandular tissue*] which resembles the runner bean [*hypophysis cerebri*] and which projects from the brain-cleft into free space and stretches to the place which is known as the 'pool' or 'cistern'. For it is convex, and it is provided with an empty cavity . . . As for these arteries, they divide themselves profusely, and are many times interwoven so that the net-like [*retiniform*] anastomosis arises from them.[20]

The details Galen provides are of a fine meshwork of interwoven small arteries derived from the carotid system.[21] As Galen describes it in *De usu partium*, each of the carotid arteries in the neck divides into two; one posteriorly (internal carotid), the other anteriorly (external carotid). The posterior branch also divides into two offshoots of unequal size, the larger, more anteriorly placed, passes through a

---

[19] In my dissections of this region, the complete exposure of the cavernous sinus was only possible when the brain had been completely removed. Thereafter, the smooth floor of the cranial fossa covered by the dura was well exhibited, and the reference point of the pituitary noted. Incisions were made on either side of it in order to peel away the dural fold and expose the cavernous sinus. The delicate retiform plexus lying within could then be teased out. See Fig. 7.

[20] XIV.3; pp. 191–192 Duckworth; square brackets Duckworth.

[21] "We cannot assume that he is trying to describe the choroid plexus, since he carefully distinguishes this from the retiform, nor can we identify it with the tela choroidea of the third ventricle which he would have regarded as part of the choroid." Harris, 1973, 357. The retiform plexus is formed usually by the occipital (directly) and via branches of the maxillary arteries (derived from the common carotid). Cf. Baldwin, 1964, 103–104, 106; Daniel, Dawes and Prichard, 1952–1954, 174ff; Sisson and Grossman, 1975, 963, 966.

canal (τρῆμα) in the "petrous [parts] of the [temporal] bone" (λιθοειδῆ τῶν ὀστῶν), and enters the retiform plexus.[22] That the retiform plexus is one of several branches of the carotid arteries is also noted by Galen in the following from *De anatomicis administrationibus*, which reinforces the seamless way in which the arterial system is fashioned by him to be an essential part of the brain:

> But from the ends of the arteries, after they have pierced the skull, immediately upon their first entry and passage into its cavity, there proceeds the so-called 'reticular network' . . . Out from the rete mirabile, again, there ascend two arteries which encircle the brain together with the delicate meninx [*pia mater*] after the fashion of a girdle, just as do the veins as we have described in the exposition of the anatomy of the brain.[23]

Galen thus defines the retiform plexus as a network of arteries derived from the carotid system, surrounding the pituitary gland on either side, the entire structure lying in the cranial fossa above the hard palate and protected by a folding of the dura over the base of the brain.

6.3 *The physiology of the retiform plexus: Galen's "paradigm" for pneumatic elaboration*

In *De placitis Hippocratis et Platonis*, Galen gives notice of the physiological importance he allocates to the retiform plexus, wedded as it is to his complex account of pneumatic elaboration:

---

[22] τῶν δ' οὖν καρωτίδων ἀρτηριῶν ἑκατέρα διχῇ σχίζεται καὶ τὸ μὲν ὀπίσω μᾶλλον ἔρχεται μόριον αὐτῆς, τὸ δὲ πρόσω· . . . ἡ δὲ λοιπὴ μοῖρα τῆς καρωτίδος ἀρτηρίας, ἣν ὀπίσω μᾶλλον ἔλεγον φέρεσθαι, σχίζεται μὲν καὶ ἥδε τὰ μέγιστα μόρια διχῇ, . . . θάτερον δ' ἐκ τῶν πρόσω τοῦδε διὰ τοῦ κατὰ τὸ λιθοειδὲς ὀστοῦν τρήματος ἀναφέρεται καὶ αὐτὸ πρὸς τὸ δικτυοειδὲς πλέγμα, *UP* II, pp. 428,6–429,2 Helmreich; IV. 333–334 K. Cf. *AA* XIV.5; p. 198 Duckworth. May, 1968, 718 n. 89, states that the description offered by Galen, "fits conditions in the pig." As evidence, May states that the "condyloid artery", is derived "not from the occipital artery as in other animals, but from the internal carotid," This is incorrect. In the pig: "at the tip of the jugular process the occipital artery releases the condylar artery," Sisson and Grossman, 1975, 1315. This artery then divides in two, with one branch passing into the cranial cavity through the hypoglossal foramen, the other entering the jugular foramen (on a point of strict nomenclature, this latter is not the "carotid canal" as stated by May, 1968, 719).

[23] XIII.9; pp. 174–175 Duckworth; square brackets Duckworth.

> Accordingly the pneuma in relation to the arteries is called vital, and that in regard to the brain is psychic, not that it exists as the substance of the soul, but rather as the first instrument of the soul which resides in the brain, whatever may be its substance. And just as vital pneuma is generated in the arteries a well as the heart, obtaining the material for its generation from inhalation and the vaporization of the humours, so the psychic pneuma is generated by a further elaboration of the vital. For it was necessary that this pneuma, by all means be changed in precisely the correct fashion. If Nature, needing to fashion semen and milk with precision, even though they are far inferior in power to psychic pneuma, nevertheless arranged for each a lengthy period in the organs of coction and for that reason provided for semen the spiral vessels for the testes and for milk the length of the vessels that go to the breasts, so naturally also that when elaborating psychic from vital pneuma in the brain it constructed close to the brain a complex labyrinth, as it were, the retiform plexus.[24]

This "complex labyrinth" (ποικίλος λαβύρινθος) is analogously related by Galen to the vasculature of the mammary gland and the testis. Like them, the rete's convoluted structure somehow makes its physiological function possible, or as Galen puts it, *reasonable* (εὔλογον), a word which, as will be seen, has particular resonance in his physiology of the brain. These convolutions, according to Galen, allow sufficient time to elapse to complete the final processing of psychic pneuma in the ventricular system. It remains to be seen if Galen can provide evidence as to why such elaborative processes take time, beyond what relates to his teleological programme.

As discussed in chapter 2, Galen's physiology of the brain is best understood by recourse to its functions being described largely—but not entirely—in terms of pneuma.[25] In this physiological scheme, the

---

[24] τὸ μὲν οὖν κατὰ τὰς ἀρτηρίας πνεῦμα ζωτικόν ἐστί τε καὶ προσαγορεύεται, τὸ δὲ κατὰ τὸν ἐγκέφαλον ψυχικόν, οὐχ ὡς οὐσία ψυχῆς ὑπάρχον, ἀλλ' ὡς ὄργανον πρῶτον αὐτῆς οἰκούσης κατὰ τὸν ἐγκέφαλον, ὁποία τις ἂν ᾖ τὴν οὐσίαν. ὥσπερ δὲ τὸ ζωτικὸν πνεῦμα κατὰ τὰς ἀρτηρίας τε καὶ τὴν καρδίαν γεννᾶται τὴν ὕλην ἔχον τῆς γενέσεως ἔκ τε τῆς εἰσπνοῆς καὶ τῆς τῶν χυμῶν ἀναθυμιάσεως, οὕτω τὸ ψυχικὸν ἐκ τοῦ ζωτικοῦ κατεργασθέντος ἐπὶ πλέον ἔχει τὴν γένεσιν· ἐχρῆν γὰρ δήπου μᾶλλον ἁπάντων αὐτὸ μεταβολῆς ἀκριβοῦς τυχεῖν. εἴπερ οὖν τό τε σπέρμα καὶ τὸ γάλα, καίτοι γ' ἀπολειπόμενα τῇ δυνάμει τοῦ ψυχικοῦ πνεύματος, ὅμως ἡ φύσις ἀκριβῶς κατεργάζεσθαι δεομένη πολυχρόνιον αὐτοῖς ἐμηχανήσατο τὴν ἐν τοῖς πεπτικοῖς ὀργάνοις διατριβὴν καὶ διὰ τοῦτο τῷ μὲν σπέρματι τὴν πρὸ τῶν ὄρχεων ἕλικα παρεσκεύασε, τῷ δὲ γάλακτι τὸ μῆκος τῶν εἰς τοὺς τιτθοὺς ἰόντων ἀγγείων, εἰκότως καὶ κατὰ τὸν ἐγκέφαλον ἐκ τοῦ ζωτικοῦ πνεύματος ἐργαζομένη τὸ ψυχικὸν οἷον λαβύρινθόν τινα ποικίλον ἐδημιούργησε πλησίον ἐγκεφάλου τὸ δικτυοειδὲς πλέγμα. pp. 444,29–446,10 De Lacy; V. 608–609 K.

[25] Galen's pathology however, is largely humoural, drawing freely from the broad

theoretical flexibility of pneuma seems paramount. Pneuma's adaptability is such that it is accepted by Galen as capable of undergoing a form of *change* (μεταβάλλειν) or *alteration* (ἀλλοίωσις). Galen's physiology is based on this distinct advantage pneuma is said to possess.[26] In utilising pneuma this way Galen cannot be called original, but in making pneuma dependent on a meticulously delineated anatomico-physiological apparatus that subjects it to what he regards as a series of *qualitative* changes, he creates an arresting and distinctive picture.[27] Galen wishes to grant pneuma an epistemological status and validity as a result of his anatomical exhibitions and physiological experiments.[28]

Galen affirms pneuma's inherent malleability by having psychic pneuma elaborated from its vital form by the retiform plexus, which begins the differentiation into psychic pneuma, with the choroid plexuses apparently completing this process. The whole conception of pneumatic elaboration is well summarised by Galen in the following citation which introduces this chapter. Here, the malleable concept of pneuma is firmly keyed into three distinct sites:

> From the outside air, pneuma is drawn in by the rough arteries and receives its first elaboration in the flesh of the lungs, its second in the heart and the arteries, especially those of the retiform plexus, and then a final elaboration in the ventricles of the brain, which completes its transformation into psychic pneuma.[29]

---

framework of Hippocratic antecedents. As noted in chapter 5.4, Galen holds that it is *reasonable* to state that epilepsy is caused by deposits of a "thick humour" impeding the flow of psychic pneuma through the ventricles (cf. chapter 5, n. 75). This is not to say that the humours play no role in Galen's physiology; humoural vaporization provides the material substrate for vital pneuma, as the above citation from *PHP* attests.

[26] Galen's concept of pneumatic elaboration rests on the notion of an "complete change of substance". *Nat. Fac.* p. 104,16 Helmreich; II. 5 K. Cf. De *plenitudine* VII. 525 K). The notion of *change* is qualitative in this instance, and is dependent on the general *faculty* or *power* (δύναμις) of the organ concerned.

[27] In the Pseudo Aristotelian *De spiritu*, p. 102 Roselli; 483b.7–8, air within the body is pneuma, which is defined as "being condensed and distributed in a certain way" (καθάπερ πυκνωθεὶς καὶ διαδοθείς πως). Galenic elaboration and differentiation of pneuma may be read as sophisticated variants of condensation.

[28] As Debru, 1996b, 147, points out: "L'anatomie (la cavité des ventricules cérébraux, le plexus réticulé) et l'expérimentation donnent au pneuma invisible un statut épistémologique beaucoup plus sûr que celui de "substance' de l'âme." Cf. Temkin, 1973, 16: "Physiological experimentation can also appear as demonstration of what is logically proved."

[29] See n. 1 above for the Greek.

The retiform plexus deals with something that is not yet psychic pneuma, just as the left ventricle of the heart deals, not with vital pneuma itself, but with something *pneuma-like* (πνευματώδης).[30] How does the retiform plexus process pneuma? The chain of changes Galen sets in train from the lungs to the brain affords one way of examining the process of pneumatic elaboration—namely, as a series of analogous accounts that occur in separate organs. When Galen says that outside air is processed or prepared in the lungs, this change is effected by a fundamental *power* (δύναμις) of the lungs that determines their status as unique organs for this particular physiological elaboration.[31] Galen first compares the process of the *elaboration* (ἐργασία) of psychic pneuma with nutrition. His handling of nutrition is not so different from that of his predecessors, but the claims he makes from using it analogously are distinct.[32] However, Galen puts himself in company with the "best physicians and philosophers" who have all *demonstrated* the principles of nutrition.[33] Second, he relies on the analogous comparison of other anatomical structures to the retiform plexus, specifically, the vascular retinaculum in the testis, which Galen holds is responsible for the concoction of semen. Each of Galen's analogous arguments will be examined in turn, beginning with his use of nutrition.

Galen underwrites the elaboration of vital to psychic pneuma (as he does with the generation of vital pneuma itself) by referring analogously to the notion of the *coction* (πέψις) of nutriment.[34] Galen refers to Aristotle who states that coction may be likened to boiling, albeit not in an exact sense.[35] Coction is readily verifiable in terms of its

---

[30] It might also be that Galen takes such pains to differentiate both vital and psychic *pneumata* in order that these elaborative steps are seen to mark the boundaries between the powers of the heart and those of the brain.

[31] According to Galen, we are governed by a number of such faculties. Each organ is a "distinct substance" (ἰδὶα οὐσία) which carries particular powers (*PHP* p. 456,11–12 De Lacy; V. 621 K; cf. *Nat. Fac.* p. 107,8–23 Helmreich; II. 9–10 K).

[32] Galen's term for elaboration reflects both the production and transformation of a substance, and the description of work performed or expended. In *AA* II. 542–543 K, for example, the term is employed in association with the process of digestion and nutrition. Galen's usage ultimately refers back to Aristotle for its validation. Cf. Lloyd, 1966, 379.

[33] ὅτι μὲν ἀλλοιοῦται τὸ τρέφον εἰς τὴν τοῦ τρεφομένου μεταβαλλόμενον οὐσίαν, ἡμῖν τε δι' ἑτέρων ἀποδέδεικται καὶ σχεδὸν ὑπὸ τῶν ἀρίστων ἰατρῶν τε καὶ φιλοσόφων ἀποδέδεικται· *PHP* pp. 408,34–410,2 De Lacy; V. 565–566 K.

[34] Cf. Peck, 1965, 34; Temkin, 1960, 85–87

[35] Cf. *Nat. Fac.* p. 222,7–10 Helmreich; II. 167 K. See also Boylan, 1982, 99;

end product. It is a process that is also seen to take time. Moreover, the familiarity of the concept of nutrition on the part of his audience allows Galen to draw the comparison between the substrates used in both pneumatic elaboration and the process of coction when he refers to "material pneuma" (ὑλικὸν πνεῦμα) as being "analogous to the dry and moist nutriment."[36] The elaboration of nutriment allows Galen to claim that the means and method of pneumatic elaboration may also be said to occur within specified organs. The familiar effects of the processing of nutriment are thereby extended analogously to render the relatively unknown elaborations of psychic pneuma more acceptable.[37] For Galen the elaboration of nutriment and the elaboration of psychic pneuma are processes to be understood in broadly similar ways; that pneuma and nutriment are both essential to life reinforces the overall concept of familiarity of both processes.[38] Galen's tools for establishing the validity of either process are, of course, extremely limited. Both sets of elaborations begin with basic substrates, food and air; both are elaborated through a number of stages until the end product is reached.[39] If the intent of the analogy of coction is to render the account of pneumatic elaboration in easily understood terms, it is also to promote it as the most likely explanation.

The second of Galen's two arguments from analogy is a more specific comparison of the elaboration of psychic pneuma with that of semen in the testicular vessels, which possess a broadly similar

---

Lloyd, 1996a, 83–103. Whether his use of coction represents a weakness in Galen's overall argumentation (as Lloyd argues it does for Aristotle) is moot. There were in any case a limited number of concepts in Antiquity which could have been used as the structural cement for binding together divergent phenomena. Even pneuma considered as air or breath is, ultimately, only an analogous comparison.

[36] τὸ δὲ ἀνάλογον τῇ ξηρᾷ καὶ ὑγρᾷ τροφῇ, τὸ προσαγορευόμενον ὑλικόν, *PHP* p. 164,17–18 De Lacy; V. 281 K.

[37] "The fact that analogies were often used uncritically . . . should not cause us to underestimate their importance as a method of discovery, especially in the investigation of problems beyond the reach of direct observation." Lloyd, 1978a, 49.

[38] Erasistratus is said to have drawn attention to the crucial importance of both processes: ὑπεβάλετο τροφήν τε καὶ πνεῦμα· δύο γὰρ πρῶτα καὶ κυριώτατά ἐστιν, οἷς διοικεῖται τὸ ζῶιον, ὥς φησιν ὁ 'Ερασίστρατος, *Anonymus Londinensis* XXII, p. 40, 50–52 Diels. Cf. Pohlenz, 1949, 51–52.

[39] Galen provides a well defined role for the "intermediate" product of pneumatic elaboration, vital pneuma. However, the "unfinished products" of nutritive elaboration are just that, and exist for the sake of what they will produce. The "initial processing" (προκατειργασμένον) of food by the stomach is sent for completion to the liver. *PHP* p. 410,21–25 De Lacy; V. 567 K.

retinacular form to that of the retiform plexus.[40] This account will lead directly to the consideration of the physiology of the retiform plexus, and, *inter alia*, of the choroid plexus. Derived from Galen's gold standard of *autopsia*, that is, "from the appearances of anatomy",[41] the testicular plexus is described as follows:

> An artery and a vein are seen to go to each of the testes, not in a straight path, as is the case everywhere else, but winding first in many shapes, in the manner of grape tendrils or ivy. Whence, I think, their own nature suggested what they are called by the anatomists themselves, some calling it ivy-shaped, others tendril-shaped. And in these many windings that they make before reaching the testes you can see the blood slowly becoming white. And when the vessel finally reaches the testis, the substance of the semen is distinctly seen in it.[42]

The testicular plexus is not precisely of the same configuration as the retiform; it is also composed of veins as well as arteries. However, its elaborative function is set out by Galen in such a way that a comparison with the working up of psychic pneuma by the retiform plexus may better be understood. Semen is made from the retinacular plexus of the testicles; it is produced "from blood" (ἐξ αἵματος).[43] That this vascular plexus is *convoluted* or *twisted* (ἑλικτός) is necessary so that blood will spend sufficient time in it in order to be elaborated into semen.[44]

The outcome of this process is semen that is of the "purest quality" (εἰλικρινεστάτη ποιότης).[45] But how does Galen draw the explicit comparison between the testicular convolution of artery and vein

---

[40] The testicular vessels consist of an artery and vein derived, according to Galen, "from the vessels along the spine" (*Sem.* p. 116,14–17 De Lacy; IV. 565 K).

[41] ἀπὸ τῶν φαινομένων ἐν ταῖς ἀνατομαῖς. *Sem.* p. 106,19–20 de Lacy; IV. 555 K.

[42] φαίνεται τοίνυν ἀρτηρία καὶ φλὲψ ἐφ᾽ ἑκάτερον ἰοῦσαι τῶν ὄρχεων, οὐκ εὐθεῖαν ὁδόν, ὥσπερ ἐπὶ τἆλλα σύμπαντα, πολυειδῶς δὲ πρότερον ἑλιττόμεναι δίκην ἑλίκων ἢ κισσῶν. ὅθεν, οἶμαι, καὶ αὐτοῖς τοῖς ἀνατομικοῖς ἀνδράσιν ἐπῆλθε κισσοειδῆ μὲν ἐνίοις, ἑλικοειδῆ δ᾽ ἄλλοις ὀνομάσαι τὴν φύσιν αὐτῶν. ἐν δὲ δὴ ταῖς πολλαῖς ἕλιξι ταύταις, ἃς ποιοῦνται πρὶν ἐπὶ τοὺς ὄρχεις ἐξικέσθαι, κατὰ βραχὺ λευκαινόμενον ἔνεστί σοι θεάσασθαι τὸ αἷμα. καὶ τέλος, ἐπειδὰν ἤδη τοῦ ὄρχεως ἅπτηται τὸ ἀγγεῖον, ἡ τοῦ σπέρματος οὐσία σαφῶς ἐν αὐτῷ φαίνεται. *Sem.* pp. 106,20–108,3 De Lacy; IV. 555–556 K. Cf. p. 118,14–22; IV. 566–567 K.

[43] *Sem.* p. 108,9 De Lacy; IV. 556 K. Galen agrees with Aristotle (and with the Pneumatist Athenaeus of Attalia) that semen is elaborated from the coction of blood: καὶ <γὰρ> ἦν ἡμῖν ὁμολόγημα κοινὸν τοῦτο πρὸς [γὰρ] ᾽Αριστοτέλην καὶ ᾽Αθήναιον, ἐκ τῆς τοῦ αἵματος πέψεως τὴν γένεσιν αὐτῷ τιθεμένους. p. 178,23–24; IV. 626 K.

[44] αὕτη γὰρ καὶ ἡ χρεία τῆς ἕλικος. ἀλλοιοῦσαι δὲ δήπου τὸ αἷμα μετέβαλον εἰς σπέρμα. *Sem.* p. 108,10–11 De Lacy; IV. 556 K. Cf. p. 114,8–21; IV. 562–563 K.

[45] *Sem.* p. 134,20–21 De Lacy; IV. 583 K.

and the retiform plexus? The importance of the elaborative abilities of a certain kind of blood vessel is highlighted in *De usu pulsuum*, where Galen explicitly compares the retiform plexus to two other broadly similar vascular structures (one of which is the testicular plexus):

> The fact that this marvellous plexus was placed by Nature, who does nothing in vain, in such a well-protected space,[46] seemed to be an indication of some great use. Since we find the vessels like those of the spiral of the intestines[47] and those that enter the testes made for the precise concoctions of the matter contained in them as well as for the abundant provisioning of their further activities, it seemed reasonable to suppose that Nature has devised this stratagem in order to elaborate for a long time the matter within the arteries, being hot and thin and air-like blood, as well as to provide abundant nourishment to the psychic pneuma in the brain.[48]

For Galen, all three vascular structures possess the right degree of anatomical complexity conferred on them by Nature in order to elaborate the appropriate *matter* (ὕλη) they contain. The above citation explicitly links the provision of nourishment to pneumatic elaboration: the retiform plexus is presented as a source of nourishment to the psychic pneuma in the brain. To be sure, the comparison between the testicular vessels and those of the retiform plexus cannot be an exact one since the retiform plexus is a more complex vascular structure. But Galen also utilises this increased complexity in the following way:

> But the retiform plexus is much more intricately coiled than the plexus with ivy-like tendrils, as the elaboration which the psychic pneuma undergoes in the brain has to be of a more precise nature than that required by semen. Thus I was correct when I demonstrated in the

---

[46] Cf: διὰ τοῦτο καὶ πάντων ἐν ἀσφαλεστάτῳ χωρίῳ κατέθηκεν ἡ φύσις αὐτό. *UP* II, p. 429,6–7 Helmreich; IV. 334 K.

[47] Galen is probably referring to the arrangement of blood vessels in the mesentery of the intestine. Cf. *AA* II. 562–567 K.

[48] τοῦτ' οὖν τὸ θαυμαστὸν πλέγμα πρὸς τῆς μηδὲν εἰκῇ ποιούσης φύσεως ἐν οὕτως ἀσφαλεῖ χώρᾳ ταχθῆναι, μεγάλης τινὸς ἐδόκει χρείας ἐνδεικτικὸν ὑπάρχειν. ἀλλ' ἐπειδὴ τὴν τῶν ἐντέρων ἕλικα καὶ τὴν τῶν εἰς τοὺς ὄρχεις ἐμφυομένων ἀγγείων ἀκριβοῦς τε πέψεως ἕνεκα τῶν περιεχομένων ὑλῶν καὶ προσέτι δαψιλοῦς παρασκευῆς ταῖς ἑξῆς ἐνεργείαις ἑωρῶμεν γεγενημένην, εὔλογον ἐδόκει κἀνταῦθα τοιοῦτόν τι μεμηχανῆσθαι τὴν φύσιν, ἅμα τε κατεργαζομένην πολλῷ χρόνῳ τὴν ἐν ταῖς ἀρτηρίαις ὕλην, αἷμα θερμὸν καὶ λεπτὸν καὶ ἀτμῶδες ὑπάρχουσαν, ἅμα τε κατασκευάζουσαν τροφὴν δαψιλῆ τῷ κατὰ τὸν ἐγκέφαλον ψυχικῷ πνεύματι. p. 200 Furley-Wilkie; V. 155–156 K.

> *Commentaries* (sc. *PHP*) that the vital pneuma drawn up through the arteries is the fitting material for the creation of psychic pneuma in the brain.[49]

The retiform plexus was constructed because psychic pneuma requires elaboration of a more precise nature. For Galen, vital pneuma is the best and most appropriate material for the physiological requirements of the retiform plexus. In what way does its elaborative change to psychic pneuma take place? As noted earlier, Galen holds that a qualitative change occurs in the substance presented to an organ.[50] This may further be illustrated by considering Galen's account of the formation of blood and bone. Blood is derived from, but is qualitatively different to ingested nutriment. For blood to be elaborated to bone, several intermediate steps are required, which are linked to the passage of time.[51] Galen states that both the production of bone and the production of blood from nutriment take place in several stages in several different organs, and each step, to be properly effective, requires an unquantified period of time. The concept of elapsed time does not have to be placed in a teleological context, but it remains for Galen a problematic notion and one which he seeks to buttress by direct appeal to a teleological agency. In *De placitis Hippocratis et Platonis* for example, Galen states that in regard to the initial preparation of blood in the liver and its final elaboration in the heart, it is a "reasonable proposition that no perfect and great work can be done at once, or from a single natural organ receive all of its fitting elaboration."[52] Galen thereby links the concepts of time and elaboration to his teleological and epistemological agendas, but goes no further beyond saying that this form of elaboration is a *reasonable* (εὔλογον) one.

---

[49] ἀλλ' ὅσῳ τὸ κατὰ τὸν ἐγκέφαλον πνεῦμα ψυχικὸν ἀκριβεστέρας ἐδεῖτο κατεργασίας τῆς τοῦ σπέρματος, τοσούτῳ καὶ τὸ δικτυοειδὲς πλέγμα τοῦ κισσοειδοῦς πολυπλοκώτερον ἐγένετο. καλῶς οὖν ἐν ἐκείνοις ἀπεδείκνυτο τοῖς ὑπομνήμασιν (sc. Περὶ τῶν Ἱπποκράτους καὶ Πλάτωνος δογμάτων) ἡ γένεσις τοῦ κατὰ τὸν ἐγκέφαλον πνεύματος ψυχικοῦ τὸ διὰ τῶν ἀρτηριῶν ἀναφερόμενον τὸ ζωτικὸν ὕλην οἰκείαν ἔχειν. *UP* II, pp. 12,20–13,2 Helmreich; III. 700 K (reading κισσοειδοῦς: De Lacy, 1992, 222, for κιρσοειδοῦς: Helmreich).

[50] Cf. Hankinson, 1987b, 82–84. To Hankinson, pneuma is a sustaining cause for Galen.

[51] ὀστοῦν δ' ἵνα γένηται, πολλοῦ μὲν δεῖται χρόνου, πολλῆς, δ' ἐργασίας καὶ μεταβολῆς τῷ αἵματι. *Nat. Fac.* p. 116,11–13 Helmreich; II. 21 K.

[52] καὶ γὰρ καὶ λόγον ἔχει μηδὲν τῶν τελέων τε καὶ μεγάλων ἔργων ἀθρόως δύνασθαι γενέσθαι μηδὲ ὑφ' ἑνὸς ὀργάνου φυσικοῦ τὸν οἰκεῖον κόσμον ἅπαντα κτήσασθαι. *PHP* p. 398,7–10 De Lacy; V. 550–551 K.

Galen is perhaps on firmer ground when he considers the functional attributes of each organ in more general terms. In Galen's scheme, an organ's physical structure is obviously essential to its proper physiological function.[53] Each organ is made up of qualities blended in a certain way,[54] which are responsible for the distinctive substance of that organ.[55] This individuation in form is also reflected by an individual *activity* (ἐνέργεια).[56] The mechanism which enables this individual activity is the particular *power* or *faculty* (δύναμις) of each organ.[57] Each faculty is absolutely distinct or peculiar to that organ. Thus the veins and the liver possess a peculiar haematopoietic faculty, the stomach a digestive, the heart a pulsatile, and the retiform plexus its own distinctive, pneumatic faculty.[58]

For Galen, the retiform plexus can function only by taking an unquantified amount of time to complete its elaborative task. In *De*

---

[53] This is not the same thing as saying that for Galen structure alone determines function, although he comes very close to doing exactly that in the case of the retiform plexus. The situation is arguably less complex in the Hippocratic Corpus, where the analogy of structure to function is broadly maintained. Cf. Gundert, 1992, 464–465.

[54] In considering the veins of the body for example, Galen declares that their function is precisely because of the way in which the four qualities have intermixed to produce them: Ἐμοὶ μὲν οὖν καὶ ἡ φλὲψ καὶ τῶν ἄλλων ἁπάντων ἕκαστον διὰ τὴν ἐκ τῶν τεττάρων ποιὰν κρᾶσιν ὡδί πως ἐνεργεῖν δοκεῖ. *Nat. Fac.* p. 106,4–6 Helmreich; II. 7 K. Cf. *Opt.Corp.Const.* IV. 737–738 K.

[55] For example, it is "by the peculiar substance of the liver" (ὑπὸ τῆς οἰκείας οὐσίας τοῦ ἥπατος), that changes the *nutritive fluid* (χυμός) from the stomach to *blood* (αἷμα). *PHP* 410,2–4 De Lacy; V. 566 K.

[56] εἶναί τινα λέγοντες ἐν ταῖς φλεψὶν αἱματοποιητικήν, ὡσαύτως δὲ κἀν τῇ κοιλίᾳ πεπτικὴν κἀν τῇ καρδίᾳ σφυγμικὴν καὶ καθ' ἕκαστον τῶν ἄλλων ἰδίαν τινὰ τῆς κατὰ τὸ μόριον ἐνεργείας. *Nat. Fac.* p. 107,16–20 Helmreich; II. 9–10 K. Cf. *PHP* p. 412,16–25 De Lacy; V. 569–570 K.

[57] αἱ γὰρ τῶν οὐσιῶν ἰδίοτητες οἰκείας ἑαυταῖς ἔχουσι καὶ τὰς τῶν δυνάμεων ἰδίοτητας, *PHP* p. 456,11–12 De Lacy; V. 621 K. On the concept of δύναμις see Von Staden, 1998.

[58] ἕκαστον γὰρ τῶν ὀργανικῶν τοῦ ζῴου μορίων, σύνθετον ὑπάρχον ἐξ ἑτέρων ἁπλουστέρων μορίων, πάντως ἕν γέ τι τοιοῦτον ὑπάρχει κατὰ τὴν οὐσίαν οἷον οὐκ ἄλλο καθ' ὅλον τοῦ ζῴου τὸ σῶμα. τοῦ γοῦν τῆς γαστρὸς σώματος τὸ ἴδιον οἷον οὐκ ἄλλο τῶν πάντων ἐστίν· ὡσαύτως δὲ καὶ τὸ τοῦ σπληνὸς ἐγκεφάλου τε καὶ νεφρῶν καὶ γλώττης καὶ ὀφθαλμῶν καὶ κύστεων καὶ μητρῶν . . . κατὰ δὲ τὸ ἴδιον εἶδος ἑκάστης οὐσίας ἰδίαν ἀναγκαῖον εἶναι τὴν ἐνέργειαν. οἷον γάρ ἐστι τὸ τοῦ πνεύμονος σῶμα κατὰ τὴν οὐσίαν οὐκ ἂν εὕροις ἕτερον οὐδ' οἷον ἐγκέφαλος οὐδὲν ἄλλο τοιοῦτον, ὥσπερ οὐδὲ ὁποῖον ἡ καρδία σῶμα κατὰ τὴν οὐσίαν ἐστὶν οὐδὲν ἀκριβῶς ἄλλο τοιοῦτον ὑπάρχει, καὶ νεφροὶ δὲ καὶ σπλήν, ἑκάτερον αὐτῶν οἷον οὐκ ἄλλο, καὶ διὰ τοῦτο ἡ ἐνέργεια καθ' ἕκαστον ἴδιος ἐπὶ τῷ τῆς οὐσίας ἰδίῳ, καὶ εἴ τις δὲ σὰρξ ἰδιότητα πολλὴν ἔχει παρὰ τὰς ἄλλας σάρκας. *PHP* p. 412,2–23 De Lacy; V. 568–570 K.

usu partium, the lungs provide a starting point from which Galen considers the importance of elapsed time in a physiological context:

> For it is reasonable that the outside air does not all at once become the nutriment of the pneuma within the body, but that this alteration takes place slowly, just as food also is, and over time it acquires the quality that is appropriate to this innate pneuma, and that the first instrument of this alteration is the flesh of the lungs . . .[59]

Galen's only way of strengthening the notion of the importance of time for pneumatic elaboration is by appeal to analogy. It therefore becomes *reasonable* that the elaboration of food cannot take place immediately. So too the outside air can only gradually be elaborated into something else. Later in the same text, Galen goes one step further, this time appealing to Nature to underwrite his notion that a certain period of time is required in each organ for elaboration to be performed correctly:

> When Nature desires matter to be most precisely elaborated, She arranges it to spend a great deal of time within those organs which concoct it . . . For our present requirements it shall be sufficient for me to cite an example of this arrangement under discussion by reminding you of the ivy-like convolutions in which blood and pneuma are rendered suitable to form the semen. For the veins and arteries there are of many-formed convolutions and in their first part contain pure blood; however, in the last part, near the testes, the humour within them is no longer perfectly red but is already whitish and requires little to complete the change into the substance of the semen, a change added by the testes themselves. For the retiform plexus is much more intricately coiled than the ivy-like plexus, as the elaboration which the psychic pneuma undergoes in the brain has to be of a more precise nature than that required by semen.[60]

[59] εὔλογον γὰρ οὐκ ἀθρόως οὐδ' ἐξαίφνης τὸν ἔξωθεν ἀέρα τοῦ κατὰ τὸ ζῷον πνεύματος γίγνεσθαι τροφήν, ἀλλὰ κατὰ βραχὺ μὲν ἀλλοιούμενον, ὥσπερ γε καὶ τὰ σιτία, δεχόμενον δὲ τὴν οἰκείαν ποιότητα τῷ συμφύτῳ πνεύματι χρόνῳ πλείονι, καὶ ταύτης τῆς ἀλλοιώσεως τὸ πρῶτον ὄργανον ὑπάρχειν τὴν τοῦ πνεύμονος σάρκα . . . *UP* I, p. 392,17–23 Helmreich; III. 540 K.

[60] ἔνθα γὰρ ἀκριβῶς κατεργάσασθαι τὴν ὕλην ἡ φύσις βούλεται, πολυχρόνιον αὐτῇ διατριβὴν ἐν τοῖς τῆς πέψεως ὀργάνοις παρασκευάζει . . . εἰς δὲ τὰ παρόντα τῆς κισσοειδοῦς ἕλικος, ἐν ᾗ τὸ πρὸς τὴν τοῦ σπέρματος γένεσιν ἐπιτήδειον αἷμα καὶ πνεῦμα παρασκευάζεται, μνημονεύσασιν ἡμῖν ἀποχρήσει παράδειγμα ποιῆσαί τι πρὸς τὰ παρόντα. φλέβες γὰρ ἐν ἐκείνῃ καὶ ἀρτηρίαι πολυειδῶς ἑλίττονται κατὰ μὲν τὰ πρῶτα μέρη τῶν ἑλίκων εἰλικρινὲς αἷμα περιέχουσαι· κατὰ δ' αὖ τὰ τελευταῖα τὰ πρὸς αὐτοῖς τοῖς ὄρχεσιν οὐκέτ' ἐρυθρὸς ἀκριβῶς, ἀλλ' ἤδη λευκότερός πώς ἐστιν ὁ ἐν αὐταῖς περιεχόμενος χυμὸς ὀλίγον ἔτι δεόμενος εἰς συμπλήρωσιν οὐσίας

Embedded in Nature's imperative, this citation provides the only "evidence" Galen offers on why the process of pneumatic elaboration should take (an unspecified amount of) time. But even this presentation is less impressive than it appears. If Galen is to be believed, then, in the case of the testicular vessels, direct observation—obtained by dissection—informs him of a change in colour of the material being processed, from red to white.[61] But all Galen does here is to combine two *separate* observations, made at *different* parts of the testicular vasculature. Galen then fashions two reasons to explain the difference in colour; namely, that this change is in the form of a physiological elaboration, and that it has taken place over time. Galen then argues that the reason the retiform plexus is more intricately coiled then the testicular is so that psychic pneuma can undergo a "more precise elaboration" (ἀκριβέστερος κατεργασία). Galen has no choice here but to suborn his skills of dissection to a necessarily speculative physiological need. Yet the blood in the testicular vessels has never been seen to change colour, either *slowly* (κατὰ βραχύ) or otherwise. It is Galen who links this change to elapsed time. For him, a convoluted vasculature is critical for the elaborative processes in the testes and at the base of the brain. Of course, Galen cannot claim that an observation of an alleged change within the retiform plexus is anatomically possible; instead he manipulates a set of observations made from the spermatic vessels in order to validate a presumed analogous activity in the retiform plexus.

The detailed vascular anatomy of the retiform plexus and its proximity to the ventricular system of the brain allow Galen to state that it, and it alone, is the most appropriate structure for the concoction of vital pneuma into its psychic form. To be sure, the validity of such an "argument from proximity" is open to question. But it cannot be denied that this elaborative scheme allows Galen to define

---

σπέρματος, ὃ παρὰ τῶν ὄρχεων αὐτῶν προσλαμβάνει. ἀλλ' ὅσῳ τὸ κατὰ τὸν ἐγκέφαλον πνεῦμα ψυχικὸν ἀκριβεστέρας ἐδεῖτο κατεργασίας τῆς τοῦ σπέρματος, τοσούτῳ καὶ τὸ δικτυοειδὲς πλέγμα τοῦ κισσοειδοῦς πολυπλοκώτερον ἐγένετο. *UP* II, p. 12,5–23 Helmreich; III. 699–700 K (reading κισσοειδοῦς: De Lacy, 1992, 222, for κιρσοειδοῦς: Helmreich).

[61] ζητῶμεν οὖν ἡμεῖς αὐτοὶ ὑποθέσεσι χρώμενοι τοῖς ἀποδειχθεῖσί τε καὶ φαινομένοις ἐναργῶς ἐν ταῖς ἀνατομαῖς. ἥ τε γὰρ ἀρτηρία καὶ ἡ φλέψ, αἱ εἰς τὴν κεφαλὴν τοῦ ὄρχεως ἐμβάλλουσαι, σπερματῶδες ὑγρὸν περιέχειν ἤδη φαίνονται, καὶ αὐτὸς ὅλος ὑγροῦ τοιούτου μεστός ἐστι. *Sem.* pp. 132,28–134,3 De Lacy; IV. 582 K. In *UP* II, p. 419,21–22 Helmreich; IV. 322 K, both semen (and milk) are made from blood which is perfectly concocted.

the vasculature not merely by what it carries, but by what it does. Yet the final elaboration of psychic pneuma occurs not in the retiform plexus but within the vascular convolutions in the ventricles, the choroid plexuses. In the account of their physiology, Galen's problems with the elaboration of psychic pneuma are further compounded.

6.4 *The anatomy and physiology of the choroid plexus*

In *De placitis Hippocratis et Platonis*, Galen states that for the production of psychic pneuma, "no small amount" (οὐκ ὀλίγον πλῆθος) of arteries and veins have their terminations in the ventricles of the brain.[62] These constitute the choroid plexuses. Perhaps because the greater part of the argumentation regarding the role of the retiform plexus can be transferred directly to the choroid plexuses, Galen does not discuss their physiology in the same detail. His description of the choroid plexuses and their relationship to the brain is as follows:

> And you will also see (sc. in the ventricles of the brain) what are known as the choroid plexuses. Those about Herophilus call them choroid-like knots (χορ<ι>οειδῆ συστρέμματα) obviously after the membranes wrapped in a circle around the foetus, these being a plexus of veins and arteries, joined together by thin membranes.[63]

If the retiform plexus was aptly likened to the twisted skeins of a fisherman's net, the choice of name for the choroid plexus is also dependent on its similarity to another structure, the *foetal membranes* or *chorion* (χόριον).[64] Galen's definition for any choroideal structure is a membranous network of veins and arteries. Galen also refers to the choroid plexuses in the ventricles as *choroid bodies* (χοριοειδῆ

---

[62] P. 444,13–14 De Lacy; V. 606–607 K.

[63] ὄψει δὲ καὶ τὰ καλούμενα χορ<ι>οειδῆ πλέγματα κατὰ ταύτας. ὀνομάζουσι δ' οἱ περὶ τὸν Ἡρόφιλον αὐτὰ χορ<ι>οειδῆ συστρέμματα, παρονομάσαντες δηλονότι τῶν χορίων, ἃ τοῖς κυουμένοις ἔξωθεν ἐν κύκλῳ περιβέβληται, φλεβῶν ὄντα καί ἀρτηριῶν πλέγματα, λεπτοῖς ὑμέσι συνεχομένων. *AA* II. 719–720 K. Cf. T 124 *Herophilus*. See also Needham, 1934, 44–45; Von Staden, 1989, 180 n. 131. Harris, 1973, 354, incorrectly states that both the choroid and retiform plexus are "networks of blood vessels containing both veins and arteries."

[64] In *De uteri dissectione* p. 50,16–18 Nickel; II. 902 K, these are described as follows: τὸ μὲν χόριον ἀγγεῖα πολλά, φλέβες τε καὶ ἀρτηρίαι, ἐγγὺς ἀλλήλων κείμεναι, τὰ μεταξὺ διαστήματα ὑμέσιν ἀναπληρούμενα. See also *AA* XII.4; pp. 116–118 Duckworth; *Foet. Form.* IV. 655 K. Cf. Kudlien, 1964, 86–88; De Witt, 1959, 363.

σώματα) or *choroid meninx* (χοριοειδὴς μῆνιγξ).[65] To emphasise structural similarity in certain organic parts, Galen compares the *pia mater* with the foetal choroid as well as with the intestinal mesentery.[66]

The arterial supply to the choroid plexus will be considered first:

> Into the brain itself pass upward a pair of arteries of no small calibre, and from where, mixing with the veins there, form the choroid plexus within the ventricles, and are interwoven with the thin meninx (pia mater).[67]

This arterial pair is derived from the internal carotid, which also supplies the retiform plexus.[68] In this context another description of the retiform plexus may be appended, which allows for a better appreciation of the way in which Galen prepares the ground for the arterial supply of the choroid plexuses. Once the internal carotid arteries have passed through the canal in the base of the skull:

> . . . in the space between it and the dura mater, they first divide up into many small and fine arteries, and next some go to the front of the head, some backwards, some to the left, and some to the right, being carried through and also interwoven with each other, and so giving the opposite appearance; namely, that they have forgotten the way to the encephalon. But this is not true either; for from all these many arteries, as roots come together to form a trunk, there arises another pair of arteries equal to those that came up in the first instance, and thus these, through the apertures in the dura mater, pass into the encephalon.[69]

---

[65] *UP* I, p. 452,1 Helmreich; III. 623 K and I, p. 479,11 Helmreich; III. 660 K, respectively. Cf. T 125 *Herophilus*. Galen also uses the phrase χοριοειδὴς μῆνιγξ to describe the nature of the relationship of the pia mater enveloping the brain (*UP* II, p. 8,22 Helmreich; III. 694 K).

[66] Ὥσπερ οὖν καὶ ἡ λεπτὴ μῆνιγξ ἅμα μὲν στηρίζει τὸν ἐγκέφαλον, ἅμα δὲ καὶ σκέπει, καὶ πρὸς τούτοις ἔτι σύνδεσμος γίγνεται τῶν κατ' αὐτὸν ἀγγείων ἁπάντων. ἔοικε μὲν γὰρ ἐμβρύου χορίῳ καὶ μεσαραίῳ ζῴου. καὶ γὰρ κἀκείνων ἑκάτερον ἐκ πολλῶν ἀρτηριῶν καὶ φλεβῶν ἐγγὺς ἀλλήλαις κειμένων ὑμένι τε λεπτῷ τὰ μεταξὺ διαστήματα συνυφασμένων ἐγένετο, *UP* I, p. 476,1–7 Helmreich; III. 656 K. Cf. Debru, 1996, 157.

[67] εἰς μὲν τὸν ἐγκέφαλον αὐτὸν ἀναφέρεται ζεῦγος οὐ σμικρὸν ἀρτηριῶν, ἐξ οὗ τό τε χοριοειδὲς ἐν ταῖς κοιλίαις αὐτοῦ γεννᾶται πλέγμα, ταῖς ἐνταῦθα φλεψὶν ἀναμιγνυμένων, ἥ τε λεπτὴ μῆνιγξ διαπλέκεται. *UP* II, p. 429,12–15 Helmreich; IV. 334–335 K.

[68] Cf. Sisson and Grossman, 1975, 956.

[69] ὑπερβᾶσαι γὰρ τὸ κρανίον ἐν τῇ μεταξὺ χώρᾳ τούτου τε καὶ τῆς παχείας μήνιγγος πρῶτα μὲν ἐσχίσθησαν εἰς πολλὰς πάνυ σμικρὰς καὶ λεπτὰς ἀρτηρίας, ἑξῆς δὲ τὸ μέν τι πρόσω τῆς κεφαλῆς, τὸ δ' ὀπίσω, τὸ δ' εἰς ἀριστερά, τὸ δ' εἰς δεξιὰ δι' ἀλλήλων φερόμεναί τε καὶ περιπλεκόμεναι πάλιν ἑτέραν δόκησιν ἐναντίαν παρέσχον, ὡς ἐπελάθοντο τῆς ἐπὶ τὸν ἐγκέφαλον ὁδοῦ. ἀλλ' οὐκ ἄρ' οὐδὲ τοῦτ'

In these descriptions, Galen establishes not only the continuity between the retiform and choroid by means of their arterial supply, but also underscores the necessity of such a source for his physiology.

As discussed in chapter 3, the venous supply to the choroid plexus derives ultimately from the venous sinuses at the torcular Herophili. From this source, branches are sent to the choroid plexuses. In the course of this distribution, the veins insinuate themselves around the *pineal* (κωνάριον), gaining the necessary support:

> . . . therefore, as (the sanguineous aqueduct, αἵματος ἀγωγή) was hard by the middle ventricle and from it was produced the large veins which themselves transmitted the veins to be distributed to the choroid plexus, [Nature] did not yet entrust so much to the pia mater alone the attachment of these veins, but crafted a gland (the pineal) to assist, and affixed it into the middle of the veins and in this way surrounded it with the pia mater, and placed around it in a circle the veins held together by the meninx, that, as long as their course was supported, and the gland at the same time accompanied them, whenever into the encephalon the veins find refuge, the gland might have its circular base propped up on the back of the encephalon. The veins dividing in this way around the gland, travel through the middle ventricle to the anterior ventricles where they are plaited together with the arteries coming from below and are combined as the choroid plexuses.[70]

It was apparently Herophilus who "not only correctly recognized the vascular nature of these plexuses, saying that they are 'plexuses of veins and arteries held together by thin membranes' . . . but, Galen seems to imply, also their projection into the ventricles."[71] These

---

ἀληθὲς ἦν. αὖθις γὰρ ἐκ τῶν πολλῶν ἐκείνων ἀρτηριῶν ὥσπερ ἐκ ῥιζῶν εἰς πρέμνα τῆς συναγωγῆς γενηθείσης ἐξέφυ ζεῦγος ἀρτηριῶν ἕτερον ἴσον τῷ κατ' ἀρχὰς ἀνιόντι καὶ οὕτως ἤδη διὰ τῶν τῆς παχείας μήνιγγος τρημάτων εἰς τὸν ἐγκέφαλον ἔδυ. *UP* II, p. 11,13–25 Helmreich; III. 698 K.

[70] ὡς ἤδη πλησίον τῆς μέσης κοιλίας προϊὼν ἐγεγένητο καὶ μεγάλας φλέβας ἐχρῆν ἀποφύειν αὐτοῦ τὰς εἰς τὰ χοριοειδῆ πλέγματα νεμηθησομένας, οὐδέπω μὲν ἐπίστευσε τῇ λεπτῇ μήνιγγι μόνῃ ξυνδεῖν τηλικαύτας φλέβας, ἀδένα δ' αὐτῇ τινα βοηθὸν ἐδημιουργήσατο καὶ μέσον ἐνδτηρίξασα ταῖς κατιούσαις φλεψὶν αὐτὸν οὕτω τήν τε λεπτὴν μήνιγγα περιέφυσε καὶ τὰς φλέβας ἐν κύκλῳ περιέστησε συνεχομένας τῇ μήνιγγι, ὅπως, μέχρις ἂν αὗται μετέωροι φέρωνται, καὶ ὁ ἀδὴν αὐταῖς συμπαρεκτείνηται, ὅταν δ' εἰς τὸν ἐγκέφαλον ἤδη καταφύωνται, τηνικαῦτα καὶ ὁ ἀδὴν τὴν ἑαυτοῦ βάσιν τὴν κυκλοτερῆ κατὰ τοῦ νώτου στηρίζῃ τοῦ ἐγκεφάλου. αἱ μὲν δὴ φλέβες οὕτως αἱ περὶ τὸν ἀδένα σχισθεῖσαι διὰ τῆς μέσης κοιλίας ἐπὶ τὰς ἔμπροσθεν ἔρχονται συμπλεκόμεναί τε κατὰ ταύτας ταῖς κάτωθεν ἀνιούσαις ἀρτηρίαις καὶ τὰ χοριοειδῆ συνιστᾶσαι πλέγματα. *UP* II, pp. 19,24–20,13 Helmreich; III. 709–710 K.

[71] Von Staden, 1989, 180.

plexuses are prominent features of the mammalian ventricular system.[72] The above citation from *De usu partium* also describes how veins pass into the anterior ventricle from the middle ventricle.[73] Although it neither confirms nor denies that a choroid plexus is to be found in the middle ventricle, its presence may be inferred from a passage in *De usu partium.* Here Galen, discussing the importance of the pia mater, states that "most anatomists" regard the part of the pia which lines the ventricles from *within* (ἔνδοθεν) as a choroid plexus (although these anatomists do not appear to make the same comparison with other parts of the pia which performs a similar function to that of the foetal choroid).[74]

From experimental and observational evidence of disruptions to the ventricular system, discussed in the previous chapter, Galen determined that these effects could be attributed to a pneumatic agency within the ventricles. To Galen, vital pneuma is processed in the retiform plexus, and apparently completed by the choroid plexuses. But, as Harris notes: "The part played by the veins (sic) of the choroid plexus . . . is very far from clear, and Galen's accounts of it appear to differ."[75] To try and establish what part the choroid plexuses play in Galen's physiology of the brain, the following from *De usu partium* may again be cited:

> From the outside air, pneuma is drawn in by the rough arteries and receives its first elaboration in the flesh of the lungs, its second in the heart and the arteries, especially those of the retiform plexus, and then a final elaboration in the ventricles of the brain which completes its transformation into psychic pneuma.[76]

Admittedly this does not explicitly mention the choroid plexuses as the site of this "final elaboration" (τελεωτάτη ἐργασία), although as noted above, Galen provides the anatomical evidence to place them

---

[72] Cf. Ariëns Kappers, Huber and Crosby, 1936, 47.

[73] In domestic animals, "the choroid plexus of the lateral ventricle is continuous, via the interventricular foramen, with the choroid plexus of the third ventricle." King, 1987, 291, see also idem, 279, fig. 21.14(b) and 282, fig. 21.15(b). For possible reasons why Galen does not refer to any mention of the choroid plexus in the *fourth* ventricle, see chapter 4, n. 225.

[74] ἀλλ' ἐνταῦθα μὲν οὐκ οἶδ' ὅπως ἀποθέμενοι τὸν ὕπνον οἱ πολλοὶ τῶν ἀνατομικῶν χοριοειδῆ πλέγματά τε καὶ συστρέμματα καλοῦσι τὸ μόριον τῆς λεπτῆς μήνιγγος, ὅσον ὑπέζωκε τὰς κοιλίας ἔνδοθεν, ἐπὶ δὲ τῶν ἄλλων μερῶν οὐκ ἐθέλουσιν οὔτ' εἰκάζειν οὔτ' ὀνομάζειν ὡσαύτως. I, p. 476,18–23 Helmreich; III. 657 K.

[75] 1973, 361.

[76] See n. 1 above for the Greek.

within the ventricles. At first glance, there seems little doubt that the choroid plexuses are involved in the elaborative process. In *De placitis Hippocratis et Platonis*, Galen links the role of the retiform plexus to that of the choroid plexuses:

> Therefore it is reasonable that this pneuma is produced in the ventricles of the brain and owing to this then no small number of arteries as well as veins terminate there, forming as they are called, the choroid plexuses ... And you would even more expect that this pneuma is produced when the vessels, especially the arteries, breathe it out into the ventricles of the brain, after you have seen the retiform plexus which is formed from the arteries that go to the head ...[77]

Anatomically at least, the retiform and choroid plexus are made to appear part of a seamless whole. However, as noted in the discussion of the elaboration of psychic pneuma in the retiform plexus in the last section, Galen qualifies its production in the choroid plexuses as no more then *reasonable* (εὔλογον); one would expect that pneuma is produced in the ventricles by the vessels of the choroid which "breathe it out". Galen's language reflects a lack of demonstrative rigour.[78] This hesitancy is also noted when, in an earlier passage from *PHP*, Galen attempts to limit the role of the veins of the choroid plexus in the production of psychic pneuma:

> Perhaps in a small way this pneuma in the ventricles is made from the veins that end in them; but the major and chief source of production is supplied from the arteries in the retiform plexus at the base of the brain.[79]

---

[77] Εὔλογον οὖν γεννᾶσθαι μὲν τουτὶ τὸ πνεῦμα κατὰ τὰς κοιλίας τοῦ ἐγκεφάλου καὶ διὰ τοῦτο ἐκεῖσε τελευτᾶν ἀρτηριῶν τε καὶ φλεβῶν οὐκ ὀλίγον πλῆθος, ἐξ οὗ τὰ καλούμενα χορ<ι>οειδῆ πλέγματα γέγονεν, ... ἔτι δ' ἂν μᾶλλον ἐλπίσαις γίγνεσθαι τὸ πνεῦμα τοῦτο τῶν ἀγγείων ἀναπνεόντων αὐτὸ καὶ μάλιστα τῶν ἀρτηριῶν εἰς τὰς κοιλίας τοῦ ἐγκεφάλου, τὸ δικτυοειδὲς ἰδὼν πλέγμα γιγνόμενον ἐκ τῶν ἐρχομένων εἰς τὴν κεφαλὴν ἀρτηριῶν, p. 444,12–18; De Lacy; V. 606–607 K.

[78] "Mais là où la démonstration est attendue, on ne trouve que l'expression d'une possibilité: "il est possible que", "il n'est pas impossible que", "comme cela peut arriver", ou "il est vraisemblable que". C'est que, dans le domaine de la physiologie, cet inachèvement de la démonstration qui ne parvient pas à dépasser le plausible est loin d'être une exception, et paraît plus souvent, chez Galien, être la règle." Debru, 1996b, 163.

[79] γένεσις δὲ τῷ πνεύματι τῷδε τῷ κατὰ τὰς κοιλίας αὐτοῦ τάχα μέν τις ὀλίγη καὶ ἐκ τῶν εἰς αὐτὰς περαιουμένων γίγνεται φλεβῶν, τὴν πλείστην δὲ καὶ κυριωτάτην αἱ κατὰ τὸ δικτυοειδὲς πλέγμα τὸ κατὰ τὴν τοῦ ἐγκεφάλου βάσιν ἀρτηρίαι χορηγοῦσιν, p. 230,19–22 De Lacy; V. 356 K.

This statement reaffirms the importance of the retiform plexus, and therefore of the arterial system as the source of psychic pneuma (that the veins may contain pneuma is another matter). But it does so at the expense of the choroid plexus.[80] There does not seem to be any reason why the arteries and veins of the choroid plexus cannot participate equally in the final stage of elaboration of psychic pneuma, but Galen never offers an account of the physiology of the choroid plexus comparable to that which he provides for the retiform. As raised at the beginning of this discussion, it is likely that Galen wishes it understood that the choroid plexus functions in a similar, albeit reduced, way to that of the retiform web. Whether the choroid plexuses can be labelled unequivocally as the site of the final elaboration of psychic pneuma remains uncertain. A fuller understanding of their function is only possible when examined in the light of certain experiments Galen performs to interrupt the supply of pneuma from heart to brain.

6.5 *The experiment of carotid ligation*

The need for Galen to consider a second source of psychic pneuma arose because of his experiments of complete ligation of the carotid arteries in animals. Did Galen perform these experiments with the notion of determining pneuma's role in his physiology? As will be discussed, it is difficult to say with complete certainty that he did. Galen noted that after carotid ligation, the animals seemed relatively undisturbed, and were able to move and even run about. In order to account for this, it seemed evident to Galen that the cerebral ventricles must have had access to another source of psychic pneuma which was not entirely dependent on that supplied by the retiform plexus. Galen concluded that external air, once inside the ventricles, could be elaborated directly into psychic pneuma to supply this need. The outside air could access the anterior ventricles directly via the

---

[80] Harris, 1973, 365–6, states that: "Galen certainly believed that the veins ending in the choroid plexus do contribute a little *pneuma* to the ventricles." (italics in original). He suggests the following as a role for the veins: "perhaps Galen considered that the arterial blood, since it had undergone so much purification, needed some elements of the humours which could only be provided from the venous blood." 362. This raises the possibility that arterial blood is in some way "exhausted" by its passages through the rete (and choroid), and stands in need of reviving nutriment.

olfactory tract. But as will be discussed below, Galen leaves unanswered the question of the mechanism by which external air receives the necessary elaboration.

To a considerable extent, Galen is a confident exponent of a pneuma-based encephalocentric physiology. Psychic pneuma is important enough to be regarded as the rational soul's "first instrument" (πρῶτον ὄργανον).[81] In a series of experiments which have been examined in chapter 5, Galen sought to establish the premise that if the brain's ventricles were disrupted, then the consequences varied from stupor to death. Galen deduced from these observations that pneuma was the mechanism behind these events, and that if some parts of the ventricular system were opened, those effects might be reversed as the supply of pneuma is renewed following closure of the affected ventricle. His investigations can be summarised as follows:

(i) Experiments have shown that the body is deprived of sensation and motion when the ventricles of the brain are opened, and the pneuma escapes.
(ii) This being so, pneuma is useful for sensation and motion.[82]

Yet although Galen has demonstrated that an agency he calls "psychic pneuma" is *present* within the ventricles of the brain, Galen is not so certain of its exact role in sensation or motion. Pneuma is deemed *useful* (χρήσιμος), but not *essential* (ἀναγκαῖος) for the two principal activities of the rational soul. By this and other similar expressions, Galen indicates his awareness of the inability of pneuma to provide a complete explanation for the functions of the brain. His ventricular experiments are also unable to determine whether the ventricles alone produce psychic pneuma; Galen only deems it *reasonable* that they should contain it. It is the experiment of carotid ligation which appears to provide Galen with an opportunity to demonstrate that psychic pneuma is produced in the ventricles. In ligating the carotid arteries in the neck, Galen was faced with what he describes as a "highly unusual" (ἀτοπώτατον) result which showed that the functions of the brain continued even if the supply of vital pneuma through the arteries was interrupted.[83] This allowed Galen to conclude that there must be a source of pneuma apart from that

[81] See the discussion in chapter 5.6.
[82] Cf. *PHP* p. 448,4–24 De Lacy; V. 611–612 K.
[83] A cardiocentrist would argue differently. See below, n. 99.

produced within the retiform plexus. Galen's accounts of these experiments are found principally in two texts, *De usu respirationis* and *De usu pulsuum*.[84] Each deals with differing physiological demands placed on the available amount of psychic pneuma. The account from *De usu respirationis* will be considered first.

In *De usu respirationis*, Galen poses the question: is it possible that psychic pneuma may be nourished from the act of breathing?[85] This inquiry is crafted to illuminate a critical point of his physiology as much as to refute an alternative, Erasistratean thesis. Although Erasistratus may be considered as an encephalocentrist, he, together with his followers, seems to have differed from Galen concerning the source of supply and site of production of psychic pneuma. The Erasistratean hypothesis posits that all pneuma comes from the heart, and thence, via the arteries, to the membranes of the brain.[86] Galen allies himself with the so-called "school of Hippocrates", according to whom pneuma enters directly through the nostrils and then into the ventricles of the brain.[87] Galen then, posits a supply of psychic pneuma directly elaborated via outside air that enters the anterior

---

[84] It is not possible to state categorically the chronological sequence of the composition of these texts. Cf. Debru, 1996, 148, n. 56. But in *De usu respirationis*, Galen does refer to *De usu pulsuum* as though it were already written. But Galen is never completely dependent on one text to validate the findings of another. And, in a wider sense, the reader is not dependent on a chronological analysis in order to understand the anatomical and physiological works. Cf. Singer, 1997b, 541 n. 58. And in what was probably Galen's last work, *De propriis placitis*, there are no expressed longings for root and branch revision or *ex post facto* tinkering.

[85] p. 120 Furley-Wilkie; IV. 501 K.

[86] p. 122 Furley-Wilkie; IVK. 502 K. Both Erasistratus and Galen maintain vital pneuma is produced in the heart and is transported to the brain via the arteries. They differ only on where in the brain psychic pneuma is elaborated. In other respects, Erasistratean and Galenic pneumatic physiologies are similar—at times, too much so for Galen's liking, as appears to be the case here. The polemics to one side, a more fundamental objection Galen has with Erasistratean physiology is its purported reliance on the so-called principle of *horror vacui* (πρὸς τὸ κενούμενον ἀκολουθία), through which, according to Galen, a mechanistic explanation may be promulgated for all physiological action. Cf. Furley and Wilkie, 1984, 32–37; Von Staden, 1997b, 199–203.

[87] p. 122 Furley-Wilkie; IV. 502 K. Cf. *Diff. Resp.* VII. 761 K; *Us. Puls.* p. 206 Furley-Wilkie; V. 161 K; *PHP* p. 164,19–20 De Lacy; V. 281 K. Galen is perhaps relying on his interpretation of *De alimento* IX. 108 L, where pneuma is allegedly nourished by air drawn from the nostrils. But there is nothing in this notoriously ambiguous Hippocratic text about air being elaborated into psychic pneuma in the brain, which may account for Galen's use of air *or* pneuma being allowed here to enter the ventricles directly through the nostrils. Cf. Jones, 1923, I, 337–341; Furley and Wilkie, 1984, 12.

ventricles from the olfactory tracts. Galen holds that this source of air is necessary to life itself.[88] Galen compares the air drawn into the nasal passages as analogous to that of air brought into the body via the pores of the skin to regulate the vital heat.[89] Does this air enter as part of the process of respiration, or does the brain itself "breathe"? In *De anatomicis administrationibus* Galen states that in a vivisected animal, "you see that the whole brain, so long as the animal does not cry out, rises and sinks slightly with a movement which resembles that of the pulsation of all beating blood vessels, that is of the arteries."[90] In *De placitis Hippocratis et Platonis*, the capacity of the brain to expand and contract is compared to the motions of the heart. The brain uses this power to send psychic pneuma from the ventricles to the nerves. The brain *itself* is able to provide its own motion, allowing it to contract and expand.[91] In *De usu partium*, Galen states that the anterior ventricles of the brain perform the actions of inspiration and expiration.[92]

---

[88] καὶ πρὸς αὐτό γε τὸ ζῆν ἀναγκαία ἡ εἰς ἐγκέφαλον εἰσπνοή, *UP* I, p. 471,25–26 Helmreich; III. 650 K.

[89] The notion of *transpiration* (διαπνοή) through passages in the skin was an established concept by Galen's time. Cf. Vallance, 1990, 51, 54–56. Although there is no convincing evidence in the Hippocratic Corpus for the notion of skin transpiration, for Galen such passages were known as "perceptible pores", apprehensible through reason alone. In his system, transpiration does not involve the lungs but is under the control of the heart and arteries. The pulsation of the arteries is responsible for attracting air into the body via the pores of the arteries on surface of the skin (*Us. Puls.* p. 208 Furley-Wilkie; V. 163 K). The arteries do not pulsate because they contain pneuma; their pulse is directed by the heart, which is the source of a faculty transmitted through the arterial tunics which makes them expand (*Art. Nat. Sang. Cont.* pp. 174–176 Furley-Wilkie; IV. 730–731 K. Cf. *Us. Puls.* p. 214 Furley-Wilkie; V. 167–169 K; *PHP* pp. 404,38–406,4 De Lacy; V. 560–561 K). This arterial motion is the cause of transpiration which helps generate psychic pneuma (cf. *PHP* p. 528,30–33 De Lacy; V. 709 K).

[90] IX.11; p. 16 Duckworth.

[91] ὥσπερ γὰρ αὐτὸ τὸ σῶμα τῆς καρδίας ἐξ ἑαυτοῦ διαστελλόμενόν τε καὶ συστελλόμενον ἐν μέρει τὰς ὕλας τε ἕλκει καὶ αὖθις ἐκπέμπει, κατὰ τὸν αὐτὸν τρόπον ὁ ἐγκέφαλος ἐπειδὰν προέληται τοῦ περιεχομένου κατὰ τὰς ἑαυτοῦ κοιλίας πνεύματος, ὃ δὴ καὶ ψυχικὸν ὀνομάζομεν, ἐπιπέμψαι τινὶ μορίῳ, τὴν εἰς τοῦτ' ἐπιτήδειον κίνησιν κινηθεὶς οὕτως ἐπιπέμπει. p. 230,14–19 De Lacy; V. 356 K. The strength of this analogy is weakened when it is recalled that, for Galen, the power of the heart is derived from the brain via the nerves. On the movement of the brain and its role in the transport of psychic pneuma see Manzoni, 2001, pp. 51–61.

[92] αἱ μὲν δὴ πρόσθιοι δύο [sc. τῶν κοιλιῶν] τήν τ' εἰσπνοὴν καὶ τὴν ἐκπνοὴν καὶ τὴν ἐκφύσησιν ἐργάζονται τὴν ἐξ ἐγκεφάλου· I, p. 481,6–8 Helmreich; III. 663 K. In *Inst. Od.* p. 44,22–29 Kollesch; II. 867–868 K, Galen is more circumspect.

One way to resolve the question of the provenance of psychic pneuma might be to deprive the brain of its arterial blood supply by ligating the carotid arteries in their course through the neck. According to Galen, the procedure is one which he has performed many times, and may well have been part of his *oeuvre* of publicly performed and privately rehearsed anatomical demonstrations. As an experimental procedure, ligation, according to Galen, is the "best as well as the clearest method of deciding the source of the [body's] activities."[93] In *De anatomicis administrationibus*, the carotid arteries in the neck are examined together with Galen's exegesis on the vagus nerve. Their exposure is as follows:

> Commence at the trachea, and continue from above downwards for a considerable distance. When you do that, then in this region portions of 'spongy flesh' (part of the thyroid gland) reveal themselves to you, which you can easily tear out and clear away with the finger-tips without using a knife. Next you can see the artery which is known as the 'artery of stupor' [*A. carotis*], one on each side. With it you can also see the deep jugular vein, and the 'nerve of the sixth pair' [*vagus*], enclosed in a common sheath. Dissect away this sheath from them. Then introduce, between the artery and the nerve, a bluntly pointed hook with an eye,[94] of which the end is moderately sharp, so that it enters easily into the fasciae without any perforation being made in the veins or arteries. And attached to this hook there should be a thread folded into two strands—either a stout thread, such as is suitable for sewing, or a thread such as is used for ligatures. This should run through the eye of the hook in order that when you draw out the nerve with the hook, you can grasp with your fingers one of the two strands of the thread, and withdraw the hook backwards together with the other strand of the thread. If you do that well, then the thread falls beneath the nerve, and when you draw both its ends outwards, the nerve will be drawn up with it, so that it can be ligatured by itself alone, without the artery.[95]

He does say that it is the action of the brain which, through the nerves connecting it to the thorax, enables it to draw something into the ventricles. Galen qualifies this to the extent that it is no more than a plausible inference, this activity of the brain being regarded as "not impossible". Cf. Debru, 1996b, 158–63.

[93] αὕτη μὲν οὖν ἀρίστη τε καὶ σαφεστάτη κρίσις ἀρχῆς ἐνεργειῶν. *Foet. Form.* IV. 679–680 K.

[94] This instrument resembles a broad needle for leather-work. Cf. Bliquez, 1994, Plate IV, Fig. 8.

[95] XI.11; p. 105 Duckworth. My gloss in round brackets; square brackets Duckworth.

Later in the same text, Galen seeks to refute claims made by others that damage to the carotid arteries alone results in a loss of voice:

> So we also find that the earlier writers had no sort of correct[96] acquaintance with the nerves lying at the sides of the arteries which they named the 'arteries of stupor', 'stupefying arteries'. These names they applied because, according to their view, it is these arteries that bring about stupor, if an injury befalls them in the way which we have described, although this is a matter which requires intensive research and investigation. For it was certainly possible that they might not know of the nerve at the side of the artery.[97] But they should not have been ignorant of the fact that, when an injury befalls the arteries and not the nerves, that causes the animal no harm which might lead to loss of sensation or mobility. For we are accustomed to see clearly that in such a condition the animal nevertheless moves and breathes without difficulty or hindrance, and nothing in it is affected except the voice alone. They may perhaps, then, have called the loss and the cutting off of the voice stupor, I mean those who endowed the artery with this name . . .[98]

Galen seeks to exclude the carotid arteries from any direct role in nervous action. He notes that following "injury" to the carotid arteries he is able to "see clearly" that movement and respiration are unaffected. It is not stated whether this "injury" includes deliberate ligation of the carotids, although it probably does, given the information quoted earlier from the same text. For Galen, all arteries have their origin in the heart and if the interruption of the arterial supply from the heart to the brain has no effect on sensation and motion, then the heart cannot be the source of psychic power.[99]

---

96 Cf. Garofalo, 1991, III, 1071 n. 132.

97 This relationship is brought out by Galen at the beginning of this chapter in *AA*, when the neck is dissected and instructions given to isolate the vagus nerve and carotid artery: "Close beside each of the two vagi there is an artery, and they are held together by a fascial sheath, common to each artery and its nerve . . . As for what concerns the two vagi, the hook normally grips the carotid artery with each nerve unless you anticipate this by dissecting and splitting the fascial sheath common to both the nerve and the artery and enveloping them." XIV.7; pp. 208–209 Duckworth.

98 XIV.7; pp. 211–212 Duckworth.

99 τὰ συνάπτοντα τὴν καρδίαν ἐγκεφάλῳ διασκέψασθαι χρὴ κατὰ τὰς τῶν ζῴων ἀνατομὰς ὁπόσα τ' ἐστὶ καὶ ὁποῖα, κἄπειτα κατὰ τὸν τράχηλον ἕκαστον αὐτῶν ἢ τέμνειν ἢ θλᾶν ἢ βρόχοις διαλαμβάνειν, εἶτ' ἐπισκέπτεσθαι τίνα καταλαμβάνει τὸ ζῷον παθήματα. συνάπτει δὲ καρδίαν ἐγκεφάλῳ τὰ τρία γένη τῶν ἀγγείων, ἅπερ δὴ καὶ παντὸς τοῦ σώματος ὑπάρχει κοινά, φλέβες καὶ ἀρτηρίαι καὶ νεῦρα, φλέβες

A nineteenth century account allows an opportunity of judging Galen's work on carotid ligation. Furley and Wilkie have drawn attention to a series of arterial ligation experiments recorded by the renowned English surgeon Astley Cooper (1768–1841). Cooper's intention was to explore the nature of potential vascular anastomoses between the carotid and vertebral arteries.[100] Here is his report on his ligation of both carotid arteries in the dog:

> In the first place, I applied a ligature to the carotid artery on each side of the neck. Little effect was produced; except, that the respiration was quickened for a few minutes, and the animal rendered dull and disinclined to eat during the day: but on the following morning it appeared lively, and ran about with its natural activity. So that it may be truly said, that these two arteries may be tied with very little change in the functions of the animal, excepting that the respiration is quickened; and this perhaps may be attributed to a greater quantity of blood being impelled through the vertebral arteries, in consequence of its interruption in the carotids.[101]

μὲν αἱ σφαγίτιδες ὀνομαζόμεναι, ἀρτηρίαι δὲ αἱ καρωτίδες, νεῦρα μὴν τὰ ταύταις ταῖς ἀρτηρίαις παραπεφυκότα. τὰς μὲν δὴ σφαγίτιδας φλέβας ἢ τὰς καρωτίδας ἀρτηρίας οὐχ ἁπλῶς χρὴ τέμνειν, ὥσπερ τὰ νεῦρα, τεθνήξεται γὰρ εὐθέως τὸ ζῷον αἱμορραγίᾳ λάβρῳ συσχεθέν, ἀλλ' ἄμεινον βρόχοις ἰσχυροῖς διαλαβόντα πρῶτον ἔν τε τοῖς ἄνω καὶ κάτω μέρεσι τοῦ τραχήλου τὸ μέσον τῶν βρόχων διατέμνειν, ὡς μηδεμίαν αἱμορραγίαν ἀκολουθῆσαι· . . . τῶν ἀρτηριῶν δὲ γε βρόχοις διαληφθεισῶν ἢ ὡς εἴρηται τμηθεισῶν ἄφωνον μὲν ἢ καρῶδες, ὡς οἱ πλεῖστοι τῶν μεθ' Ἱπποκράτην κακῶς ἀνατεμόντων ἔγραψαν, οὐκ ἔσται τὸ ζῷον, *PHP* p. 148,14–33 De Lacy; V. 263–264 K. A cardiocentrist could say this proves nothing: the nerves about the heart could provide the necessary psychic power to the rest of the body (including the brain).

[100] For a discussion of these experiments, which formed part of Cooper's own research programme on aneurysmal surgery, see Brock, 1952, 49–50.

[101] 1836, 463. Wilkie (in Furley and Wilkie, 1984, 48–50) does not discuss this experiment but another which Cooper performed where both carotid arteries as well as the vertebral arteries were ligated in the dog. In this experiment, performed on 28 January, 1831, the animal was at first rendered "insensible". Then, as Cooper, 1836, 458–459, records, "After an hour and a half, however, it was able to stand, and, although with difficulty, to stagger around a small room." Only from 1 February did it begin to fully recover. Wilkie, 1984, 50, comments on the significance of this experiment for interpreting Galen's results: "in Galen's dog the circulation of the brain could well have been maintained by such anastomoses, together with the vertebral arteries. The tying of the arteries would have diminished the circulation and this diminution could well have been exposed by forcing the animal to exert itself, exactly in accordance with Galen's description." But these anastomoses take several months to develop, as Cooper attests (cf. 1836, 459–460 and Plate 1, Fig.1, which is a beautiful hand-coloured drawing of the casts of the anastomoses). Debru, 1996b, 149 n. 57, follows Wilkie. Galen's experiments could not have benefited from these anastomoses not just because his animals are killed within a day (from *Usu Resp.* it

Although Galen does not mention which animal he used to perform his experiments, Cooper's account allows us to infer that Galen employed a dog for his ligation experiments.[102] Cooper records no surprise from his results; Galen, in *De usu respirationis*, professes himself amazed by what he has seen:

> . . . this phenomenon should be examined, of which we have often made experiment. For when the arteries throughout the neck are secured with ligatures[103] the animal suffers nothing, either immediately or later, as we have several times separated and ligated them in our experiments. Having seen the animal respiring and breathing out and moving unhindered the whole day besides into the depth of the night, we killed it, no longer supposing that the experiment, nearly completed, to be trustworthy: for it was possible that the principal organs would be sympathetically affected from the ligatures being maintained for so long a time.[104] So we were amazed to discover that the arteries that extend from the most important of the vital organs, the heart, to the most important of the psychic, the encephalon, bearing in this manner the damage without harm to life as a whole. But this however will be investigated more fully in *On the use of the pulse*: but from it one thing may be repeated as being useful, when, for the sake of our argument, I have called to mind once more the carotid arteries; that the brain does not at all need the pneuma from the heart. At any rate it is admitted that either the vapour which arises from the blood to the brain is sufficient or that which is breathed in through the nostrils. But it is likely the vapour becomes not even abundant when the arteries are grasped and ligatured: and this also was shown

---

can be roughly estimated that they were killed between 12 to 24 hours after ligation), but because Galen did not perform this particular experiment: the carotids are the only arteries Galen ligates.

[102] Wilkie, 1984, 50, states that in *De usu pulsuum*, Galen's experimental animal is a dog. Debru, 1996b, 149 n. 57, points out that: "Galien ne précise pas de quel animal il s'agit", and goes on to state that a dog can be inferred, based on Cooper's 1836 experiments on carotid ligation (cf. idem, 1994, 1744). This does not exclude the possibility of Galen using other animals. However, in the sheep and goat, virtually no blood reaches the brain via the vertebral arteries, and therefore these animals could not have provided Galen with the results he has claimed to have obtained. Cf. Andersson and Jewell, 1956, 462–474; King, 1987, 2–6.

[103] These were probably made of *silk* (σηρικόν), a material, says Galen, available to wealthy women in cities under Roman rule (*Meth. Med.* X. 942 K). Galen also used ligatures made of dried gut. It is least subject to putrefaction, according to Galen, but tends to come away easily. Cf. Majno, 1975, 403–404.

[104] That is, if the experiment had run its course, the animal, according to Galen, would have died. But, as Cooper, 1836, 463, showed, the animal, provided it survives the procedure and any complications, can survive indefinitely the ligation of the carotids alone.

> in *On the use of the pulse.* So it necessarily follows that for the most part the nourishment for psychic pneuma is from breathing in through the nostrils.[105]

Contrary to Galen's expectation (at least, that is what he tells us), the animals whose carotid arteries were ligated did not perish but suffered minimal locomotor disturbance. Why does he profess to find this result unexpected? The demands of the argument in *De usu respirationis,* where Galen seeks to account for *all* sources of nourishment of psychic pneuma, mean that he must allow psychic pneuma to be nourished via external air through the nostrils, seemingly at the expense of the contribution from the retiform plexus.[106] But Galen qualifies this as a reasonable supposition only: it is only *likely* (εἰκός) that the supply of pneuma from the retiform plexus is decreased when the carotids are ligated. Galen further qualifies the amount of psychic pneuma involved: the supply from the retiform plexus is "not even abundant" (οὐδὲ δαψιλής). Granted that this external source of nourishment provides Galen with an explanation for his experimental result, it can do so only by reducing the role of the retiform plexus. In an attempt to redress this imbalance, Galen, in *De usu pulsuum,* turns to a different interpretation of the carotid ligation experiment, which will now be examined.

In the first chapter of *De usu pulsuum,* as in *De usu respirationis,* Galen finds it "highly unusual" (ἀτοπώτατον) that ligation of the

---

[105] . . . ἐπισκεψώμεθα τουτὶ τὸ φαινόμενον, οὗ πολλάκις ἐπειράθημεν. ἐν γὰρ τῷ βρόχοις διαλαμβάνεσθαι τὰς κατὰ τὸν τράχηλον ἀρτηρίας οὐδὲν πάσχει τὸ ζῷον οὔτ' εὐθὺς οὔθ' ὕστερον, ὡς ἡμεῖς ἐνίοτε πειρώμενοι ἐν βρόχοις αὐτὰς διελαμβάνομεν. ὅλην δὲ ἡμέραν τὸ ζῷον εἰσπνέον τε καὶ ἐκπνέον καὶ κινούμενον ἀκωλύτως ἰδόντες νυκτὸς ἤδη βαθείας ἐσφάξαμεν, οὐκέθ' ἡγούμενοι τὴν ἐπὶ πλέον πεῖραν πιστὴν ὑπάρχειν· δύνασθαι γὰρ ἐν τοσούτῳ χρόνῳ διὰ τοὺς περικειμένους βρόχους συμπαθῆσαί τι τῶν κυριωτάτων μορίων. ἐθαυμάζομεν οὖν τὰς ἐκ τοῦ κυριωτάτου τῶν ζωτικῶν ὀργάνων, [τῶν] τῆς καρδίας, εἰς τὸ κυριώτατον τῶν ψυχικῶν, τὸν ἐγκέφαλον, τεταμένας ἀρτηρίας οὕτως ἄλυπον τῷ παντὶ βίῳ τὴν βλάβην ἐχούσας εὑρίσκοντες. ἀλλὰ περὶ μὲν τούτων κἀν τῷ Περὶ χρείας σφυγμῶν ἐπὶ πλέον ζητήσομεν· ὅπερ δ' ἐξ αὐτῶν ἐστιν εἰς τὰ παρόντα χρήσιμον, οὗ χάριν ἐμνημόνευσα τῶν καρωτίδων ἀρτηριῶν, τοῦτ' αὖθις λεγέσθω, τὸ μὴ πάνυ τι δεῖσθαι τοῦ παρὰ τῆς καρδίας πνεύματος τὸν ἐγκέφαλον. ἀπολείπεται γοῦν ἤτοι τὴν ἀναθυμίασιν αὐτῷ τὴν ἐκ τοῦ αἵματος ἱκανὴν ὑπάρχειν <ἢ τὴν> διὰ τῶν ῥινῶν εἰσπνοήν. ἀλλ' οὐδὲ τὴν ἀναθυμίασιν εἰκὸς γίνεσθαι δαψιλῆ βρόχῳ διαληφθεισῶν τῶν ἀρτηριῶν· δείκνυται <δὲ καὶ> τοῦτ' ἐν τῷ Περὶ χρείας σφυγμῶν. ἀναγκαῖον <οὖν> ἐκ τῆς διὰ τῶν ῥινῶν εἰσπνοῆς τὴν πλείστην εἶναι τροφὴν τῷ ψυχικῷ πνεύματι. pp. 122–124 Furley-Wilkie; IV. 502–504.

[106] Breathing nourishes the substance of the psychic pneuma. *Caus. Resp.* p. 240 Furley-Wilkie; IV. 466 K. Cf. Debru, 1996b, 150.

carotids produces "no evident harm" to the animal, in spite of the arterial connection of the heart to the brain.[107] A little later in the same text, Galen recounts these experiments:

> Since it has been demonstrated by us elsewhere that the ventricles in the brain are full of psychic pneuma, requiring continuous nourishment, and that it has also been demonstrated that it is thus consumed during voluntary motion, we resolved to ligate the arteries of an animal, and then compelled it to run. And it did so for a good while, but could not keep this up for ever, so we decided to inquire why it ran for a long while, since it seemed that it ought not to endure even for a while, but should be exhausted forthwith, since the psychic pneuma was expended.[108]

If this account is interpreted solely with reference to *De usu respirationis*, then the answer to Galen's inquiry lies in pneuma being provided to the ventricles via the external air. However, as noted above, that solution carries with it the difficulty of appearing to deny the physiological value of the retiform plexus as an elaborative pneumatic agent. Therefore, Galen's answer in *De usu pulsuum* must involve the retiform plexus in some way. Yet this will create a further tension in Galen's schemata of pneumatic elaboration by downplaying or even gainsaying the alleged second source of psychic pneuma, the ventricles of the brain. It may be that, in order to avoid this circular argument, Galen introduces the concept of voluntary motion into the account of carotid ligation in *De usu pulsuum*. This is employed in the following way. The twin hallmarks of the rational soul for Galen are that it is responsible for sensation and voluntary motion.[109]

---

[107] τὸ δὲ δὴ πάντων ἀτοπώτατον, εἰ τὰς κατὰ τὸν τράχηλον ἀρτηρίας βρόχοις διαλάβοις, οὐδὲν σαφὲς βλάψεις τὸ ζῷον, καίτοι συνάπτουσιν αὗται καρδίαν ἐγκεφάλῳ, τὸ κυριώτατον τῶν ζωτικῶν ὄργανον τῷ κυριωτάτῳ τῶν ψυχικῶν. p. 194 Furley-Wilkie; V. 150 K.

[108] ἐπεὶ γὰρ ἐν ἑτέροις ἡμῖν ἀποδέδεικται, τὰς κατὰ τὸν ἐγκέφαλον κοιλίας πνεύματος εἶναι ψυχικοῦ μεστάς, τροφῆς χρῄζοντος συνεχοῦς, ἀποδέδεικται δὲ καὶ ὡς δαπανᾶται τοῦτο ἐν ταῖς καθ' ὁρμὴν κινήσεσιν, ἄμεινον ἐδόκει τρέχειν ἀναγκάζειν τὸ ζῷον, ᾧ τὰς ἀρτηρίας βρόχοις διελάβομεν. ἐπεὶ δὲ μέχρι μὲν πολλοῦ καλῶς ἔτρεχε, μέχρι παντὸς δ' οὐκ ηδύνατο, ζητεῖν ἐδόκει τὴν αἰτίαν δι' ἥνπερ μέχρι πολλοῦ ἔτρεχε, χρῆναι γὰρ οὐδὲ μέχρι πολλοῦ διαρκεῖν, ἀλλ' εὐθέως ἐκλύεσθαι, δαπανωμένου τοῦ ψυχικοῦ πνεύματος. *Us. Puls.* pp. 198–200 Furley-Wilkie; V. 154–155 K.

[109] Involuntary motions, being natural like those of the arteries and the heart, do not require significant neural involvement (cf. *Mot. Musc.* IV. 442 K). The heart is an organ of nature by Galen's definition of involuntary motion (IV. 455 K). It is not a muscle (IV. 377 K). To move is to initiate muscular action, an activity of the rational soul and therefore a voluntary action. Although in *Plen.* VII. 534 K,

These are mediated in some way by the actions of psychic pneuma in the nerves. The experimental animal in *De usu pulsuum* is made to run in order to expend its supply of psychic pneuma which is presumably sent to the muscles via the nerves. Eventually, the animal stops, and Galen concludes—in the sentence immediately following the above citation—that the cause of the animal running at all is pneuma in the retiform plexus.[110] To accommodate the second source of psychic pneuma, Galen qualifies the extent of the role of the retiform plexus by stating that in this situation—where the brain is deprived of its connection with the heart—the retiform plexus supplies the brain for a "long while", especially if the animal is not expending extra psychic pneuma in voluntary activity.[111] The retiform plexus is able to supply the experimental animal in a resting situation, and probably for a period after the animal commences running. But it is necessarily a finite supply: for the animal to exert itself as long as Galen says it does, then psychic pneuma must be replenished, and this source comes from the only other source available to Galen which has access to the ventricles, the outside air. Galen, it is true, does not say exactly this in so many words. But, in *De usu pulsuum*, he does say that the brain benefits from breathing through the nostrils, which provides nourishment to psychic pneuma.[112] Nourishment implies replenishment. If this is added to the role of the retiform plexus as outlined here, it is possible to accommodate both sources of supply of psychic pneuma without necessarily creating too great a tension in Galen's interpretations of his experiments of carotid ligation both in *De usu respirationis* and *De usu pulsuum*. The tension is exacerbated, however, when the role of the choroid plexus is once more considered.

Galen does say that it is the pneuma flowing through the nerves which is responsible for muscular motion, this is denied later in the same text (VII. 563 K; cf. *PHP* p. 452,8–11 De Lacy; V. 616 K). In *Mot. Musc.* IV. 371 K, this theory seems replaced by the arguably simpler one of a δύναμις flowing from the brain through the nerves and to the muscles. This power is not innate to the nerves; they are simply carriers of a δύναμις from the brain: οὕτως ἄρα μεγάλη τις δύναμις ἐν τοῖς νεύροις ἐστὶν, ἄνωθεν ἀπὸ τῆς μεγάλης ἀρχῆς ἐπιρρέουσα· οὐ γὰρ δὴ ἐξ αὐτῶν γε οὐδὲ σύμφυτον αὐτὴν ἔχει . . . λόγον οὖν ὀχετῶν ἔχοντα τὰ νεῦρα καθάπερ ἔκ τινος πηγῆς τοῦ ἐγκεφάλου τοῖς μυσὶ παράγονται τὰς δυνάμεις.

[110] Cf. Harris, 1973, 354.

[111] καὶ διὰ τοῦτο, κἂν στερηθῇ τῆς πρὸς τὴν καρδίαν συνεχείας ὁ ἐγκέφαλος, ἐξαρκεῖν αὐτῷ τὸ δικτυοειδὲς πλέγμα μέχρι πολλοῦ, καὶ μάλισθ' ὅταν ἀτρεμῇ τὸ ζῷον, ὡς ἂν μὴ δαπανωμένου τηνικαῦτα τοῦ ψυχικοῦ πνεύματος εἰς τὴν καθ' ὁρμὴν ἐνέργειαν. p. 200 Furley-Wilkie; V. 156 K.

[112] p. 198 Furley-Wilkie; V. 154 K.

6.6. *The limits of Galenic pneumatic physiology*

In *De usu pulsuum*, the physiology of the arterial vasculature is the focal point of Galen's argumentation, and the carotid ligation experiment is made to show the necessity of the retiform plexus as a reservoir for psychic pneuma. In that text, both breathing and pulsation are common to the coction of psychic pneuma, but the role of the retiform plexus is of particular importance.[113] In *De usu respirationis*, on the other hand, a second source of psychic pneuma must be fitted into the discussion of Galen's physiology of respiration. It is only when both narratives are considered together that it is possible to account for the retiform as well as the choroid plexus. The experiment of carotid ligation provides Galen with an opportunity to fully integrate the retiform and choroid plexus into his physiology of the brain. But is there any evidence that Galen elucidated the matter in this way?

Given the results of the experiment of carotid ligation, it might be expected that Galen would try to integrate it with the existence of the choroid plexus in the ventricles (especially the anterior). Galen, after all, must explain a second source of psychic pneuma in such a way so as to avoid compromising his overall physiology of the brain, which is principally dependent on vital pneuma from the heart entering the retiform plexus. How does Galen consider the role of the choroid plexus in the elaboration of a second source of psychic pneuma? In *De methodo medendi*, Galen states:

> Of the psychic pneuma, we have clearly demonstrated that for instance, the brain is its well-head, and it is watered and nourished both by inspiration and the supply of the retiform plexus.[114]

If "nourished . . . by inspiration" is an indirect reference to the role of the ventricles, then the choroid plexuses cannot be excluded from pneumatic elaboration. In the following from *De usu partium*, which recapitulates the general features of the vasculature of the brain, Galen provides a further clue concerning the choroid plexus as an organ of elaboration:

---

[113] εἶναι δὲ καὶ τὸ τῆς πέψεως τοῦ ψυχικοῦ πνεύματος κοινὸν μὲν ἀμφοῖν, ἀλλ' ἰδιαίτατον τῶν ἀρτηριῶν, . . . τοῦ δικτυοειδοῦς πλέγματος. *Us. Puls.* p. 206 Furley-Wilkie; V. 161 K.

[114] τοῦ μὲν δὴ ψυχικοῦ πνεύματος ἐναργῶς ἐδείξαμεν οἷον πηγήν τινα οὖσαν τὸν ἐγκέφαλον, ἀρδομένου καὶ τρεφομένου διά τε τῆς εἰσπνοῆς καὶ τῆς ἐκ τοῦ δικτυοειδοῦς πλέγματος χορηγίας. X. 839 K.

> ... and a large amount of psychic pneuma is contained (in the brain), where it obtains its distinct quality by elaboration ... Further, here it is observed that both the retiform plexus together with the rest of the fabric of the brain are in marvellous agreement with those correct demonstrations. For the entire brain is interwoven with these manifold divided arteries and many of these split and terminate in the ventricles, just as indeed do those veins which come from the crown of the head. For from this opposite direction they encounter the arteries, unite with them, and are distributed to all the parts (of the brain), including the ventricles themselves.[115]

If an earlier passage from the same text is recalled, Galen states that only within the ventricles of the brain does the final elaboration of psychic pneuma take place.[116] Yet the Galen of *De placitis Hippocratis et Platonis*, discussed in section 6.4, claims only that it is *reasonable* that this pneuma is produced within the ventricles of the brain, and that for this reason "no small amount" of arteries and veins end there.[117] If the role of the retiform plexus as an organ of elaboration is granted, Galen is still left to address the question of how external air obtains the necessary elaboration from the choroid plexus. Yet a depiction of a seamless set of Galenic elaborative organs in the brain working in harmony is misleading, even if it were considered desirable. Each set of elaborations must be taken in context. In *De usu pulsuum*, psychic pneuma is cited as if it is entirely produced within the retiform plexus, and in a completed state of elaboration since it is expended in voluntary motion. In *De usu respirationis*, on the other hand, Galen's attention is directed to formulating a role for the outside air entering the brain. As far as interpreting these accounts within the context of carotid ligation, Galen requires a second source of psychic pneuma, and concluded that external air, taken into the ventricles via the olfactory tracts, must be its source. But is such a direct elaboration of psychic pneuma by the choroid

---

[115] καὶ πνεῦμα ψυχικὸν ἐν αὐτῷ περιέχεσθαι πάμπολυ, τὴν ἰδιότητα τῆς ποιότητος ἐκ τῆς ἐν ἑαυτῷ κατεργασίας κτώμενον, ... ἐνταυθοῖ δὲ τά τ' ἄλλα τῆς κατασκευῆς αὐτοῦ καὶ τὸ δικτυοειδὲς πλέγμα θαυμαστῶς ὁμολογεῖν φαίνεται τοῖς ὀρθῶς ἀποδεδειγμένοις. ὅ τε γὰρ ὅλος ἐγκέφαλος ὑπὸ τούτων τῶν ἀρτηριῶν διαπλέκεται πολυειδῶς σχισθεισῶν καὶ πολλαὶ τῶν ἀποσχίδων εἰς τὰς κοιλίας αὐτοῦ τελευτῶσιν, ὥσπερ οὖν καὶ τῶν ἐκ τῆς κορυφῆς κατιουσῶν φλεβῶν. ἐξ ἐναντίων μὲν γὰρ τόπων ἐμβάλλουσι ταῖς ἀρτηρίαις, εἰς ἅπαντα δ' ὡσαύτως αὐτοῦ τὰ μόρια διανέμονται, τά τ' ἄλλα και [κατ'] αὐτὰς τὰς κοιλίας. II, p. 13,8–20 Helmreich; III. 700–701 K.

[116] ἔπειτα τὴν τελεωτάτην ἐν ταῖς τοῦ ἐγκεφάλου κοιλίαις, ἔνθα δὴ καὶ ψυχικὸν ἀκριβῶς γίγνεται [πρότερον]. I, p. 394,4–6 Helmreich; III. 541–542 K.

[117] P. 444,13–14 De Lacy; V. 606–607 K.

plexus possible? Although Galen states that the anterior ventricles contain a "vaporous pneuma" (ἀτμώδης πνεῦμα), and not simply external *air* (ἀέρος),[118] he does not offer any information on how this one-stage differentiation from external air to pneuma is effected. But since, according to Galenic physiology, the lungs elaborate outside air into the "pneuma-like" (πνευματώδης) substrate—and he is silent on the mechanism of this process, too—which is then altered to vital pneuma by the heart, then it might well be that the anterior ventricles, like the lungs, prepare outside air into a substrate (*qua* ἀτμώδης πνεῦμα) suitable for elaboration by the choroid plexuses.[119]

6.7 *Conclusion*

Galen's physiology of the brain is dependent on an anatomically verifiable vascular system of which the retiform plexus is his best argued and definitive example. In the case of the retiform plexus, and to a lesser extent the choroid, Galen attempts to construct a set of internally consistent functions for vascular structures he has taken some care to describe. That their function is problematic when considered together is only to be expected, given the pressures placed on them by the requirements Galen demands, as well as his choice of physiological agent. The consequences of explaining the results of the experiment of carotid ligation creates an additional set of interpretations to which pneuma must be fitted with varying degrees of success.

---

[118] *UP* I, p. 469,17 Helmreich; III. 647 K.

[119] According to Eastwood, 1981, 269 n. 3, "It is unclear whether Galen understands a pneumatic substance, drawn from the outer air, to be the only thing actually reaching the ventricles of the brain, or if he conceives of the outer air as such to reach the brain . . . On balance, I am inclined to believe that Galen conceives the air (*aer*) itself to be drawn into the brain, and he may understand *pneuma* as the effective part of the air for purposes of sensation and systematic cooling." (italics in text). This assumes pneuma to be a part of the outside air, but Eastwood does not mention if the ventricles themselves *extract* pneuma from the air.

CHAPTER SEVEN

CONCLUSION

By the second century AD, there existed something akin to a broad consensus among certain physicians and philosophers that the governing agency of the body resided in one of two locations. Although there were no formal groupings of what this book has referred to as cardiocentrists and encephalocentrists, these contemporary labels are a convenient way to approach the arguments of each position. To Galen, the controlling organ of the body was the brain, and, as these chapters have sought to document, he devoted considerable time, effort and ingenuity to try and prove his thesis that the brain was responsible for the two defining qualities of the *hegemonikon* of the rational soul, sensation and voluntary motion. Others held that these functions should be invested in the heart, and their arguments possessed a valid currency. When, in *De locis affectis* VIII. 159 K, Galen sardonically comments that the proofs of the location of the *hegemonikon* are self-evident to all except the best physicians and philosophers (who place it in the heart), he is also acknowledging the strength of the debate regarding its supposed nature and location. That Galen devotes much of his time to attacking the cardiocentric arguments of Chrysippus, for example, illustrates Galen awareness that such a position is important and probably influential, and must be addressed and defeated. At the very least, Galen's assault on the Chrysippean viewpoint shows, as chapters 1 and 2 have discussed, that he is aware of a potential weakness in his encephalocentric argumentation in regard to the origin of the nerves. Galen overcomes this deficiency to a significant extent by his relentless creation of a detailed architecture of the brain.

Galen espoused a formal scientific methodology of anatomical dissection and experimentation based on Aristotelian first principles and on the Aristotelian quest to comprehend nature. To this theoretical bedrock, Galen, after a process of selective filtration, admits the legacy of certain of his predecessors, especially Herophilus, Erasistratus, and Marinus. The immense contribution of Herophilus and Erasistratus in anatomically and experimentally revealing the brain for the first time as a structure worthy of consideration cannot be denied. If, in the account of the meningeal coverings of the brain, its divisions into cerebrum and cerebellum, and the general configuration of the ventricular system, Galen simply repeated (or even transmitted) the work of Herophilus and Erasistratus, then he would have performed a valuable service to the history of anatomical science (and at least this much might be said of Marinus of Alexandria). But Galen also constructs a dynamic extension to Alexandrian anatomical knowledge, especially in regard to the ventricles of the brain. This anatomical genealogy has one additional factor in common: all of these men practiced a form of dissection and experimentation which not only expanded the biological methodology of Aristotle, but allocated considerable research to the study of the brain. In Galen's assessment, his Alexandrian predecessors, in their examination of the brain and nerves, correctly accorded these structures an hegemonic status. This view was by no means persuasive or pervasive. Witness that in Galen's own time, others neither recognised such a status for the brain nor, as the Empiricists and Methodists espoused, saw the need for formal anatomical investigation in the everyday practice of medicine.

Galen's anatomical epistemology is impressive. As a consistent account of the structure of the brain, it was not bettered until Thomas Willis. Galen's chief epistemological problems, as far as the brain is concerned, are that the anatomy under discussion is conceptually difficult and the physiology problematic. The brain is a far more complex structure than the heart, even if both are accorded a broadly similar set of functional criteria. Galenic brain anatomy, apart from any other single factor, demanded an intense degree of participation from his pupils in ways which reflected Galen's awareness of the epistemological challenges of his anatomical interpretation of the brain. Galen repeatedly reveals what may be described as a keenness in making his pupils see for themselves the details being discussed rather than taking what he has to say at face value. On the one hand, this shows Galen as being thorough (one might say, *Galenic*), and a good teacher who makes the effort to root out and correct the errors of his predecessors and their perpetuation in his students. On the other, Galen's employment of anatomical observation and interpretation, conveyed by the term *autopsia* (αὐτοψία), seems to be necessary psychologically for the proofs required. Autopsia is also linked to a sense of *wonder* (θαῦμα). Galen's anatomical demonstrations are designed to draw attention to Nature's Grand Design, highlighting a Nature which famously "does nothing in vain", but also does it *skilfully* (τεχνικῶς). Galen's own *skill* (τέχνη) in elucidating the architecture of the brain is framed from a similar perspective and is structured in such a way as to be equally as incontrovertible and to elicit the same sense of wonder. In *De anatomicis administrationibus* II. 727–728 K, Galen states that "those who perform a dissection *badly* (μὴ καλῶς) not only make such mistakes in anatomy itself, but introduce those errors into their physiological explanations. For it is inevitable that, just as the uses of those things properly observed in anatomy provoke amazement, so too the account of the use of things mis-observed is impossible." Galen's study of the brain is designed to encourage this feeling of amazement. The interactive process between teacher, pupil and anatomical material gives *autopsia* its full meaning. It is obvious that the dissector's ability must be of a sufficiently high standard or else an *autopsia* is incomplete and mistakes will be made. This applies more to the study of the brain than for any other part of the body. The need to generate a sense of awe in the dissector lies behind Galen's comments on the shortcomings of his predecessors: they lacked that requisite sense of wonder. To Galen, these doctors miss the complete picture, not because they are poor anatomists, but because they have not completely understood Nature. This argument is encapsulated in Galen's discussion of the optic nerves and their relationship to the anterior ventricles, as detailed in chapter 4. Here, Galen links an extremely strong teleology with his epistemology. Or, to put it in perhaps a better way, Galen uses his teleological standpoint both to justify his epistemological claims and to neutralise his opponents.

Galen's efforts to construct a detailed anatomy and physiology of the ventricular system represent his specific contribution to the encephalocentric debate (the elucidation of the function of the spinal cord and nerves is another and requires a separate study). As noted above, Galen's anatomical expositions also serve to draw attention to the skill of a provident Nature and to reflect his own technical skills in the most favourable light. Nowhere is that skill better revealed than in Galen's comprehensive depiction of the ventricles of the brain. The magnitude of this is such that it is easy to run the risk of either ignoring or minimising what Galen has to say about the rest of the brain. But, as chapter 3 has shown, Galen's account of the parts of the brain which surround the ventricles forms an essential prerequisite to an understanding of their anatomy and physiology.

It has therefore been important to highlight Galen's handling of the anatomy of the brain *qua* substance, not merely because it illustrates his technical expertise as an anatomist (which in fact it does). Rather, the need to understand clearly the substance of the brain lies in the fact that without its support the ventricles cannot exist. In this respect, the substance of the brain has a crucial role to play in Galen's physiological thinking.

To an appreciable extent, it may be that, for Galen, the brain functions because it has ventricles, just as in the same way the heart operates because of its cavities (although any parallel between the two organs should not be pressed too far). But Galen's choice of the ventricles as the functional unit of the *hegemonikon* also means that their physiology must be cogently presented. Since Galen places *psychic pneuma,* his chosen effector agent of the soul, within the ventricular cavities, he can best show that the brain is the hegemonic organ of sensation and voluntary motion by experimenting on these cavities. It is essential for Galen's enterprise that the results from such experiments are rendered comprehensible; in other words, that such observations may be related to function. To this end, Galen devotes considerable effort in seeking to establish that the ventricles of the brain are in possession of a set of distinct characteristics which eminently fit that organ, and no other, for hegemonic status. It then requires a body of observational and experimental data for Galen to make his case for a ventricular-based encephalocentrism. In a series of experiments which have been examined in chapter 5, Galen noted that it is only when the ventricles are incised does the animal lose sensation and motion, and that similar symptoms are sometimes seen in trepanation. Moreover, Galen attempts to establish a localisation of effect: least harm occurs to the animal if either of the anterior ventricles is incised, the most if the posterior ventricle is opened, with sensation and motion being lost (the middle ventricle seems a half-way house of affected modalities). Galen concluded that loss of psychic pneuma was the mechanism behind these events, but that these effects were sometimes reversible provided the supply of pneuma could be renewed. Galen then, sought to address the question of how these experimental results are consistent with his own theoretical standpoint. Unanswered is a more general query: by what theory or theories are these observations best explained? In the absence of any contemporary reactions to Galen's hegemonic claims at this level (outside the wider one of why the brain and not the heart was chosen), this question would not be adequately addressed for more than fifteen centuries.

Galen's physiology of the brain relies on pneuma within the ventricles and on its elaboration by two vascular structures, the retiform and choroid plexuses. His account of the retiform plexus in the ox brain is based on a series of meticulous observations. That it is also a complex structure is no more than fitting from Galen's point of view since it deals with the beginning of the elaboration of psychic pneuma. For Galen, elaboration requires a certain degree of anatomical complexity. Galen bases his understanding of the function of the retiform plexus by an analogous comparison to a broadly similar vascular network in the testes. Galen's account of the structure of the choroid plexus is also a model of descriptive anatomy. However, its function can only be inferred after considering some disparate evidence gathered from his experiments of carotid ligation and the contexts in which they have been framed. Chapter 6 has examined how Galen's emphasis on the importance of the results of one experimental account is necessarily at the expense of the other and thereby increases the tension within Galen's pneuma-based physiology. This is inevitable because of Galen's choice of pneuma. However, in general terms, pneuma is an appropriate physiological candidate since much of Greek speculation on the cause of life centred on this spatially mobile and fine substance. For those who maintain that the heart commands the body, the physiological process involved seems straightforward: from the outside air, pneuma goes directly to the heart via the lungs. Galen has to explain not only a much greater distance (from outside air to lungs and then from heart to brain) but to justify a more complex system. The arteries are Galen's pneumatic carriers. To this he adds the physiological imperative (perhaps first adumbrated by Diocles of Carystus) that pneuma must be progressively elaborated into a qualitatively finer form. Hence the preliminary steps take place in the lungs (the "pneuma-like" substance) and the heart ("vital pneuma"), leading to a final set of elaborations in the retiform and choroid plexuses of the brain ("psychic pneuma").

For Galen, the descriptive anatomies of the retiform and choroid plexus appear to serve as a paradigm of how complex anatomical structures can be made to bear the weight of physiological speculation. But it is too much to expect that his empirical methodology would be capable of entirely resolving the tensions implicit in the elaboration of psychic pneuma. As far as his experiments on carotid ligation are concerned, Galen demands something different from each of these vascular structures. In *De usu pulsuum*, the importance of the retiform plexus is stressed in a text which, after all, deals with the importance of the arterial system as a whole. In *De usu respirationis*, the role of the entire cycle of breathing is Galen's first concern, and an established place for the communication between the outside air and the ventricular system must be found. However, the choroid plexus is not explicitly mentioned in this text, although its role in pneumatic elaboration is implied. Although Galen goes to considerable lengths to resolve the physiological status of psychic pneuma in his system, all that his carotid ligation experiments can show is an incomplete record of some of the effects attributed to his chosen physiological agent.

Galen owed his reputation and influence to three, interlinked factors. The first was his undeniable ability, the product of a remarkable and exceptionally thorough education and his own innate intelligence. The second was his choice of Rome as the city in which to develop and promote his career and which gave him a stable research platform. Third, his aggressive ability to demonstrate and promulgate his ideas both in writing as well as in private investigations and public demonstrations of his anatomical discoveries. Galen was able to manipulate each of these factors to promote a particular image of himself which can impress as well as mislead. But it is to Galen's credit that he permits a window into his uncertainties in the field of physiological experimentation of the brain. Here he makes no real (or, at least, convincing) attempt to dress speculation as dogma. Yet within the operational restraints of his era, Galen made a significant contribution to the history of the anatomy and physiology of the brain. If Galen sometimes gives a picture of himself at the centre of a medical maelstrom, and *Galenus contra mundum* doubtless formed an attractive concept for him, it is because he felt it necessary to defend the importance of anatomical epistemology and a research methodology based, however imperfectly, on dissection and vivisection. Galen's work in this field can therefore never be as disinterested as he might claim. His thirst for anatomical knowledge had its own agenda, not simply as a quest for science for its own sake, but as the means by which he could take his place as the legitimate head of a hierarchy of anatomical investigation. Apart from anything else, Galen's study of the brain is a hallmark of this strategy, a measure of his success, and an index to a number of unresolved physiological questions.

APPENDIX ONE

THE DOCTRINE OF VENTRICULAR LOCALISATION

One of the longest themes in the history of knowledge of the brain is that of ventricular localisation.[1] The ventricles, in varying descriptive and pictorial forms, developed into one of the dominant factors in theories of brain function.[2] This representation has been described as part of a two-stage process: "firstly, the localization of psychological functions within the ventricles was advocated, and secondly a method of representing this system pictorially was devised."[3] While this was not devised by Galen, his handling of ventricular function led inexorably to the later conceptualisation of ventricular localisation, the doctrine of "cell theory."[4] Todd states that a "division ... into the imaginative, the ratiocinative, and the commemorative, and their location in respectively the front ventricles, the middle ventricle, and the back ventricle of the brain, is a doctrine that Galen hints at rather than specifies."[5] For Galen the rational soul is responsible for sensation and voluntary motion,[6] and resides somewhere in the brain substance.[7] The activities of the rational soul also encompass imagination, reason and memory, but these too are not placed in any specific part of the brain.[8] Nevertheless, Galen's comprehensive

---

[1] See Clarke and O'Malley, 1968, chapter IX. See also Koelbing, 1985; Manzoni, 1998; Soury, 1899, 322–327.

[2] Even a cursory examination of the wealth of drawings from the medieval to the Renaissance shows, with varying degrees of ingenuity, various depictions of the ventricular system, with each ventricle labelled and given responsibility for a specific function. Cf. Choulant, 1945; Roberts and Tomlinson, 1992, 84–5.

[3] Clarke, 1962, 86.

[4] Cf. Clarke and Dewhurst, 1996, chapter 3; Leyacker, 1927; Putscher, 1973, 17–19, 48–55, 132–142; Sudhoff, 1913.

[5] 1984, 107.

[6] ἡ μὲν οὖν αἰσθητικὴ τῆς ψυχῆς ἐνέργεια πέντε τὰς πάσας ἔχει διαφορὰς ὁρατὰς, καὶ ὀσφρητὰς, καὶ γευστὰς, καὶ ἀκουστὰς, καὶ ἁπτάς· ἡ δὲ κιηντικὴ τὸ μὲν προσεχὲς ὄργανον ἓν ἔχει καὶ τὸν τρόπον αὐτοῦ τῆς κινήσεως ἕνα, (δέδεικται γὰρ οὕτως ἐν τοῖς περὶ μυῶν κινήσεως) ποικίλλεται δὲ ἐν τοῖς κατὰ μέρος ὀργάνοις. ὡς φαίνεσθαι πολυειδής. *Symp. Diff.* VII. 55–56 K. Cf. Soury, 1899, 279.

[7] Cf. Clarke and Jacyna, 1987, 218. And see chapter 5.6.

[8] ἡ λοιπὴ δὲ ἐνέργεια τῆς ψυχῆς ἡ κατ' αὐτὸ τὸ ἡγεμονικὸν εἴς τε τὸ φανταστικὸν καὶ διανοητικὸν καὶ μνημονευτικὸν διαιρεῖται. *Symp. Diff.* VII. 56 K. Galen, according

delineation of the ventricular system provided a suitable template from which to elaborate an increasingly formalised and specialised role for each ventricle. In this regard, one of the chief characteristics of Galen's ventricular physiology was not so much its all-encompassing scope but its plasticity,[9] affording considerable scope for speculative interpretation. Later commentators, medical, philosophical or theological, removed Galen's distinction between the physiology of the ventricles and the activities of the rational soul, laying the groundwork for the placement of imagination, reason, and memory within a separate ventricle.

Although the concept of a formal *ventricular* localisation is a significant feature of the developing dogma of *Galenism*, it is difficult to determine exactly when and by whom this localisation came about.[10] According to Aëtius of Amida (ca. 530–560 AD), who is our only source for this, the physician Posidonius of Byzantium (fl. end of the fourth century AD) apparently placed imagination in the forepart of the brain, reason in the middle cavity and memory in the hind part of the brain.[11] A more explicit localisation theory is to be found in Nemesius of Emesa (fl. 400 AD).[12] Nemesius placed the faculty of *imagination* (φανταστικόν) in the front of the brain, as did Posidonius, but Nemesius localised it to the anterior ventricles.[13] Nemesius fixed *intelligence* (διανοητικόν) within the middle ventricle.[14] *Memory* (μνημονευτικόν) was placed in the posterior ventricle.[15]

---

to *De moribus*, which survives only in the Arabic, states that: "Understanding resides only in the rational soul, and is a faculty that perceives agreement and disagreement in all things." XXXVIII, 212; p. 247 Mattock.

[9] Cf. Temkin, 1973, chapters 2 and 3.

[10] It has been suggested that Porphyry (232/3–c. 305 AD) invented the first theory of this type. Cf. Leyacker, 1927.

[11] τοῦ μὲν οὖν ἐμπροσθίου μέρους τοῦ ἐγκεφάλου βλαβέντος τὸ φανταστικὸν μόνον ἠδίκηται, τῆς δὲ μέσης κοιλίας τοῦ ἐγκεφάλου βλαβείσης παρατροπὴ γίγνεται τοῦ λογιστικοῦ, τοῦ δὲ κατὰ τὸ ἰνίον ὀπισθίου ἐγκεφάλου βλαβέντος ἀπόλλυται τὸ μνημονευτικόν, σὺν αὐτῷ δὲ ὡς ἐπίπαν καὶ τὰ ἕτερα δύο. *Iatrica* VI. 2, p. 125,16–20 Oliveri. Cf. Bloch, 1902a, 490; idem, 1902b, 492–568. See also Hunger, 1978, 287–320; Scarborough, 1984, ix–xvi; Temkin, 1962, 97–115.

[12] Cf. Todd, 1984, 107. Telfer's translation of Nemesius is unreliable from an anatomical point of view. The differentiation of faculties of the soul into each of the ventricles is also noted in the medical writer, Theophilus Protospatharius (seventh or ninth-tenth centuries AD), but there is no essential difference between his account and that of Nemesius. Cf. Soury, 1899, 324–327; *ODB*, 2067.

[13] *De natura hominis* p. 56,2 Morani. Cf. Leyacker, 1927, 254 n. 3.

[14] P. 68,11–12 Morani.

[15] P. 69,18–19 Morani. This codification is unchanged in Hunayn ibn Ishaq. Cf. Meyerhof, 1928, 17.

Galen's exposition of the ventricular system set in train the concept of "ventricular dominance" which was not fully reversed until the substance of the brain was explored in the seventeenth century by Thomas Willis, who transferred the functions of the ventricles into the brain substance.[16] Until this step was taken, the refinement of ventricular localisation remained the only path, apparently sanctioned by Galen, that could be followed concerning theories of the physiology of the brain and the *hegemonikon* of the rational soul.

---

[16] Cf. Clarke and O'Malley, 1968, 469, 472–474.

APPENDIX TWO

GALENISM AND THE STATUS OF THE RETIFORM PLEXUS

Although it held no such position for Galen, the retiform plexus has aptly been described as a "fascinating mythological organ."[1] This ascription is not because Galen was incorrect in his description of what he saw; rather, the tensions inherent in later accounts arose partly because Galen used a vascular structure found in several mammals and transferred this analogously to man.[2] That it does not exist in humans makes it all the more important to elucidate why it was held in such regard and given an important place in the arterial vasculature of man, from the first presumed citing by Herophilus (or his followers), its detailed elucidation by Galen, its passionate avowal amongst the majority of later anatomists until the first seeds of doubt sown by Berengario Da Carpi, and its unequivocal denial in the *Fabrica* of Vesalius. In its fashion, the doctrine of the retiform plexus reflects one aspect of the historical development of *Galenism*, and the authority of his *ipse dixit*. It is also because of the function Galen ascribes to the retiform plexus that its denial as an allegedly crucial part of the human cranial vasculature was made difficult. Galen is of course in some measure responsible for the almost mystical devotion many later anatomists had for this plexus by referring to it as the "most marvellous" (μέγιστον θαῦμα) of the structures in the region around the pituitary gland at the base of the brain. But even after Galen's pneumatic physiology had been abandoned, the physiological requirement for a retiform plexus remained.[3]

---

[1] De Gutiérrez-Mahoney and Schechter, 1972, 141. This is an invaluable guide to the anatomy of the retiform plexus as well as an analysis of its misapplication to any comparable structure in the human cranial vascular system. For an overview see Clarke and Dewhurst, 1996, 59–64.

[2] Cf. Dobson, 1925, 20; Meyer and Hierons, 1962, 123.

[3] Siegel, 1968, 109, maintains that, "a structure similar to the rete exists also in man." This is incorrect. On the possible function of the retiform plexus, which is by no means a settled question, see De Gutiérrez-Mahoney and Schechter, 1972, 156–157; King, 1987, 7–8; Forrester, 2002, 215–216.

Few anatomical structures in the Galenic Corpus have been the subject of such disputation as the retiform plexus.[4] Harris has stated that "this net-like complex is something of a mystery."[5] Yet Galen grounds the beginning of the elaboration of psychic pneuma in a well-defined vascular structure established by meticulous dissection, and given his unambiguous anatomical depiction of the retiform plexus, its importance in his pneumatic physiology, and Galen's own formidable authority, it is little wonder that this structure was referred to as a *rete mirabile* and was regarded as being part of the human brain. This process was part of the Western assimilation and codification of Galen's works and was also driven by the realisation—implicit or otherwise—that there were a limited number of physiological theories available.[6] Western anatomists refer to a *rete mirabile,* beginning with Bartholomaeus Anglicus (fl. c. 1220–1240).[7] Europe's first exponent of human anatomical demonstration, Mondino de' Liuzzi (c. 1270–1326), also uses the term *rete mirabile* in his *Anathomia* (1316).[8] There, it is stated that the rete is found in humans, but that it is composed of both arteries and veins.[9] On the whole, however, the retiform plexus was more usually referred to as an arterial structure alone. Yet the affirmation of the retiform plexus in human vascular

---

[4] The retiform plexus is cited only once in the Pseudo-Galenic Corpus (*Praesagitio omnino vera expertaque* XIX. 514 K), where its description is essentially similar to that given in *De usu partium.*

[5] 1973, 354.

[6] Renaissance anatomists, like their Classical predecessors, had little in the way of choice: "The Aristotelian alternative was not really viable: Galen had amply demonstrated the anatomical impossibility of the heart's being the origin of sensation and hence the producer of the spirits of sensation. There was no other established theory to which the anatomists could turn; they had either to create a new theory or to cover up the unfortunate discovery that the *rete mirabile* was not to be found in man. In this case, the underlying teaching of Galen was preserved and even the denial of a detail—the *rete mirabile*—was by no means clear-cut; there was compromise, prevarication and a general 'fudging' of the picture." Wear, 1981, 237. On the development of the later doctrine of animal spirits see Bono, 1984.

[7] "ex corde procedunt arteriae, sicut dicit Haly, ex quibus componitur rete mirabile, in quo quasi inuoluitur cerebrum, et in illo reti digeritur spiritus animalis." *De proprietatibus rerum* V, ch. 3. I am grateful to the anonymous reader who supplied this reference.

[8] On Mondino's importance to the development of Western anatomy see Cunningham, 1997, 42–56. On the rediscovery of Galenic anatomical texts from the end of the thirteenth century, together with an analysis of philological development and skills in handling the material, see Nutton, 1988b, 112–116. See also Siraisi, 1990, chapter four.

[9] Cf. Clarke and O'Malley, 1968, 24, 763.

physiology rests upon a falsification, whether accidental or contrived. The anatomist Nicolo Massa (1485–1569), could "declare that I have found and dissected this *rete* many times in the presence of others, sometimes so widespread that not even an idiot might deny it; but sometimes I found it so slight that I was unable to remove it."[10] Massa's case is instructive, nor is it isolated.[11] His statement does not mean that he was deliberately engaged in an act of fabrication. It may be that Massa was looking at some other vascular complex in the base of the brain—in all likelihood the veins of the cavernous sinus—or, less likely, was engaged in some form of comparative anatomical work, examining different animals in which the size of the retiform plexus varies.[12] These possibilities aside, the existence of the retiform plexus in humans was rarely the subject of doubt.[13]

The first seeds of uncertainty were sown by Berengario da Carpi, in his 1521 *Commentary* on the anatomy of Mondino.[14] In Berengario's text, the retiform plexus is given a section of its own (*De rhete mirabile*) which runs in part as follows:

> Reader be aware that I worked very hard to become acquainted with this *rete* and its position. I anatomized (sic) more than 100 human heads with regard only to this *rete* and I am well acquainted with the

---

[10] *Liber Introductorius Anatomiae* 1536 (quoted in Clarke and O'Malley, 1968, 764).

[11] Caspar Bauhin, 1605, 609, and therefore after Vesalius, states that he demonstrated the rete in man, but that in the calf and ox it is much larger. Cf. Wear, 1981, 235–236.

[12] Cf. De Gutiérrez-Mahoney and Schechter, 1972, 148. On the anatomy of the cavernous sinus, see Sisson and Grossman, 1975, 956; Dyce, Sack, and Wensing, 1996, 305.

[13] Leonardo Da Vinci attempted to depict the retiform plexus in the human brain, although the first representation seems to have been made by Magnus Hundt (*Antropologium* 1501). Cf. De Gutiérrez- Mahoney and Schechter, 1972, 145, 147. Considering Galenic physiology was still the prevalent medical orthodoxy, Leonardo's efforts could be seen as making a contribution to visually depicting this orthodoxy (even though his pictorial representations were never distributed in his lifetime). He depicts the base of an ox brain, denuded of its vasculature, but then draws upon it a fine tracery of vessels which he claims is the *rete mirabile* (Q V 7r; O'Malley and Saunders, §147, fig. 5). If Leonardo actually dissected an ungulate's brain for the purpose of locating the *rete*, it is not reflected in his illustration, which seems an imagined structure (cf. Woollam, 1957, 100). This is not to say that Leonardo wanted to misinterpret Galen, but his desire to give pictorial representation to Galen's descriptions seems to have resulted in a distortion that also serves as a reminder of the powerful influence of the Galenic legacy. Cf. Kemp, 1971, 116 n. 7; O'Malley and Saunders, 1952, 27.

[14] Cf. Lind, 1959; French, 1985, 19–21.

> matter. It is indeed true that at the sides of the glands which are under the lake of Mundino (sic) on both the right and left side and toward the back I often touched something in the shape of a net which was very entangled and this can be considered to be this *rete mirabile* and also under the dura mater behind the foramen luchae and partly toward the laude bone over the os basilaris, I felt something very entangled as I described before but I am not able to judge whether this is the *rete* or something entangled and of some other nature. That is, if the *rete* exists as was described by Galen, I maintain that such entanglements are the *rete mirabile*. But, since Galen, as I said before, states that the *rete mirabile* is in the substance of the dura mater I believe that it is not possible to see the whole *rete mirabile* since the dura mater is entangled with it in such a way that it is not possible to distinguish either one or the other. But since Galen says that the *rete mirabile* occupies a great part of the os basilaris under the skull I believe that Galen has more than anything else imagined that the *rete mirabile* is in the place above mentioned because even I have good eyes and hands and instruments suitable for separating the dura mater from the skull . . . and have never found such *retes* (sic) other than as I have described above . . . After considering all this, I maintain that this *rete* does not exist but I believe that the above mentioned ascending arteries which divide into very small arteries in the pia mater are sufficient for the diffusion of the vital spirit which then passes in a very thin nature through the substance of the brain . . . So I believe that Galen has imagined the *rete mirabile* and he never saw it and I believe that other men after Galen believe in the *rete mirabile* more because of the opinion of Galen than because of fact . . .[15]

According to Lind, the Circle of Willis is "Berengario's unseen *rete mirabile*."[16] This is a reasonable deduction given that Berengario only mentions a structure he has felt in a dissection, not what he has seen. Berengario's is also the first explicit mention that human anatomies were performed to solve the question of the existence of a structure regarded since Galen as part of the anatomy of the human brain. Berengario mentions both how many times he per-

---

[15] Quoted in Lind, 1959, 458–460. The *Commentary* appeared before Berengario knew of the existence of the Greek text of Galen's anatomical work. Until the publication of the Aldine edition, physicians who wished to study Galenic anatomy and physiology could only make the acquaintance of an abridged Latin version of *De usu partium*, the so-called *De juvamentis membrorum*, heavily and clumsily edited, and replete with errors (cf. French, 1979, 96–109). Berengario later compiled an edition of *De anatomicis administrationibus*, based on the Greek text and published in 1529. Cf. Nutton, 1987c, 28.

[16] 1959, 147 n. 51.

formed his dissections and the nature of his anatomical material. As far as the retiform plexus is concerned, the opinion of Berengario is that Galen was led astray (*imaginatum fuisse*). To be sure, for Berengario, the acceptance of Galenic anatomy and physiology was absolute; his solution to the non-appearance of the rete in humans was to transfer its function to the pial arteries, a stratagem he repeats in his *Isagoge*, published the following year:

> In my judgement it is in these very small branches of the arteries dispersed everywhere in the pia mater that the blood or vital spirit is rendered subtle and prepared so that in the substance of the brain and in its ventricles it may be made into animal spirit.[17]

That this approach was found useful can be seen in its adoption by other anatomists, especially after the formal disavowal of the retiform plexus in humans by Vesalius, and the rediscovery and promulgation (in varying degrees and versions) of Galen's key anatomical and physiological works from the 1520s onward.[18] The most likely vascular structure in the human brain that could assume the role of the rete was another, similar vascular structure.[19]

Even though he formally denied the existence of the retiform plexus in humans in his 1543 *De Corporis Humani Fabrica*,[20] Vesalius made a "straightforward replacement of the *rete mirabile* by the cerebral arteries."[21] In a different anatomical location, the doctrine of a *rete mirabile* was as strong as ever, reinforced by new claims to anatomical authority promulgated by Berengario and Vesalius.

---

[17] Ibid., 141.

[18] Cf. Mani, 1956, 29–52; Nutton, 1993c, 16–24; Potter, 1998, 259–261.

[19] Laurentius, 1600, 146, 555, places the attributes and name of the *rete mirabile* in the *choroid plexus*. Colombo, 1559, 191, does the same thing. Cf. Wear, 1981, 234–235.

[20] The Vesalius of the *Tabulae Anatomicae Sex* 1538, however, still held to the existence of the *rete*. But by 1540, as Demonstrator to Curtius' lectures on the *Anatomia Mundini*, Vesalius was well aware that he had previously affirmed the existence of the *rete* in man, and his demonstrations after 1543 were always given with a sheep's head as well as a human's, to demonstrate the *rete* in the former, but to show its absence in the latter. Cf. Eriksson, 1959, 25–26, 324 n. 13.

[21] Wear, 1981, 234. Cf. Clarke and O'Malley, 1968, 768–769; Cunningham, 1997, chapter 4.; De Gutiérrez- Mahoney and Schechter, 1972, 149–151; Temkin, 1973, 139–142.

GLOSSARY OF ANATOMICAL TERMS

The following cites only those terms which have appeared or are referred to in this book. It is not intended as a comprehensive list of Greek anatomical terminology.

ἀγγεῖον. A general expression for a *blood vessel*, either artery or vein. Cf. φλέψ.

ἀδήν (ἀδένος). A general term for *gland* (for example, the pituitary).

αἵματος ἀγωγή. *Aqueduct of blood* (the confluence of the straight and sagittal dural venous sinuses).

αἰσθήσεων (ἡ) ὀσφρητική. *The olfactory sense* (for Galen not a nerve but an outgrowth, ἔκφυσις (q.v.) of the brain).

αἰσθητὴ κοιλία. *Perceptible cavity*. The infundibular recess.

αἰσθητοὶ πόροι. *Perceptible pores* (usually in reference to the optic nerves).

ἀορτή. *Aorta*. Cf. ἀρτηρία μεγάλη, ἀρτηρία παχεία.

ἀπόφυσις. *Process* or *outgrowth* (usually of a nerve, but also a bony process). Cf. ἔκφυσις.

ἀρτηρία μεγάλη. *Great artery*, the *aorta*. Cf. ἀορτή, ἀρτηρία παχεία.

ἀρτηρία παχεία. *Thick artery*. Another term for *aorta*. Cf. ἀορτή, ἀρτηρία μεγάλη.

ἀρτηρίαι καρωτίδες. *Carotid arteries*. Cf. καρωτίδες.

ἀστράγαλος. *Spine* or joints of the *spinal column* (or ankle joint).

αὐχήν. *Neck* or *nape of the neck*. Cf. δειρή, τράχηλος.

βρέγμα. *Bregma* (anterior part of the skull or head).

γλουτία. The first pair of bodies making up the *corpora quadrigemina* (*tectum*). Cf. διδύμια, ὄρχεις.

γομφίος. Saw-like nature of the sutural arrangement of the cranial bones.

δειρή. Another term for the *neck* or *nape of the neck*. Cf. αὐχήν, τράχηλος.

δεξαμένη. *Cistern*. A confluence of the dural venous sinuses. Also another term for *infundibulum*, πύελος (q.v.). Cf. χοάνη.

δερματώδης μῆνιγξ. *Skin-like meninx*. The dura mater. Cf. παχεῖα μῆνιγξ, σκληρὰ μῆνιγξ.

δεσμοί. *Bands*. These consist of *thin membranes* (λεπτοὶ ὑμένες, q.v.) that project from the dura mater to the sutures of the skull.

διάτρησις. *Perforation*.

διαφράττον. *Septum lucidum* (*pellucidum*). The translucent membrane between the anterior ventricles.

διδύμια. Pair of bodies which, together with the γλουτία (q.v.) make up the *corpora quadrigemina*. Also a term for the *testes*, ὄρχεις (q.v.).

διέξοδος. *Outlet* or *passage*.

δικτυοειδὲς πλέγμα. *Retiform plexus* (*rete mirabile*). Cf. ποικίλος λαβύρινθυς.

διπλῆ μήνιγξ. *Double fold of meninx* (the double fold (δίπλωσις, q.v.) of dura mater which forms the venous sinuses).

διπλόη. The space between the inner and outer table of the skull.

δίπλωσις. The *double fold* of the dura mater which forms the *tentorium cerebelli*. Cf. διπλῆ μήνιγξ.

ἐγκέφαλος. The *contents of the head*. In general, the *brain*.

ἐγκράνιον. *Cerebellum*. Cf. ἐπεγκρανίς, παρεγκεφαλίς.

ἔκφυσις. An *outgrowth*. Usually of a nerve. Cf. ἀπόφυσις.

ἔμπροσθεν κοιλιῶν. A term for both *anterior and middle ventricles*.

ἐνδιπλομένη ἡ σκληρὰ μῆνιγξ. *Tentorium cerebelli*. (double fold of dura mater separating the cerebral hemispheres from the cerebellum).

ἐντεριώνη. *Core* of a nerve.

ἐπεγκρανίς. *Cerebellum*. Cf. ἐγκράνιον, παρεγκεφαλίς.

ἐπίφυσις. *Epiphysis* (outgrowth between two bones formed for the sake of articulation).

ἠθμοειδὲς ὀστοῦν. *Ethmoid bone* (on its nature, see ἠθμός, σπογγοειδής).

ἠθμός. *Sieve* or *colander* (a description of the nature of the ethmoid bone).

θαλάμη. *Optic thalamus* (terminal part of the inferior horn of the anterior ventricle).

κάλαμος. *Calamus scriptorius* of the fourth ventricle.

καμάρα. *Vaulted chamber* (fornix).

καμάριον. *Fornix*. Diminutive of καμάρα. A more general term than ψαλιδοειδής (ψαλιδοειδές, q.v.).

καρωτίδες. *Carotid arteries*. Cf. ἀρτηρίαι καρωτίδες.

κέρεβρον (Latin origin). *Cerebrum* or *brain*, excluding the cerebellum. The *cerebral hemispheres*. Cf. πρόσθεν ἐγκέφαλον.

κεφαλή. The *head*.

κιρσοειδὴς (κισσοειδὴς) παραστάτης. *Seminal duct*.

κιρσοειδὲς (κισσοειδής) πλέγμα (ἕλιξ). *Varicose plexus* (testicular vessels).

κοιλία. Any *cavity* in the body.

κοιλίαι τοῦ ἐγκεφάλου. *Ventricles of the brain*.

κοιλία τῆς παρεγκεφαλίδος. *Ventricle of the cerebellum* (fourth ventricle). Cf. μικρὰ κοιλία, ὀπίσω (ὄπισθεν) κοιλία, τετάρτη κοιλία.

κόρη. *Pupil* (of the eye).

κορωνὰ τῆς κεφαλῆς. *Condyles of the head* (occipital bone).
κρανίον. *Cranium* or *skull.*
κωνάριον. *Pineal* (cf. σῶμα κωνοειδές).
κωνοειδὲς σῶμα. *Pineal body.* Another term for κωνάριον (q.v.).
λαμβδοειδὴς (λαβδοειδὴς) ῥαφή. *Lambdoid suture* of the skull.
λεπιδοειδὴς ῥαφή. *Squamiform suture* of the skull.
λεπιδοειδῆ προσκολλήματα (ῥαφαί). *Squamous agglutinations* (the squamosal and sphenofrontal sutures).
λεπτὴ μῆνιγξ. *Thin meninx* (pia mater). Cf. μαλακὴ μῆνιγξ, ὑμενώδης μῆνιγξ.
λεπτοὶ ὑμένες. *Thin (fine) membranes* (in any part of the body). Cf. δεσμοί.
λεπτὸς ὑμήν. *Thin (fine) membrane* which binds the veins on the outside of the brain (part of the pia mater).
ληνός. *Treading (pressing) floor (calcatorium).* The *torcular Herophili* (a confluence of veins of the cerebral venous sinus).
λιθοειδῆ τῶν ὀστῶν. The *petrous [parts]* of the *[temporal] bones.*
μαλακὴ μῆνιγξ. *Soft meninx* (pia mater). Cf. λεπτὴ μῆνιγξ, ὑμενώδης μῆνιγξ.
μεγάλη φλέψ. *Great vein* (the cerebral vein of Galen).
μέση (τρίτη) κοιλία. *Middle* or *third ventricle* (of the brain).
μέση χώρα. *Middle space* (another term for the third ventricle).
μῆκος εὐθεῖα ῥαφή. *Sagittal suture* (of the skull).
μῆνιγξ. *Meninx* or *membrane* (covering layers of the brain; covering layer of any organ or structure).
μικρὰ κοιλία. *Small ventricle* (ventricle of the cerebellum or fourth ventricle). Cf. κοιλία τῆς παρεγκεφαλίδος, ὄπισθεν κοιλία, τετάρτη κοιλία.
νεῦρον. *Nerve* (also *tendon* or *ligament*).
νευρῶδες. *Nerve-like* (also a description of very fine arteries, attributed to Praxagoras.
νευρώδη σώματα. *Nerve-like bodies* (the *cordae tendineae* and *trabeculae carnae* of the heart).
νωτιαῖος μυελός. *Spinal marrow* (spinal cord).
ὁδός. Portmanteau term for *passage, channel,* or *vessel.*
ὀπή. *Foramen* or *communication.*
ὀπίσω (ὄπισθεν) κοιλία. *Fourth* (posterior) *ventricle.* Cf. μικρὰ κοιλία, κοιλία τῆς παρεγκεφαλίδος, τετάρτη κοιλία.
ὄρχεις. *Testes* (also the second pair of bodies making up the *corpora quadrigemina.* Cf. διδύμια.
ὀστᾶ τοῦ βρέγματος. *Bones of the bregma* (the parietal bones).

ὀφθαλμοῖς τὰ νεῦρα. *The optic nerves.*

παρεγκεφαλίς. *Cerebellum.* Cf. ἐπεγκρανίς, ἐγκράνιον.

παχεῖα μῆνιγξ. *Thick meninx.* The *dura mater.* Cf. δερματώδης μῆνιγξ, σκληρὰ μῆνιγξ.

περικράνιον. A structure, composed of *bands* (δεσμοί, q.v.) of *thin membranes,* λεπτοὶ ὑμένες (q.v.), which attaches the *dura mater* to the cranium.

ποικίλος λαβύρινθυς. The *complex labyrinth* (retiform web). Cf. δικτυοειδὲς πλέγμα.

πόρος. Any *passage* or *channel* in the body. Cf. τρῆμα.

πόρος αἰσθητικός. *Perceptible channel.* Alleged channel in the optic nerve.

πρόσθεν ἐγκέφαλον. *Anterior brain* (cerebral hemispheres). Cf. κέρεβρον.

προσθίαι κοιλίας. *Anterior ventricles* (s. πρόσθεν κοιλία).

πύελος. *Infundibulum* (or infundibular region; also a term for the orbital cavity). Cf. χοάνη, δεξαμένη.

ῥαφή. *Suture.* Cf. συνάρθρωσις.

σκληρὰ μῆνιγξ. *Hard meninx.* The *dura mater.* Cf. παχεῖα μῆνιγξ, δερματώδης μῆνιγξ.

σκωληκοειδὴς ἐπίφυσις. *Vermis* of the cerebellum.

σπογγοειδής. *Sponge-like* (a better term, according to Galen, to describe the nature of the *ethmoid bones* (ἠθμοειδῆ ὀστᾶ). Cf. ἠθμός.

στεφανιαία ῥαφή. *Coronal suture* (of the skull).

στόματα. *Openings.* For example, the alleged pores of the arteries on the surface of the skin.

συμβολή. *Junction* (in the brain, a confluence of small veins which are part of the *torcular Herophili* (q.v.)).

σύμφυσις. *Union.* For example, between cerebrum and cerebellum.

συνάρθρωσις. *Immovable articulation* (another term for suture). Cf. ῥαφή.

σύνδεσμοι. Membrane-like *bands of union* (also a description of the *brachium conjunctivum* or superior cerebellar peduncles). Cf. τένοντες.

συνθέσεις. *Articulations* between bones.

σύντρησις. *Opening, passage* or *channel.* According to Galen, a word also used by some (unnamed) anatomists to refer to the *third ventricle* (q.v.).

σῶμα κωνοειδές. The *pineal.* Cf. κωνάριον.

σῶμα τοῦ ἐγκεφάλου. *Body of the brain* (used by Galen to refer to the brain excluding the ventricles and the cranial nerves).

τένοντες. *Tendons* (*brachium conjunctivum* or superior cerebellar peduncles). Cf. σύνδεσμοι.

τετάρτη κοιλία. *Fourth ventricle* (of the brain). Cf. ὄπισθεν (ὀπίσω) κοιλία, μικρὰ κοιλία, κοιλία τῆς παρεγκεφαλίδος.

τράχηλος. *Neck.* Cf. δειρή, αὐχήν.

τρῆμα. *Foramen, passage* or *perforation*, usually associated with a nerve. Cf. πόρος.

τρήματα τοῦ μέσου διαφράγματος. *Perforations of the middle diaphragm.* Alleged perforations in the interventricular septum of the heart.

τρίτη κοιλία. *Third ventricle* (of the brain). Cf. μέση κοιλία.

τυλοειδής (τυλώδεις). *Callus-like.* The description of the *corpus callosum.*

ὑαλοειδὲς ὑγρόν. *Vitreous humour* (of the eye).

ὑμενώδεις ἀποφύσεις. *Membranous outgrowths* (for example, those connecting the brain to the ethmoid).

ὑμενώδης (ὑμενοειδής). *Membrane-like.*

ὑμενώδης (ὑμενοειδὴς) μῆνιγξ. *Membrane-like meninx* (pia mater). Cf. λεπτὴ μῆνιγξ, μαλακὴ μῆνιγξ.

ὑμήν. The general term for *membrane.*

φλέβες. *Veins* (older term for *blood vessels*).

φλέβες σφαγίτιδες. *Jugular veins.*

φλέψ. *Vein.* A blood vessel in general. Cf. ἀγγεῖον.

φλέψ κοίλη. *Hollow vein.* Vena cava (superior and inferior). Cf. φλέψ παχεία.

φλέψ παχεία. *Thick vein.* Another term for vena cava. Cf. φλέψ κοίλη.

φρένες (φρήν). Older (Homeric) term for *diaphragm.*

χιτών. *Tunic.* For example, the venous tunic or coat.

χοάνη. *Infundibulum* (q.v.). Cf. δεξαμένη, πύελος.

χοριοειδὴς μῆνιγξ. *Choroid meninx* (also a term for the *choroid plexus* of the ventricles of the brain).

χοριοειδῆ πλέγματα. *Choroid plexuses.* Cf. χοριοειδῆ συστρέμματα, χοροειδῆ σώματα.

χοριοειδῆ συστρέμματα. *Choroid-like bands.* Older (Herophilean) term for the *choroid plexus* (infrequently used by Galen).

χοριοειδῆ σώματα. *Choroid bodies* (choroid plexuses). Cf. χοριοειδῆ πλέγματα, χοριοειδῆ συστρέμματα.

χοριοειδής. *Choroid-like.*

χόριον. *Chorion.* The foetal membrane. Cf. χοριοειδὴς μῆνιγξ.

χώρα. *Cavity* (Pre-Galenic term for the orbital cavity). Cf. πύελος, δεξαμένη.

ψαλιδοειδὲς (ψαλιδοειδής) σῶμα (ψαλίς). *Fornix.* Cf. καμάριον (καμάρα).

BIBLIOGRAPHY

For editions of the source-texts, see Index of Sources and List of Editions.

Adams, F. (1849), *The Genuine Works of Hippocrates*, London.

Algra, K.A.; Van der Horst, P.W.; Runia, D.T. (eds.) (1996), *Polyhistor. Studies in the History and Historiography of Ancient Philosophy*, Leiden.

Allbutt, T.C. (1921), *Greek Medicine in Rome*, London.

Amouretti, M.-C.; Brun, J.-P. (eds.) (1993), *La production du vin et de l'huile en Méditerranée, Bulletin De Correspondance Hellénique*, Supplement XXVI.

Andersson, B.; Jewell, P.A. (1956), "The distribution of carotid and vertebral blood in the brain and spinal cord of the goat", *Journal of Experimental Physiology*, 41, 462–474.

André, J. (1987), *Être médecin à Rome*, Paris.

Annas, J.E. (1981), *An Introduction to Plato's Republic*, Oxford.

—— (1992), *Hellenistic Philosophy of Mind*, Berkeley.

Annoni, J.-M.; Barras, V. (1993), "La découpe du corps humain et ses justifications dans l'antiquité", *Canadian Bulletin of Medical History*, 10, 185–227.

Argoud, G.; Guillaumin, J.-Y. (eds.) (1996), *Sciences exactes et sciences appliquées à Alexandrie*, St Étienne.

Ariëns Kappers, C.U.; Huber, G.C.; Crosby, E.C. (1936), *The Comparative Anatomy of the Nervous System of Vertebrates, Including Man*, Two Vols., New York.

Arnott, R.; Finger, S.; Smith, C. (eds.) (2002), *Trepanation—History, Discovery, Theory*, Lisse.

Arnott, R.; Stuckey, E. (forthcoming), "Finding the Hydra: the search for malaria in the prehistoric Aegean", in Tzedakis and Martlew (eds.)

Asmis, E. (1993), "Asclepiades of Bithynia rediscovered?" *CPhil.*, 88, 2, 145–156.

Baldwin, B. (1991), "Nemesius", *ODB*, 1452–1453.

Baldwin, B.A. (1964), "The anatomy of the arterial supply to the cranial regions of the sheep and ox", *American Journal of Anatomy*, 115, 101–118.

Baldwin, B.A.; Bell, F.R. (1963), "The anatomy of the cerebral circulation of the sheep and ox. The dynamic distribution of the blood supplied by the carotid and vertebral arteries to cranial regions", *Journal of Anatomy*, 97, 203–215.

Balme, D. (1965), *Aristotle's Use of the Teleological Explanation*, London.

—— (1972), *Aristotle's De Partibus Animalium I and De Generatione Animalium I (with passages from II.1–3), Translated with Notes*, Oxford.

—— (1987), "Teleology and necessity", in Gotthelf and Lennox (eds.), 275–285.

Balsdon, J.P.V.D. (1974), *Roman Women*, London.

Barnett, C.H.; Harrison, R.J.; Tomlinson, J.D.W. (1958), "Variations in the venous system of mammals", *Biological reviews of the Cambridge Philosophical Society*, 33, 442–487.

Barnes, J. (1991), "Galen on logic and therapy", in Kudlien and Durling (eds.), 50–102.

—— (1993), "Galen and the utility of logic", in Kollesch and Nickel (eds.), 33–52.

Barnes, J.; Brunschwig, J.; Burnyeat, M.; Schofield, M. (eds.) (1982), *Science and Speculation. Studies in Hellenistic Theory and Practice*, Cambridge.

Barnes, J.; Griffin, M. (eds.) (1997), *Philosophia Togata II*, Oxford.

Barnes, T.D. (ed.) (1994), *The Sciences in Greco-Roman Society, Apeiron*, XXVII.

Bastholm, E. (1950), *The History of Muscle Physiology. From the natural philosophers to Albrecht von Haller*, Translated by W.E. Calveron, Copenhagen.

Bates, D. (ed.) (1995), *Knowledge and the Scholarly Medical Traditions*, Cambridge.
Bergsträsser, G. (1925), *Hunain ibn Ishâq, Über die syrischen und arabischen Galen-Übersetzungen*, Leipzig.
Bicknell, P. (1968), "The seat of the mind in Democritus", *Eranos*, 66, 10–23.
Bliquez, L.J. (1994), *Roman Surgical Instruments and Other Minor Objects in the National Archaeological Museum of Naples*, Mainz.
Bliquez, L.J.; Kazhdan, A. (1984), "Four testimonia to human dissection in Byzantine times", *BHM*, 58, 554–557.
Bloch, I. (1902), "Byzantininsche Medizin", in Neuburger and Pagel (eds.), 492–568.
—— (1902), "Griechische Aerzte des dritten und vierten (nachchristlichen) Jahrhunderts", in Neuburger and Pagel (eds.), 481–491.
Bono, J.J. (1984), "Medical spirits and the medieval language of life", *Traditio*, 40, 91–130.
Bossier, F. et al. (eds.) (1976), *Symbolae. Images of Man in Ancient and Medieval Thought, Studia Gerardo Verbeke*, Leuven.
Boswell, J. (1988), *The Kindness of Strangers*, London.
Boudon, V. (1994), "Les oeuvres de Galien pour les débutants ('De sectis', 'De pulsibus ad tirones', 'De ossibus ad tirones', 'Ad Glauconem de methodo medendi', et 'Ars medica'): médecine et pédagogie au IIe siècle ap. J.-C.", *ANRW* II. 37.2, 1421–1467.
—— (2002), "Galen's *On my own books.* New material from Meshed, Rida, tibb. 5223", in Nutton (ed.), 9–18.
Boudouris, K.J. (ed.), (1998), *Philosophy and Medicine. II. Studies in Greek Philosophy, 29*, Athens.
Bourgey, L.; Jouanna, J. (eds.) (1975), *La Collection hippocratique et son rôle dans l'histoire de la médecine*, Leiden.
Bowen, A.C. (ed.) (1991), *Science and Philosophy in Classical Greece*, New York—London.
Bowersock, G.W. (1969), *Greek Sophists in the Roman Empire*, Oxford.
—— (ed.) (1994), *Approaches to the Second Sophistic*, Pennsylvania.
Bowie, E.L. (1974), "Greeks and their past in the Second Sophistic", in Finley (ed.), 166–209.
—— (1982), "The importance of sophists", *YClS*, XXVII, 29–59.
Boylan, M. (1982), "The digestive and 'circulatory' systems in Aristotle's biology", *Journal of the History of Biology*, 15, 89–118.
Brain, P. (1986), *Galen on Bloodletting. A study of the origins, development and validity of his opinions, with a translation of the three works*, Cambridge.
Breasted, J.H. (1930), *The Edwin Smith Surgical Papyrus*, Two Vols., Chicago.
Briggs, E.A. (1939), *Anatomy of the Sheep's Brain*, Sydney.
Brock, A.J. (1916), *Galen. On the Natural Faculties*, Cambridge Mass.—London.
Brock, R.C. (1952), *The Life and Work of Astley Cooper*, Edinburgh—London.
Browning, H. (1953), "The confluence of the dural venous sinuses", *American Journal of Anatomy*, 93, 307–329.
Browning, R. (1985), "A further testimony to human dissection in the Byzantine world", *BHM*, 59, 518–520.
Brunschwig, J.; Lloyd, G.E.R.; Pellegrin, P. (eds.) (1997), *Le Savoir Grec*, Paris.
Brunschwig, J.; Nussbaum, M.C. (eds.) (1993), *Passions and Perceptions. Studies in Hellenistic Philosophy of Mind*, Cambridge, 150–183.
Brunt, P.A. (1994), "The bubble of the Second Sophistic", *BICS*, 25–52.
Burke, P.F. (1996), "Malaria in the Greco-Roman world. A historical and epidemiological survey", *ANRW* II. 37.3, 2252–2281.
Burkert, W. (1996), *Creation of the Sacred*, Harvard, 1996.
Burnet, J. (1955), *Greek Philosophy*, London, 1955.
Bussemaker, U.C.; Daremberg, C.; Molinier A. (1851, 1854, 1858, 1862, 1873, 1876), *Œuvres d'Oribase*, Vols. I–VI, Paris.

Bylebyl, J.J. (1977), "Nutrition, quantification and circulation", *BHM*, 51, 369–385.
Bynum, W.F.; Porter, R. (eds.) (1993), *Medicine and the Five Senses*, Cambridge.
Cambiano, G. (ed.) (1986), *Storiografia e dossografia nella filosofia antica*, Turin, 1986.
Carrick, P. (1985), *Medical Ethics in Antiquity*, Dordrecht.
Casadei, E. (1997), "La dottrina corpuscolare di Asclepiade e i suoi rapporti con la tradizione atomista", *Elenchos*, XVIII, 77–106.
Caton, R. (1914), "Notes on a group of medical and surgical instruments found near Kolophon", *JHS*, 34, 114–118.
Caujolle-Zaslawsky, F. (1984), "Dionysios (Denys) D' Égée (Aigéai)", in Goulet (ed.), 862–864.
Charlton, W. (1987), "Aristotle on the place of mind in nature", in Gotthelf and Lennox (eds.), 408–423.
Chauveau, A. (1891), *The Comparative Anatomy of the Domesticated Animals*, Second English Edition Translated and Edited by G. Fleming, London.
Cherniss, H. (1933), "Galen and Posidonius' theory of vision", *AJP*, 54, 154–161.
Choulant, L. (1945), *History and Bibliography of Anatomical Illustration*, Translated and Annotated by M. Frank, New York.
Clarke, E. (1962), "The early history of the cerebral ventricles", *Transactions and Studies of the College of Physicians of Philadelphia*, 30 (2), 85–89.
—— (1963), "Aristotelian concepts of the form and function of the brain", *BHM*, 37, 1–14.
Clarke, E.; Dewhurst, K. (1996), *An Illustrated History of Brain Function*, Second Edition, San Francisco.
Clarke, E.; Jacyna, L.S. (1987), *Nineteenth-Century Origins of Neuroscientific Concepts*, California.
Clarke, E.; O'Malley, C.D. (1968), *The Human Brain and Spinal Cord*, California.
Clarke, E.; Stannard, J. (1963), "Aristotle on the anatomy of the brain", *J.Hist.Med.*, 18, 130–148.
Cleary, J.J (ed.) (1991), *Boston Area Colloquium in Ancient Philosophy V*, Lanham.
Cleary, J.J.; Shartin, D.C. (eds.) (1992), *Boston Area Colloquium in Ancient Philosophy VI*, Lanham.
Cohen, M.R.; Drabkin, I.E. (1948), *A Sourcebook in Greek Science*, Harvard.
Cole, F.J. (1944), *A History of Comparative Anatomy. From Aristotle to the Eighteenth Century*, London.
Como, J. (1925), "Das Grab eines römischen Arztes in Bingen", *Germania*, IX, 152–162.
Conrad, L.I.; Neve, M.; Nutton, V.; Porter, R.; Wear, A. (eds.) (1995), *The Western Medical Tradition*, Cambridge.
Cooper, A. (1836), "Some experiments and observations on tying the carotid and vertebral arteries, and the pneumo-gastric, phrenic, & sympathetic nerves", *Guy's Hospital Reports*, 1, 457–475.
Cosans, C.E. (1994), "Anatomy, metaphysics, and values: The ape brain debate reconsidered", *Philosophy and Biology*, 9 (2), 129–165.
—— (1995), "The Platonic origins of anatomy", *Perspectives in Biology and Medicine*, 38 (4), 581–596.
—— (1997), "Galen's critique of Rationalist and Empiricist anatomy", *Journal of the History of Biology*, 30, 35–54.
Craik, E.M. (1998b), "The Hippocratic treatise *On anatomy*", *CQ*, 48, 135–167.
Crombie, A.C. (1994), *Styles of Scientific Thinking in the European Tradition. I*, London.
Cunliffe, R.J. (1963), *A Lexicon of the Homeric Dialect*, New Edition, Oklahoma.
Cunningham, A. (1997), *The Anatomical Renaissance. The Resurrection of the Anatomical Projects of the Ancients*, Aldershot—Brookfield.
Daniel, P.M.; Dawes, J.D.K.; Prichard, M.M.L. (1952–54), "Studies of the carotid rete and its associated arteries", *Philosophical Transactions of the Royal Society (Biological Sciences)*, 237, 173–208.

Daremberg, C. (1841), *Exposition des connaissances de Galien sur l'anatomie, la physiologie et la pathologie, du système nerveux*, Paris.
—— (1854), *Œuvres anatomiques, physiologiques et médicales de Galien*, Two Vols., Paris.
—— (1870), *Histoire des sciences médicales*, Paris (reprinted: Graz, 1974).
Davies, R.W. (1970a), Review of: Scarborough (1969), *JRS*, 60, 224–225.
—— (1970b), "Some Roman medicine", *Med.Hist.*, 14, 101–106.
—— (1970c), "The Roman military medical service", *Saalburg Jahrbuch*, 27, 84–104.
Dawson, W.R. (1932), "Studies in the Egyptian medical texts", *JEg.Arch.*, 18, 150–154.
Debru, A. (1993), "Expérience, plausibilité et certitude chez Galien", In López-Férez (ed.), 31–40.
—— (1994), "L'expérimentation chez Galien", *ANRW* II. 37.2, 1718–1756.
—— (1995), "Les démonstrations médicales à Rome au temps de Galien", in Van der Eijk, Horstmanshoff and Schrijvers (eds.), 69–81.
—— (1996a), "L'Anonyme de Bruxelles: un témoin Latin de l'hippocratisme tardif", in Wittern and Pellegrin (eds.), 311–327.
—— (1996b), *Le Corps Respirant. La pensée physiologique chez Galien*, Leiden.
—— (ed.) (1997a), *Galen on Pharmacology*, Leiden.
—— (1997b), "Galien", in Brunschwig, Lloyd and Pellegrin (eds.), 677–688.
Deichgräber, K. (1930), "Marinos", *RE* 14, col. 1796.
Devereux, D.; Pellegrin, P. (eds.) (1990), *Biologie, logique et métaphysique chez Aristote*, Paris.
Diels, H. (1879), *Doxographi Graeci*, Berlin. (reprinted: 1965).
Diller, A. (1955), "The authors named Pausanias", *TAPA*, LXXXVI, 268–279.
Dillon, J. (1977), *The Middle Platonists*, Cornell.
—— (1982), "Self-definition in later Platonism", in Meyer and Saunders (eds.), 60–75.
Dillon, J.; Long, A.A. (eds.) (1988), *The Question of "Eclecticism"*, California.
Dionisotti, A.C.; Grafton, A.; Kraye, J. (eds.) (1988), *The Uses of Greek and Latin. Historical Essays*, Warburg Institute Surveys and Texts XVI, London. 111–126.
Dobson, J.F. (1925), "Herophilus of Alexandria", *Proceedings of the Royal Society of Medicine*, XVIII, 19–32.
—— (1927), "Erasistratus", *Proceedings of the Royal Society of Medicine*, XX, 825–832.
Dockrill, D.W.; Tanner, R.G. (eds.) (1985), *The Concept of Spirit. Prudentia* Supplementary Number, Auckland.
Dodds, E.R. (1965), *Pagan and Christian in an Age of Anxiety*, Cambridge.
Donagan, A.; Perovich, A.N.; Wedin, M.V. (eds.) (1986), *Human Nature and Natural Knowledge*, Dordrecht.
Donini, P. (1988), "Tipologia degli errori e loro correzione secondo Galeno", in Manuli and Vegetti (eds.), 65–116.
—— (1992), "Galeno e la filosofia", *ANRW* II. 36.5, 3484–3505.
Drabkin, I.E. (1944), "On medical education in Greece and Rome", *BHM*, 15, 333–351.
—— (1950), *Caelius Aurelianus on Acute and Chronic Diseases*, Baltimore.
Drachmann, A.G. (1932), *Ancient Oil Mills and Presses*, Copenhagen.
—— (1973), *The Mechanical Technology of Greek and Roman Antiquity*, Oxford.
Duckworth, W.L.H. (1949), *Some Notes on Galen's Anatomy*, Cambridge.
Duminil, M.-P. (1983), *Le sang, les vaisseaux, le cœur dans la Collection hippocratique. Physiologie et anatomie*, Paris.
Durling, R.J. (1988), "The innate heat in Galen", *Medizinhistorisches Journal*, 23, 2/3, 210–212.
—— (1993), *A Dictionary of Medical Terms in Galen*, Leiden.
Dyce, K.M.; Sack, W.O.; Wensing, C.J.G. (1996), *Textbook of Veterinary Anatomy*, Second Edition, Philadelphia.
Easterling, P.E.; Knox, B.M.W. (eds.) (1989), *The Cambridge History of Classical Literature. Volume I part 4. The Hellenistic Period and the Empire*, Cambridge.

Eastwood, B.S. (1981), "Galen on the elements of olfactory sensation", *Rh.Mus.*, 124, 268–290.
Edelstein, L. (1932/1967), Die Geschichte der Sektion in der Antike", *Quellen und Studien zur Geschichte der Naturwissenschaften und der Medizin*, 3 (2), 100–156 (reprinted as: "The history of anatomy in antiquity" in Temkin and Temkin (eds.), 247–301).
—— (1935), "The development of Greek anatomy", *BHM*, 3, 235–248.
Edlow, R.B. (1977), *Galen on Language and Ambiguity. An English Translation of Galen's 'De Captionibus (On Fallacies)'*, Introduction, Text, and Commentary, Leiden.
Eichholz, D.E. (1951), "Galen and his environment", *Greece & Rome*, XX, 60–71.
Van der Eijk, P. (1995), Review of: Kollesch and Nickel (eds.) (1993) *CR*, 45 (ii), 405–407.
—— (ed.) (1999a), *Ancient Histories of Medicine. Essays in Medical Doxography and Historiography in Classical Antiquity*, Leiden.
—— (1999b), "Historical awareness, historiography and doxography in Greek and Roman medicine", in Van der Eijk (ed.), 1–31.
—— (1999c), "The Anonymus Parisinus and the doctrines of "the Ancients'", in Van der Eijk (ed.), 295–331.
—— (2001), *Diocles of Carystus*. A Collection of the Fragments with Translation and Commentary. Volume Two: Commentary, Leiden.
Van der Eijk, P.J.; Horstmanshoff, H.F.J.; Schrijvers, P.H. (eds.) (1995), *Ancient Medicine in its Socio-Cultural Context*, Two Vols., Amsterdam.
Eriksson, R. (1959), *Andreas Vesalius' First Public Anatomy at Bologna, 1540. An Eyewitness report by Baldasar Heseler together with his notes on Matthaeus Curtius' Lectures on Anatomia Mundini*, Edited, with an Introduction, Translation and Notes, Uppsala.
Erskine, A. (1990), *The Hellenistic Stoa*, London.
Everson, S. (ed.) (1990), *Epistemology. Companions to Ancient Thought: 1*, Cambridge.
—— (ed.) (1991), *Psychology. Companions to Ancient Thought: 2*, Cambridge.
—— (ed.) (1994), *Language. Companions to Ancient Thought: 3*, Cambridge.
Eyben, E. (1980/81), "Family planning in Graeco-Roman antiquity", *Ancient Society*, 11/12, 5–82.
Feindel, W. (ed.) (1965), *The Anatomy of the Brain* . . . Tercentenary Edition, Two Vols., McGill.
Fenik, B. (1968), "Typical battle scenes in the *Iliad*", *Hermes Einzelschriften*, Stuttgart, 21.
Ferngren, G.B. (1982), "A Roman declamation on vivisection", *Transactions and Studies of the College of Physicians of Philadelphia*, 5 (4), 272–290.
—— (1985), "Roman lay attitudes towards medical experimentation", *BHM*, 59, 495–505.
Field, J.V.; James, F.A.J.L. (eds.) (1993), *Renaissance and Revolution*, Cambridge.
Finlayson, J. (1893), *Herophilus and Erasistratus: A bibliographical demonstration*, Glasgow, (reprinted from: *The Glasgow Medical Journal*, 39.5, 321–352).
Finley, M.I. (ed.) (1974), *Studies in Ancient Society*, London.
—— (ed.) (1981), *The Legacy of Greece*, Oxford.
Fischer, K.-D.; Nickel, D.; Potter, P. (eds.) (1998), *Text and Tradition. Studies in Ancient Medicine and its Transmission*, Leiden.
Fitzgerald, T.C. (1961), "Anatomy of cerebral ventricles of domestic animals", *Veterinary Medicine*, 56, 38–45.
Flashar, H. (ed.) (1971), *Antike Medizin*, Darmstadt.
Fleck, L. (1935/79), *Genesis and Development of a Scientific Fact*, edited by T.J. Trenn and R.K. Merton. Translated by F. Bradley and T.J. Trenn, Chicago.
Forbes, R.J. (1955), *Studies in Ancient Technology*, III, Leiden.
—— (1964), *Studies in Ancient Technology*, IX, Leiden.
Forrester, J.M. (2002), "The marvellous network and the history of enquiry into its function", *J.Hist.Med.*, 57, 198–217.

Frandson, R.D. (1965), *Anatomy and Physiology of Farm Animals*, London.
Frankel, R. (1993), "Screw weights from Israel", in Amouretti and Brun (eds.), 107–118.
Fraser, P.M. (1972), *Ptolemaic Alexandria*. Three Vols., Oxford.
Frede, M. (1981), "On Galen's epistemology", in Nutton (ed.), 65–98 (reprinted: 1987, 279–298).
—— (1982), "The method of the so-called Methodical school of medicine", in Barnes, Brunschwig, Burnyeat, Schofield (eds.), 1–23 (reprinted: 1987, 261–278).
—— (1985), *Galen. Three Treatises on the Nature of Science*, Indianapolis.
—— (1986), "Philosophy and medicine in antiquity", in Donagan, Perovich and Wedin (eds.), 211–232 (reprinted: 1987, 225–242).
—— (1987), *Essays in Ancient Philosophy*, Minnesota.
—— (1990), "An empiricist view of knowledge: memorism", in Everson (ed.), 225–250.
Frede, M.; Striker, G. (eds.) (1996), *Rationality in Greek Thought*, Oxford.
French, R.K. (1979), "*De juvamentis membrorum* and the reception of Galenic physiological anatomy", *Isis*, 70, 96–109.
—— (1985), "Berengario da Carpi and the use of the commentary in anatomical teaching", in Wear, French and Lonie (eds.), 42–74.
—— (1994), *William Harvey's natural philosophy*, Cambridge.
French, R.K.; Greenaway, F. (1986), *Science in the Early Roman Empire: Pliny the Elder, His Sources and Influence*, London.
Freudenthal, G. (1995), *Aristotle's Theory of Material Substance*, Oxford.
Fuchs, R. (1894), "Anecdota Medica Graeca", *Rh.Mus.*, 49, 532–558.
—— (1902), "Geschichte der Heilkunde bei den Griechen: Galenos", in Neuburger and Pagel (eds.), 373–402.
Furley, D.J. (1989), *Cosmic Problems. Essays on Greek and Roman Philosophy of Nature*, Cambridge.
Galdston, I. (ed.) (1960), *Human Nutrition. Historical and Scientific*, New York.
García-Ballester, L. (1994), "Galen as a clinician: his methods in diagnosis", *ANRW* II. 37.2, 1636–1671.
Garnsey, P. (1988), *Famine and Food Supply in the Graeco-Roman World*, Cambridge.
Garnsey, P.; Saller, R. (1987), *The Roman Empire. Economy, Society and Culture*, London.
Garnsey, P.; Whittaker, C.R. (eds.) (1978), *Imperialism in the Ancient World*, 1978.
Garofalo, I. (1991), *Galeno. Procedimenti anatomici*, Three Vols., Milan.
—— (1992), "Galeno e l'anatomia di Ippocrate", in López-Férez (ed.), 609–622.
—— (1993), "The six classes of animals in Galen", in López-Férez (ed.), 73–87.
—— (2000), *Anatomicarum administrationum libri qui supersunt novem, earundem interpretatio Arabica Hunaino Isaaci filio ascripta, tomus alter libros V–IX continens*, Naples.
Giannantoni, G.; Vegetti, M. (eds.) (1984), "La scienza ellenistica", *Elenchos*, IX, Naples.
Gibbs, E.L.; Gibbs, F.A. (1934), "The cross section areas of the vessels that form the torcular and the manner in which flow is distributed to the right and left lateral sinus", *Anatomical Record*, 59, 419–425.
Gippert, J. (1997), *Index Galenicus. Wortformenindex zu den Schriften Galens*, Two Vols., Dettelbach.
Gladigow, B. (1995), "Anatomia sacra", in Van der Eijk, Horstmanshoff, Schrijvers (eds.), 345–361.
Glucker, J. (1978), *Antiochus and the Late Academy*, Göttingen.
Gotthelf, A. (ed.) (1985), *Aristotle on Nature and Living Things*, Pittsburgh—Bristol.
Gotthelf, A.; Lennox, J.G. (eds.) (1987), *Philosophical Issues in Aristotle's Biology*, Cambridge.
Gottschalk, H.B. (1987), "Aristotelian philosophy in the Roman world from the time of Cicero to the end of the second century AD", *ANRW* II. 36.2, 1079–1174.
Goulet, R. (ed.) (1994), *Dictionnaire des philosophes antiques*, Vol. 2, Paris.

Gourevitch, D. (1987), "L'esthetique médicale de Galien", *LEC*, LV, 267–289.
—— (1998), "The paths of knowledge: medicine in the Roman world", in Grmek (ed.), 104–138.
Gracia, D. (1978), "The structure of medical knowledge in Aristotle's philosophy", *ZWG*, 62, 1–36.
Grapow, H.; Von Deines, H.; Westendorf, W. (1954–62), *Grundriss der Medizin der alten Ägypten*, Berlin.
Greenhill, W.A. (1843), "Professor Marx's *Herophilus*", *British and Foreign Medical Review*, XV, 106–114.
—— (1864), "Adversaria medico-philologica", *British and Foreign Medico-chirurgical Review*, Part I, n.s. 34, 555–560.
—— (1865), Parts II & III, n.s. 35, 36, 1865, 275–281; 556–560.
—— (1866), Parts IV & V, n.s. 37, 38, 1866, 266–279; 551–560.
—— (1869), Part VII, n.s. 44, 1869, 555–560.
—— (1870), Parts VIII & IX, n.s. 45, 46, 547–554; 548–557.
—— (1871), Parts X & XI, n.s. 47, 48, 552–557; 554–557.
—— (1872), Parts XII & XIII, n.s. 49, 50, 275–281; 538–541.
Grmek, M.D. (1983), "Ancienneté de la chirurgie hippocratique", in Laserre and Mudry (eds.), 285–295.
—— (1990), *La première révolution biologique. Réflexions sur la physiologie et la médecine du XVIIIe siècle*, Paris.
—— (1996), *Il calderone di Medea: La sperimentazione sul vivente nell'antichità*, Rome.
—— (ed.) (1980), *Hippocratica*, Paris.
—— (ed.) (1993), *Storia del pensiero medico occidentale 1. Antichità e medioevo*, Rome.
—— (ed.) (1998), *Western Medical Thought from Antiquity to the Middle Ages*, translated by A. Shugaar, Harvard.
Grmek, M.D.; Gourevitch, D. (1988), "L'école médicale de Quintus et de Numisianus", in Sabbah (ed.), 43–60.
—— (1994), "Aus sources de la doctrine médicale de Galien: l'enseignement de Marinus, Quintus et Numisianus", *ANRW* II. 37.2, 1491–1528.
Gundert, B. (1992), "Parts and their roles in Hippocratic medicine", *Isis*, 83, 453–465.
Gurlt, E. (1898), *Geschichte der Chirurgie*, I, Berlin.
Guthrie, L.G. (1921), *The History of Neurology*, London.
De Gutiérrez-Mahoney, C.G., and Schechter, M.M. (1972), "The myth of the *rete mirabile* in man", *Neuroradiology*, 4, 141–158.
Haase, W. (ed.) (1993–), *Aufstieg und Niedergang der Römischen Welt (ANRW)*, Berlin.
Hahm, D.E. (1977), *The Origins of Stoic Cosmology*, Ohio.
Hall, T.S. (1969), *Ideas of Life and Matter. Studies in the History of General Physiology*, Volume I, Chicago.
Hamma, K (ed.) (1996), *Alexandria and Alexandrianism*, California.
Hankinson, R.J. (1987a), "Causes and empiricism: A problem in the interpretation of later Greek medical method", *Phronesis*, XXXII (3), 329–348.
—— (1987b), "Evidence, externality and antecedence: inquiries into later Greek causal concepts", *Phronesis*, XXXII (1), 80–100.
—— (1988), "Galen explains the elephant", *Canadian Journal of Philosophy*, Supplement 14, 135–157.
—— (1989), "Galen and the best of all possible worlds", *CQ*, 39, 206–227.
—— (1991a), "A purely verbal dispute? Galen on Stoic and Academic epistemology", *Revue Internationale de Philosophie*, 3 (178), 267–300.
—— (1991b), "Galen's anatomy of the soul", *Phronesis*, XXXVI (2), 197–233.
—— (1991c), *Galen on the Therapeutic Method. Books I and II*, Oxford.
—— (1991d), "Greek medical models of mind", in Everson (ed.), 194–217.
—— (1992), "Galen's philosophical eclecticism", *ANRW*, II. 36.5, 3505–3522.
—— (1993), "Galen on the foundations of science", in López-Férez (ed.), 15–29.

—— (1994a), "Galen and the logic of relations", in Schrenk (ed.), 57–75.
—— (1994b), "Galen's anatomical procedures: A second-century debate in medical epistemology", *ANRW* II. 37.2, 1834–1855.
—— (1994c), "Usage and abusage: Galen on language", in Everson (ed.), 166–187.
—— (1995), "The growth of medical empiricism", in Bates (ed.), 60–83.
—— (1997), "Natural criteria and the transparency of judgement. Antiochus, Philo and Galen on epistemological justification", in Inwood and Mansfeld (eds.), 161–216.
—— (1998), *Galen on Antecedent Causes*, Cambridge.
—— (ed.) (forthcoming), *The Cambridge Companion to Galen*, Cambridge.
Hanson, N.R. (1958), *Patterns of Discovery*, Cambridge.
Hanson, V.D. (1989), *The Western Way of War*, New York.
Harger, P. (1982), "Chrysippus' theory of pneuma", *Prudentia*, XIV, (2), 97–108.
Harig, G.; Kollesch, J. (1975), "Galen und Hippokrates", in Bourgey and Jouanna (eds.), 257–274.
Harris, C.R.S. (1973), *The Heart and the Vascular System in Ancient Greek Medicine. From Alcmaeon to Galen*, Oxford.
Harris, W.V. (1989), *Ancient Literacy*, Harvard.
Hartman, C.G.; Straus, W.-L. (eds.) (1933), *The Anatomy of the Rhesus Monkey*, London.
Healy, J.F. (1978), *Mining and Metallurgy in the Greek and Roman World*, London.
Herrlinger, R. (1970), *History of Medical Illustrations from Antiquity to 1600*, London.
Hill, W.C.O. (1953), *Primates. Comparative Anatomy and Taxonomy. I. Strepsirhini*, Edinburgh.
—— (1966), *VI. Catarrhini. Carcopithecoidea. Carcopithecinæ*, Edinburgh.
—— (1970), *VIII. Cynopithecinæ. Papio, Mandrillus, Theropithecus*, Edinburgh.
—— (1974), *VII. Cynopithecinæ. Cercocebus, Macaca, Cynopithecus*, Edinburgh.
Holford-Strevens, L. (1997), "Favorinus: the man of paradoxes", in Barnes and Griffin (eds.), 188–217.
Hope, V.M. (2000), "Contempt and respect. The treatment of the corpse in ancient Rome", in Hope and Marshall (eds.), 104–127.
Hope, V.M.; Marshall, E. (2000), *Death and Disease in the Ancient City*, London—New York.
Hopkins, K. (1983), *Death and Renewal*, Cambridge.
Horne, J.F. (1894), *Trephining in Ancient and Modern Aspect*, London.
Horstmanshoff, H.F.J. (1990), "The ancient physician: craftsman or scientist?", *J.Hist.Med.*, 45, 176–197.
—— (1995), "Galen and his patients", in Van der Eijk, Horstmanshoff, Schrijvers (eds.), 83–99.
Huby, P.; Neal, G. (eds.) (1989), *The Criterion of Truth*, Liverpool.
Hülser, K. (1992), "Galen und die Logik", *ANRW* II. 36.5, 3523–3554.
Hunger, H. (1978), *Byzantinisches Handbuch. Die Hochsprachliche Profane Literatur Der Byzantiner, II*, Munich.
Hyrtl, J. (1880), *Onomatologia Anatomica* (reprinted: 1970, Hildesheim).
Ieraci Bio, A.-M. (1993), "Sulla concezione del medico peripaideuménos in Galeno e nel tardoantico", in López-Férez (ed.), 259–272.
Ierodiakonou, K. (1995), "Alexander of Aphrodisias on medicine as a stochastic art", in Van der Eijk, Horstmanshoff, Schrijvers (eds.), 473–485.
Igarashi, S.; Kamiya, T. (1972), *Atlas of the Vertebrate Brain*, Tokyo.
Ilberg, J. (1899), "Ueber die Schriftstellerei des Klaudios Galenos", *Rh.Mus.*, n.f. 44, 207–239; 47,
—— (1892), II, *Rh.Mus.* n.f. 47, 489–514.
—— (1896), III, *Rh.Mus.* n.f. 51, 165–196.
—— (1897), IV, *Rh.Mus.* n.f. 52, 591–623.
—— (1905) "Aus Galens Praxis", *Neue Jahrbücher*, 15, 276–312 (reprinted: Flashar (ed.), 1971, 361–416).

—— (1930), "Wann ist Galenos geboren?", *AGM*, 23, 289–92.
Inwood, B. (1985), *Ethics and Human Action in Early Stoicism*, Oxford.
Inwood, B.; Mansfeld, J. (eds.) (1997), *Assent and Argument. Studies in Cicero's Academic Books*, Leiden.
Ioannidi, H. (1983), "Les notions de partie du corps et d'organe", in Lasserre and Mudry (eds.), 327–330.
—— (1992), "Les sensation-perception dans le *corpus hippocratique*", in López-Férez (ed.), 69–73.
Ioppolo, A.M. (1993), "The Academic position of Favorinus of Arelate", *Phronesis*, 38 (2), 183–213.
Ireland, S.; Steel, F.L.D. (1975), "Φρένες" as an anatomical organ in the works of Homer", *Glotta*, 53, 183–195.
Irigoin, J. (1980), "La formation du vocabulaire de l'anatomie en grec: du mycénien aux principaux traités de la Collection hippocratique", in Grmek (ed.), 247–257.
Irwin T.H. (1988), *Aristotle's First Principles*, Oxford.
Iverson, E. (1947), "Some remarks on the terms *mm* and *ais*", *JEg.Arch.*, 33, 47–51.
Jackson, R. (1986), "A set of Roman medical instruments from Italy", *Britannia*, XVII, 119–167.
—— (1988), *Doctors and Diseases in the Roman Empire*, London.
—— (1990), "Roman doctors and their instruments: recent research into ancient practice", *Journal of Roman Archaeology*, 3, 5–27.
—— (1995), "The composition of Roman medical *instrumentaria* as an indicator of medical practice: a provisional assessment", in Van der Eijk, Horstmanshoff, Schrijvers (eds.), 189–205.
Jennison, G. (1937), *Animals for Show and Pleasure in Ancient Rome*, Manchester.
Johansen, T.K. (1998), *Aristotle on the Sense-Organs*, Cambridge.
Johnstone, P. (ed.) (1984), *Studies in Medieval Arabic Medicine. Theory and Practice*, London.
Joly, R. (ed.) (1977), *Corpus Hippocraticum*, Mons.
Jones, C.P. (1967), "A friend of Galen", *CQ*, 17, 311–12.
Jones, W.H.S. (1923–1931), *Hippocrates*, Vols., I, II, IV, Cambridge Mass.—London.
—— (1947), *The Medical Writings of Anonymus Londinensis*, Cambridge.
Jouanna, J. (1998), *Hippocrate*, Fayard (reprinted: (1999) *Hippocrates*, translated by M.B. De Bevoise, 1999, Baltimore).
Kaplan, H.A.; Ford, D.H. (1966), *The Brain Vascular System*, Amsterdam.
Katz, O. (1890), *Die Augenheilkunde des Galens*, Berlin.
Keele, K.D. (1964), "Leonardo da Vinci's influence on Renaissance anatomy", *Med.Hist.*, 8, 360–370.
Keller, O. (1909), *Die Antike Tierwelt*, I, Leipzig.
Kemp, M. (1971), "'Il concetto dell'anima' in Leonardo's early skull studies", *Journal of the Warburg and Courtauld Instititute*, 34, 115–134.
Kieffer, J.S. (1964), *Galen's Institutio Logica*, English Translation, Introduction, and Commentary, Baltimore.
Kier, E.L. (1977), "The cerebral ventricles: a phylogenetic and ontogenetic study", in Newton and Potts (eds.), 2787–2914.
King, A.S. (1987), *Physiological and Clinical Anatomy of the Domestic Mammals, Vol. I, Central Nervous System*, Oxford.
King, H. (1998), *Hippocrates' Women*, London.
Kirk, G.S.; Raven, J.E.; Schofield, M. (1991), *The Presocratic Philosophers*, Second Edition, Cambridge.
Koelbing, H.M. (1985), "Remarques à propos de l'évolution des connaissances sur le cerveau", *Gesnerus*, 42, 315–328.
Kollesch, J. (1965a), "Aus Galens Praxis am römischen Kaiserhof", in Welskopf (ed.), 57–61.
—— (1965b), "Galen und seine ärztlichen Kollegen", *Alterum*, 11, 47–53.

—— (1967), "René Chartier-Herausgeber und Fälscher der Werke Galens", *Klio*, 48, 183–198.
—— (1981), "Galen und die Zweite Sophistik", in Nutton (ed.), 1–11.
Kollesch, J.; Nickel, D. (eds.) (1993), *Galen und das hellenistische Erbe. Verhandlungen des IV. Internationalen Galen-Symposiums, Sudhoffs Archiv Beihefte 32*, Stuttgart.
De Koning, P. (1903), *Trois traités d'anatomie Arabes*, Leiden.
Kudlien, F. (1964), "Nomenklaturgeschichtliche Addenda zur 'Nachgeburt'", *AGM*, 48, 86–88.
—— (1968), "Pneumatische Ärzte", *RE*, Supplementband XI, cols. 1097–1108.
—— (1969), "Antike Anatomie und menschlicher Leichnam", *Hermes*, 97, 78–94.
—— (1976), "Medicine as a 'liberal art' and the question of the physician's income", *J.Hist.Med.*, 31, 448–459.
—— (1981), "Galen's religious belief", in Nutton (ed.), 117–130.
—— (1991), " 'Endeixis' as a scientific term" A) Galen's usage of the word (in medicine and logic)", in Kudlien and Durling (eds.), 103–111.
Kudlien, F.; Durling, R.J. (eds.) (1991), *Galen's Method of Healing*, Leiden.
Kullmann, W.; Föllinger, S. (eds.) (1997), *Aristotelische Biologie*, Stuttgart.
Künzl, E. (1982), "Medizinische Instrumente aus Sepulkralfunden der römischen kaiserzeit", *Bonner Jahrbücher*, 182, 1–131.
—— (1996), "Forschungsbericht zu den antiken medizinischen Instrumenten", *ANRW* II.37.3, 2433–2639.
De Lacy P. (1966), "Galen and the Greek poets", *GRBS*, 57, 259–266.
—— (1972), "Galen's Platonism", *AJP*, XCIII, 27–39.
—— (1974), "Plato and the intellectual life of the second century AD", in Bowersock (ed.), 4–10.
—— (1988), "The third part of the soul", in Manuli and Vegetti (eds.), 43–63.
—— (1991), "Galen's response to skepticism", *ICS*, XVI, 283–306.
De Lahunta, A. (1983), *Veterinary Neuroanatomy and Clinical Neurology*, Second Edition, Philadelphia.
Laks, A. (1999), "Soul, sensation, and thought", in Long (ed.), 250–270.
Langholf, V. (1990), *Medical Theories in Hippocrates. Early Texts and the 'Epidemics'*, Berlin.
Langslow, D. (1991), "The development of Latin medical terminology: some working hypotheses", *PCPS*, 37, 106–130.
—— (2000), *Medical Latin in the Roman Empire*, Oxford.
Lasserre, F.; Mudry, P. (eds.) (1983), *Formes de pensée dans la Collection hippocratique*, Geneva.
Last, R.J.; Tompsett, D.H. (1953), "Casts of the cerebral ventricles", *British Journal of Surgery*, 40, 525–543.
Lennox, J.G. (1994), "The disappearance of Aristotle's biology: A Hellenistic mystery", in Barnes, T.D (ed.), 7–24.
Leyacker, J. (1927), "Zur Entstehung der Lehre von den Hirnventrikeln als Sitz psychischer Vermögen", *AGM*, 19, 253–286.
Lind, R.L. (1959), *Jacopo Berengario Da Carpi. A Short Introduction to Anatomy (Isagogae Breves)*, Chicago.
—— (1969), *The Epitome of Andreas Vesalius*, Massachusetts.
Lloyd, G.E.R. (1966), *Polarity and Analogy. Two Types of Argumentation in Early Greek Thought*, Cambridge.
—— (1968), "Ancient medicine and modern controversies". Review of: Temkin and Temkin (eds.) (1967), *History of Science*, 7, 125–129.
—— (1975a), "Alcmaeon and the early history of dissection", *ZWG*, 59, 113–147 (reprinted: 1991a, 164–193).
—— (1975b), "A note on Erasistratus of Ceos", *JHS*, 95, 172–175.
—— (1978a), *Hippocratic Writings*. Edited with an Introduction, London.

—— (1978b), "The empirical basis of the physiology of the *Parva naturalia*", in Lloyd and Owen (eds.), 215–239 (reprinted: 1991a, 224–247).
—— (1979), *Magic, Reason and Experience. Studies in the Origin and Development of Greek Science*, Cambridge.
—— (1980), "Hippocratic problems". Review of: Smith (1979),*CR*, xxx, (i), 186–189.
—— (1983a), "Plato on mathematics and nature, myth and science" (reprinted: 1991a, 338–351).
—— (1983b), *Science, Folklore and Ideology. Studies in the Life Sciences in Ancient Greece*, Cambridge.
—— (1985), *Science and morality in Greco-roman antiquity*, Cambridge, 1985 (reprinted: 1991a, 352–371).
—— (1987a), "Empirical research in Aristotle's biology", in Gotthelf and Lennox (eds.), 53–63.
—— (1987b), *The Revolutions of Wisdom. Studies in the Claims and Practice of Ancient Greek Science*, California.
—— (1988), "Scholarship, authority and argument in Galen's *Quod animi mores*", in Manuli and Vegetti (eds.), 11–42.
—— (1990), "Aristotle's zoology and his metaphysics. The *status questionis*. A critical review of some recent theories", in Devereux and Pellegrin (eds.), 7–35.
—— (1991a), *Methods and Problems in Greek Science. Selected Papers*, Cambridge.
—— (1991b), "The definition, status, and methods of the medical τέχνη in the fifth and fourth centuries", in Bowen (ed.), 249–260.
—— (1992a), "The theories and practices of demonstration", in Cleary and Shartin (eds.), 371–401 (reprinted: 1996a, 7–37).
—— (1992b), "The transformations of ancient medicine", *BHM*, 66, 114–132.
—— (1993), "Galen on Hellenistics and Hippocrateans: contemporary battles and past authorities", in Kollesch and Nickel (eds.), 125–143 (first published: 1991a, 398–416).
—— (1996a), *Aristotelian Explorations*, Cambridge.
—— (1996b), "Theories and practices of demonstration in Galen", in Frede and Striker (eds.), 255–277.
Lloyd, G.E.R.; Owen, G.E.L. (eds.) (1978), *Aristotle on mind and the senses*, Cambridge.
Lokhorst, G.-J.C. (1982), "An ancient Greek theory of hemispheric specialisation", *Clio Medica*, 17, 33–38.
—— (1996), "The first theory about hemispheric specialization: new light on an old codex", *J.Hist.Med.*, 51, 293–312.
Lones, T.E. (1912), *Aristotle's Researches in Natural Science*, London.
Long, A.A. (1982), "Soul and body in Stoicism", *Phronesis*, 27, 34–57.
—— (1986), *Hellenistic Philosophy*, Second Edition, London.
—— (1988), Ptolemy *On the Criterion*: an epistemology for the practicing scientist", in Dillon and Long (eds.), 176–207.
—— (ed.) (1999), *The Cambridge Companion to Early Greek Philosophy*, Cambridge.
Long, A.A.; Sedley, D.N. (1987), *The Hellenistic Philosophers*. Two Vols., Cambridge.
Longrigg, J. (1981), "Superlative achievement and comparative neglect: Alexandrian medical science and modern historical research", *History of Science*, XIX, 155–200.
—— (1985), "A seminal 'debate' in the fifth century BC?", in Gotthelf (ed.), 277–287.
—— (1988), "Anatomy in Alexandria in the third century BC", *British Journal of the History of Science*, 21, 455–488.
—— (1993), *Greek Rational Medicine. Philosophy and Medicine from Alcmaeon to the Alexandrians*, London.
—— (1998), *Greek Medicine. From the Heroic to the Hellenistic Age. A Source Book*, London.
Lonie, I.M. (1964), "Erasistratus, the Erasistrateans, and Aristotle, *BHM*, 38, 426–443.
—— (1965), "Medical theory in Heraclides of Pontus", *Mnemosyne*, 18, 126–143.

—— (1973), "The paradoxical text 'On the heart'", *Med.Hist.*, 17, 1–15; 136–153.
López-Férez, J.A. (1991), "Le temoignage de Galien sur les Methodiques à Rome", in Mudry and Pigeaud (eds.), 188–201.
—— (ed.) (1992), *Tratados Hippocráticos*, Madrid.
—— (ed.) (1993), *Galeno. Obra, Pensamiento, E Influencia*, Madrid.
Macgregor, R. (1980), *The Structure of the Meat Animals. A guide to their anatomy and physiology*, Third Edition. Revised by F. Gerrard, Oxford.
Machamer, P.K.; Turnbull, R.G. (eds.) (1978), *Studies in Perception*, Ohio.
Maehle, A.-H.; Tröhler, U. (1987), "Animal experimentation from antiquity to the end of the nineteenth century: attitudes and arguments", in Rupke (ed.), 14–47.
Majno, G. (1975), *The Healing Hand. Man and Wound in the Ancient World*, Harvard.
Mani, N. (1956), "Die griechische Editio princeps des Galenos (1525), ihre Enstehung und uhre Wirkung", *Gesnerus*, 13, 29–52.
—— (1991), "Die wissenschaftlichen Grundlagen der Chirurgie bei Galen (mit besonderer Berücksichtigung der *MM*)", in Kudlien and Durling (eds.), 26–49.
Mansfeld, J. (1971), *The Pseudo-Hippocratic Text ΠΕΡΙ 'ΕΒΔΟΜΑΔΩΝ Ch. 11 and Greek Philosophy*, Assen.
—— (1975), "Alcmaeon: 'physikos or physician? With some remarks on Calcidius' 'on vision' compared to Galen's *Plac.Hipp.Plat.* VII", in Mansfeld and de Rijk (eds.), 26–38.
—— (1990a), "Chrysippus and the Placita", *Phronesis*, 34, 311–342.
—— (1990b), "Doxography and dialectic: the *Sitz im Leben* of the *Placita*", *ANRW* II. 36.4, 3056–3229.
—— (1991), "The idea of the will in Chrysippus, Posidonius, and Galen", in Cleary (ed.), 107–145.
—— (1994), *Prolegomena. Questions to be Settled Before the Study of an Author, or a Text*, Leiden.
Mansfeld, J.; de Rijk, L.M. (eds.) (1975), *Kephalaion: Essays offered to C.J. de Vogel*, Assen.
Mansfeld, J.; Runia, D.T. (1996), *Aëtiana. The Method and Intellectual Context of a Doxographer. Volume I, The Sources*, Leiden.
Mansion, S. (ed.) (1961), *Aristote et les problèmes de méthode*, Louvain.
Manuli, P. (1977), "La *techne* medica nella tradizione encephalocentrica e cardiocentrica", in Joly (ed.), 182–195.
—— (1986), "Traducibilità e molteplicità dei linguaggi nel *De placitis* di Galeno", in Cambiano (ed.), 245–265.
—— (1988), "La passione nel *De placitis Hippocratis et Platonis*", in Manuli and Vegetti (eds.), 185–214.
—— (1993), "Galen and Stoicism", in Kollesch and Nickel (eds.), 53–61.
Manuli, P.; Vegetti, M. (1977), *Cuore, sangue e cervello: biologica e antropologica nel pensiero antico*, Milan.
Manuli, P.; Vegetti, M. (eds.) (1988), *Le opere psicologiche di Galeno*, Naples.
Manzoni, T. (1998), "The cerebral ventricles, the animal spirits and the dawn of brain localization of function", *Archives Italiennes de Biologie*, 136, 103–152.
—— (2001), *Il cervello secondo Galeno*, Ancona.
Marasco, G. (1995), "L'introduction de la médecine grecque à Rome: une dissension politique et idéologique", in Van der Eijk, Horstmanshoff, Schrijvers (eds.), 35–48.
Marganne, M.-H. (1986), "La *Cirurgia Eliodori* et le P. Genève Inv. III", *Études de Lettres*, 65–73.
—— (1988), "Le chirurgien Héliodore: tradition directe et indirecte", in Sabbah (ed.), 107–111.
—— (1998), *La chirurgie dans l'Egypte gréco-romaine d'après les papyrus littéraires grecs*, Leiden.

Marmelzat, W.L. (1991), "Galen on 'glands' and 'spongy flesh': with special reference to an ancient mechanical-biophysical theory for superficial lymph node function", *J.Hist.Med.*, 46, 419–439.
De Martini, U. (1964), "Considerazioni sulla dottrina dello pneuma in Galeno", *Pagine di Storia della Medicina*, 8, 41–47.
Matthen, M. (1988), "Empiricism and ontology in ancient medicine", in *Method, Medicine and Metaphysics. Studies in the Philosophy of Ancient Science, Apeiron*, XXI (2), 99–121.
May, M.T. (1958), "Galen on human dissection", *J.Hist.Med.*, 13, 409–410.
—— (1968), *Galen. On the Usefulness of the Parts of the Body*, Two Vols., Cornell.
—— (1970), "On translating Galen", *J.Hist.Med.*, 25, 168–176.
McDermott, W.C. (1938), *The Ape in Antiquity*, Baltimore.
McMinn, R.M.H.; Hutchings, R.T.; Logan, B.M. (1981), *A Colour Atlas of Head and Neck Anatomy*, London.
Mewaldt, J. (1912), "Galenos", *RE*, cols. 578–591.
Meyer, A. (1971), *Historical Aspects of Cerebral Anatomy*, Oxford.
Meyer, A.; Hierons, R. (1962), "Observations on the history of the 'Circle of Willis'", *Med.Hist.*, 6, 113–130.
Meyer, B.E.; Saunders, E.P. (eds.) (1982), *Jewish and Christian Self-Definition III. Self-Definition in the Greco-Roman World*, Philadelphia.
Meyer, H. (1954), "Macroscopic brain dissection in veterinary anatomy", *American Journal of Veterinary Research*, 15, 143–146.
Meyerhof, M. (1926), "New light on Hunain Ibn Ishâq and his period", *Isis*, 8, 685–724 (reprinted (1984), *Studies in Medieval Arabic Medicine. Theory and Practice*, edited by P. Johnstone, London).
—— (1928), *The Book of the Ten Treatises on the Eye Ascribed to Hunain Ibn Is-Haq (809–877 AD), The Arabic Text edited from the only two known Manuscripts, with an English Translation and Glossary*, Cairo.
Meyer-Steineg, T. (1912/13), "Studien zur Physiologie des Galenos", *AGM*, V, 172–224; VI, 417–448.
Michler, M. (1968), *Die Alexandrinischen Chirurgen. Eine Sammlung und Auswertung ihrer Fragmente*, Wiesbaden.
Millen, J.W.; Woollam, D.H.M. (1962), *The Anatomy of the Cerebrospinal Fluid*, Oxford.
Miller, H.W. (1948), "A medical theory of cognition", *TAPA*, 78, 168–183.
Miller, T.S. (1985), *The Birth of the Hospital in the Byzantine Empire*, Baltimore.
—— (1999), "Review of: G. Weber, *Areteo di Cappadocia*", *BHM*, 73, 141–2.
Milne, J.S. (1907), *Surgical Instruments in Greek and Roman Times*, Oxford.
Modrak, D.K.W. (1987), *Aristotle. The Power of Perception*, Chicago.
Moraux, P. (1976), "Galien et Aristote", in Bossier et al. (eds.), 127–146.
—— (1981), "Galien comme philosophe: la philosophie de la nature", in Nutton (ed.), 87–116.
—— (1984), *Der Aristotelismus bei den Griechen. Von Andronikos bis Alexander von Aprodisias, I*, Berlin.
—— (1985a), "Galen and Aristotle's *De partibus animalium*" (tr. A. Preus), in Gotthelf (ed.), 327–344.
—— (1985b), *Galien de Pergame. Souvenirs d'un médecin*, Paris.
Mudry, P. (1982), *La préface du 'De medicina' de Celse*, Lausanne.
—— (1993), "Le 'De medicina' de Celse. Rapport bibliographique", *ANRW* II. 37.1, 787–799.
Mudry, P.; Pigeaud, J. (eds.) (1991), *Les Écoles Médicales à Rome*, Geneva.
Møller-Christensen, V. (1938), *The History of the Forceps*, London.
Needham, J. (1934), *A History of Embryology*, Cambridge.
Neuburger, M. (1897/1981), *The Historical Development of Experimental Brain and Spinal Cord Physiology before Flourens.* Translated and edited, with additional material by E. Clarke, Baltimore.

Neuburger, M.; Pagel, J. (eds.) (1902), *Handbuch der Geschichte der Medizin, I*, Jena.
Newton, T.H.; Potts, D.G. (eds.) (1977), *Radiology of the Skull and Brain. Volume III. Anatomy and Pathology*, St. Louis.
Nickel, D. (1989), *Untersuchungen zur Embryologie Galens*, Berlin.
Nijhuis, K. (1995), "Greek doctors and Roman patients: a medical anthropological approach", in Van der Eijk, Horstmanshoff, Schrijvers (eds.), 49–67.
Nunn, J.F. (1996), *Ancient Egyptian Medicine*, Oklahoma—London.
Nutton, V. (1969), "Medicine and the Roman army: a further reconsideration", *Med.Hist.*, 13, 260–270.
—— (1970), *The Medical profession in the Roman Empire from Augustus to Justinian*, Ph.D. Diss. Cambridge.
—— (1971), "L. Gellius Maximus, physician and procurator", *CR*, xxi, 262–272.
—— (1972a), "Ammianus and Alexandria", *Clio Medica*, 3, 165–176.
—— (1972b), "Galen and medical autobiography", *PCPS*, n.s. 18, 50–62 (reprinted: 1988a, I).
—— (1973), "The chronology of Galen's early career", *CQ*, 23, 158–171 (reprinted: 1988a, II).
—— (1974), "Greek medicine dissected". Review of: Phillips (1973) and Harris (1973), *History of Science*, xii, 59–69.
—— (1975), "Museums and medical schools in classical antiquity", *History of Education*, 4, no. 1, 3–15.
—— (1977), "*Archiatri* and the medical profession in antiquity, *PBSR*, 45, 191–226 (reprinted: 1988a, V).
—— (1978), "The beneficial ideology", in Garnsey and Whittaker (eds.), 209–221.
—— (ed.) (1981), *Galen: Problems and Prospects*, London.
—— (1984a), "From Galen to Alexander, aspects of medicine and medical practice in late Antiquity", *Dumbarton Oaks Papers*, 38, 1–14 (reprinted: 1988a, X).
—— (1984b), "Galen in the eyes of his contemporaries", *BHM*, 58, 315–324 (reprinted: 1988a, III).
—— (1985), "Murders and miracles: lay attitudes towards medicine in classical antiquity", in Porter (ed.), 23–53 (reprinted: 1988a, VIII).
—— (1986), "The Perils of patriotism: Pliny and Roman medicine", in French and Greenaway (eds.), 30–58 (reprinted: 1988a, VII).
—— (1987a), "Galen's philosophical testament: 'On my own opinions'", in Wiesner (ed.), 27–51.
—— (1987b), "Numisianus and Galen", *ZPG*, 71, 235–9.
—— (1987c), *John Caius and the Manuscripts of Galen*, Cambridge Philological Society, Supplement 13.
—— (1988a), *From Democedes to Harvey: Studies in the History of Medicine*, London.
—— (1988b), "'Prisci dissectionum professores'", in Dionisotti, Grafton and Kraye (eds.), 111–126.
—— (1990), "The patient's choice: a new treatise by Galen", *CQ*, 40, 236–257.
—— (1992), "Healers in the medical market place: towards a social history of Graeco-Roman medicine", in Wear (ed.), 15–58.
—— (1993a), "Galen and Egypt", in Kollesch and Nickel (eds.), 11–31.
—— (1993b), "Galen at the bedside: the methods of a medical detective", in Bynum and Porter (eds.), 7–16.
—— (1993c), "Greek science in the sixteenth-century Renaissance", in Field and James (eds.), 15–28.
—— (1993d), "Roman medicine: tradition, confrontation, assimilation", *ANRW* II. 37.1, 49–78.
—— (1995a), "Galen *ad multos annos*", *Dynamis*, 15, 25–40.
—— (1995b), "Roman medicine, 250 BC to AD 200", in Conrad, Neve, Nutton, Porter, Wear (eds.), 39–70.

—— (1995c), "The medical meeting place", in Van der Eijk, Horstmanshoff, Schrijvers (eds.), 3–25.
—— (1997), "Galen on theriac: problems of authenticity", in Debru (ed.), 133–151.
—— (1998), "Galen", *DNP*, cols., 747–756.
—— (2000), "Medicine", in CAH², 943–965.
—— (ed.) (2002), *The Unknown Galen: Galen beyond Kühn, BICS* Supplement 77.
Oberhelman, S.M. (1983), "Galen, *On Diagnosis from Dreams*", *J.Hist.Med.*, 38, 36–47.
O'Malley, C.D.; Saunders, J.B., de C.M. (1952), *Leonardo da Vinci on the Human Body*, New York.
Onians, R.B. (1951), *The Origins of European Thought about the Body, the Mind, the Soul, the World, Time and Fate*, Cambridge.
Ormos, I. (1993), "Bemerkungen zur editorischen Bearbeitung der Galenschrift *Über die Sektion toter Lebewesen*", in Kollesch and Nickel (ed.), 165–172.
Owen, G.E.L. (1961), "τιθέναι τὰ φαινόμενα", in Mansion (ed.), 83–103 (reprinted: (1986), *Logic, Science and Dialectic*, London, 239–251).
Owen, R. (1837/1992), *The Hunterian Lectures in Comparative Anatomy*, Edited and with an Introductory Essay and Commentary by P.R. Sloan, London.
—— (1868), *On the Anatomy of Vertebrates, Vol. III. Mammals*, London.
Parente, I.M. (1961), "Technê", *PP*, XVI, 257–296.
Passmore, J. (1975), "The treatment of animals", *Journal of the History of Ideas*, 36, 195–218.
Pearcy, L.T. (1983), "Galen and Stoic rhetoric", *GRBS*, 24, 259–272.
—— (1993), "Medicine and rhetoric in the period of the Second Sophistic", *ANRW* II. 37.1, 445–456.
Peck, A.L. (1965), *Aristotle. Generation of Animals*, Cambridge Mass.—London.
Pellegrin, P. (1986), *Aristotle's Classification of Animals. Biology and the conceptual unity of the Aristotelian Corpus*, translated by A. Preus, California.
Pendrick, G. (1994), "A note on Galen and Asclepiades", *Mnemosyne*, XLVII, (2), 226–229.
Penfield, W. (1958), "Hippocratic preamble: the brain and intelligence", in Poynter (ed.), 1–4.
Phillips, E.D. (1973), *Greek Medicine*, London (= (1987), *Aspects of Greek Medicine*, Philadelphia).
Pigeaud, J. (1980), "Quelques aspects du rapport de l'âme et du corps dans le Corpus Hippocratique", in Grmek (ed.), 417–432.
—— (1993a), "Les problèmes de la création chez Galien", in Kollesch and Nickel (eds.), 87–103.
—— (1993b), "L'introduction du Méthodisme à Rome", *ANRW* II. 37.1, 565–599.
Pleket, H.W. (1995), "The social status of physicians in the Graeco-Roman world", in Van der Eijk, Horstmanshoff, Schrijvers (eds.), 27–34.
Pohlenz, M. (1949), *Die Stoa.* Two Vols., Göttingen.
Popesko, P. (1977), *Atlas of Topographical Anatomy of the Domestic Animals, I*, Second Edition, Philadelphia.
Porter, R. (ed.) (1985), *Patients and Practitioners*, Cambridge.
Pott, P. (1768), *Observations on the Nature and Consequences of those Injuries to which the Head is liable from External violence*, London.
Potter, P. (1976), "Herophilus of Chalcedon: An assessment of his place in the history of anatomy", *BHM*, 50, 45–60.
—— (1988, 1995), *Hippocrates*, Vols. V, VI, VIII, Cambridge Mass.—London.
—— (1998), "The *editiones princepes* of Galen and Hippocrates and their relationship", in Fischer, Nickel, Potter (eds.), 243–261.
Poynter, F.L.N. (ed.) (1958), *The History and Philosophy of Knowledge of the Brain and its Functions*, Springfield Ill.—Oxford.

Prendergast, J. (1930), "The background of Galen's life and activities, and its influence on his achievements", *Proceedings of the Royal Society of Medicine*, 23, 1131–1148.
Preus, A. (1975), *Science and Philosophy in Aristotle's Biological Works*, New York.
Prioreschi, P. (1992), "Supernatural elements in Hippocratic medicine", *J.Hist.Med.*, 47, 389–404.
Putscher, M. (1973), *Pneuma, Spiritus, Geist*, Wiesbaden.
Ragep, F.J.; Ragep, S.P. (eds.) (1996), *Tradition, Transmission, Transformation*, Leiden.
Rawson, E. (1982), "The life and death of Asclepiades of Bithynia", *CQ*, 32, 358–370.
—— (1985), *Intellectual Life in the Late Roman Republic*, London.
Riddle, J.M. (1993), "High medicine and low medicine in the Roman Empire", *ANRW* II. 37.1, 102–120.
Roberts, C.; Manchester, K. (1995), *The Archaeology of Disease*, Cornell.
Roberts, K.B.; Tomlinson, J.D.W. (1992), *The Fabric of the Body. European Traditions of Anatomical Illustration*, Oxford.
Rocca, J. (1997), "Galen and the ventricular system", *Journal of the History of the Neurosciences*, 6 (3), 227–239.
—— (1998a), "A note on the term *fornix*", *Journal of the History of the Neurosciences*, 7 (3), 243.
—— (1998b), "Galen and Greek neuroscience (notes towards a preliminary survey)", *Early Science and Medicine*, 3 (3), 216–240.
—— (1998c), "Galen and Greek pneuma theory. The limitations of physiological explanation", in Boudouris (ed.), 171–197.
—— (2002a), "Galen and the uses of trepanation", in Arnott, R.; Finger, S.; Smith, C. (eds.), 253–271.
—— (2002b), "The brain beyond Kühn: reflections on *Anatomical procedures*, book IX", in Nutton (ed.), 87–100.
—— (forthcoming), "Anatomy", in Hankinson (ed.).
Rogers, L. (1947), "The function of the circulus arteriosus of Willis", *Brain*, 70, 171–178.
Roselli, A. (1999), "Notes on the *doxai* of doctors in Galen's commentaries on Hippocrates", in Van der Eijk (ed.), 359–381.
Rubinstein, H.S. (1953), *The Study of the Brain*, New York.
Runia, D.T. (1999a), "The *placita* ascribed to doctors in Aëtius' doxography on physics", in Van der Eijk (ed.), 189–250.
—— (1999b), "What is doxography?", in Van der Eijk (ed.), 33–55.
Rupke, N.A. (ed.) (1987), *Vivisection in Historical Perspective*, London.
Ryberg, I.S. (1955), *Rites of the State Religion in Roman Art*, Rome.
Sabbah, G. (ed.) (1988), *Études de médecine romaine*, St Étienne.
Salazar, C.F. (2000), *The Treatment of War Wounds in Graeco-Roman Antiquity*, Leiden.
Saller, R., P. (1982), *Personal Patronage under the Early Empire*, Cambridge.
Sambursky, S. (1956), *The Physical World of the Greeks*, London.
—— (1959), *Physics of the Stoics*, London.
—— (1962), *The Physical World of Late Antiquity*, London.
Saunders, K.B. (1999), "The wounds in *Iliad* 13–16", *CQ*, 49, 345–363.
Savage-Smith, E. (1969), *Galen on Nerves, Veins, and Arteries: A Critical Edition and Translation from the Arabic with Notes, Glossary and an Introductory Essay*, Ph.D. Diss. Wisconsin.
—— (1971), "Galen's account of the cranial nerves and the autonomic nervous system", Parts 1 and 2, *Clio Medica*, 6, 77–98; 173–194.
Scarborough, J. (1968), "Roman medicine and the legions: a reconsideration", *Med.Hist.*, 12, 254–261.
—— (1969), *Roman Medicine*, London.
—— (1971), "Galen and the gladiators", *Episteme*, 5, 98–111.
—— (1976a), "Celsus on human vivisection at Ptolemaic Alexandria", *Clio Medica*, 11, 25–38.

—— (1976b), "Galen's investigations of the kidney", *Clio Medica 11*, 171–177.
—— (1981), "The Galenic question", *ZWG*, 65, 1–31.
—— (1984), "Symposium on Byzantine Medicine. Introduction", *Dumbarton Oaks Papers*, 38, ix–xvi.
—— (1985), "Galen's dissection of the elephant", *Koroth*, 8, 123–134.
—— (1991), "Aetios", *ODB*, 30–31.
—— (1992), *Medical and Biological Terminologies. Classical Origins*, Oklahoma.
—— (1993), "Roman medicine to Galen", *ANRW* II. 37.1, 3–48.
Scarborough, J.S.; Talbot, A.-M. (1991), "Alexander of Tralles", *ODB*, 58.
Scarborough, J.S.; Vikan, G.V. (1991), "Theophilos", *ODB*, 2067.
Scheidel, W. (1999), "Emperors, aristocrats, and the grim reaper: towards a demographic profile of the Roman élite", *CQ*, 49, 254–281.
Schrenk, L.P. (ed.) (1994), *Aristotle in Late Antiquity*, Washington.
Schürmann, A. (1991), *Griechische Mechanik und Antike Gesellschaft*, Stuttgart.
Sedley, D.N. (1982), "On signs", in Barnes et al., 239–272.
—— (1993), "Chrysippus on psychological causality", in Brunschwig and Nussbaum (eds.), 313–331.
Sharp, J.A. (1961), "Alexander Monro *secundus* and the interventricular foramen", *Med.Hist.*, 5, 83–89.
Shaw, J.R. (1972), "Models for cardiac structure and function in Aristotle", *Journal of the History of Biology*, 5, 355–388.
—— (1974), Review of: Harris (1973), *Isis*, 65, 533–534.
Skoda, F. (1988), *Médicine ancienne et métaphore. La vocabulaire de l'anatomie et de la pathologie en grec ancien*, Paris.
Shryock, R.H. (1961), "The history of quantification in medical science", *Isis*, 52, 215–237.
Sideras, A. (1994), "Rufus von Ephesos und sein Werk im Rahmen der antiken Medizin", *ANRW* II. 37.2, 1077–1253.
Siegel, R.E. (1968), *Galen's System of Physiology and Medicine. An analysis of his doctrines and observations on bloodflow, respiration, humors and internal disease*, Basel.
—— (1970), *Galen on Sense Perception. His doctrines, observations and experiments on vision, hearing, smell, taste, touch and pain, and their historical sources*, Basel.
—— (1973), *Galen on Psychology, Psychopathology, and Function and Diseases of the Nervous System. An analysis of his doctrines, observations and experiments*, Basel.
—— (1976), *Galen. On the Affected Parts*, Basel.
Sigerist, H. (1920), "Die 'Cirurgia Eliodori'", *AGM*, 12, 1–9.
Simon, M. (1906), *Sieben Bücher Anatomie des Galen*, Two Vols., Leipzig.
Singer, C. (1952), *Vesalius on the Human Brain*, Oxford—Wellcome.
—— (1956), *Galen. On Anatomical Procedures*, Oxford—Wellcome.
Singer, P.N. (1992), "Some Hippocratic mind-body problems", in López-Férez (ed.), 131–143.
—— (1993), "Aspects of Galen's Platonism", in López-Férez (ed.), 41–55.
—— (1997a), *Galen. Selected Works*, Oxford.
—— (1997b), "Levels of explanation in Galen", *CQ*, 47, 525–542.
Siraisi, N.G. (1990), *Medieval and Early Renaissance Medicine. An Introduction to Knowledge and Practice*, Chicago.
—— (1995), "Early anatomy in comparative perspective: introduction", *J.Hist.Med.*, 50, 3–10.
Sisson, S. (1930), *The Anatomy of the Domestic Animals*, Second Edition, Philadelphia.
Sisson, S.; Grossman, J.D. (1975), *The Anatomy of the Domestic Animals*, Fifth Edition, Two Vols., Philadelphia.
Smallwood, J.E. (1992), *A Guided Tour of Veterinary Anatomy*, Philadelphia.
Smith, W.D. (1979), *The Hippocratic Tradition*, Cornell.
—— (1994), *Hippocrates*, Vol. VII, Cambridge Mass.—London.

Solin, H. (1995), "Die sogenannten Berufsnamen antiker Ärzte", in Van der Eijk, Horstmanshoff, Schrijvers (eds.), 119–142

Solly, S. (1847), *The Human Brain: its Structure, Physiology and Diseases*, Second Edition, London.

Solmsen, F. (1957), "The vital heat, the inborn pneuma and the aether", *JHS*, 77, 1957, 118–123 (reprinted: (1968), *Kleine Schriften, I*, Hildesheim, 605–611).

—— (1961), "Greek philosophy and the discovery of the nerves", *MH*, 19, 1961, 150–197 (reprinted: (1968), *Kleine Schriften, I*, 536–582; reprinted in Flashar (ed.) (1971), 202–279).

Sorabji, R. (1988), *Matter, Space and Motion. Theories in Antiquity and their Sequel*, London.

—— (ed.) (1997), *Aristotle and After, BICS* Supplement 68, London.

Souques, A. (1933), "Les connaissances neurologiques de Galien", *Revue Neurologique*, 3, 297–340.

—— (1936), *Étapes de la neurologie dans l'antiquité grecque (d'Homerè à Galien)*, Paris.

Soury, J. (1899), *Le système nerveux central. Structure et fonctions. Histoire critique de théories et des doctrines*, Two Vols., Paris.

Spillane, J.D. (1981), *The Doctrine of the Nerves*, New York—Oxford.

Von Staden, H. (1975), "Experiment and experience in Hellenistic medicine", *BICS*, 22, 178–199.

—— (1978), "The Stoic theory of perception and its 'Platonic' critics", in Machamer and Turnbull (eds.), 96–136.

—— (1992), "The Discovery of the body: human dissection and its cultural contexts in Ancient Greece", *Yale Journal of Biology and Medicine*, 6, 223–241.

—— (1993), "Galen as historian: his use of sources on the Herophileans", in López-Férez (ed.), 205–222.

—— (1995a), "Anatomy as rhetoric: Galen on dissection and persuasion", *J.Hist.Med.*, 50, 47–66.

—— (1995b), "Science as text, science as history: Galen on metaphor", in Van der Eijk, Horstmanshoff, Schrijvers (eds.), 499–518.

—— (1996a), "Andréas de Caryste et Philon de Byzance: médecine et mécanique à Alexandrie", in Argoud and Guillaumin (eds.), 147–172.

—— (1996b), "Body and machine: interactions between medicine, mechanics, and philosophy in early Alexandria", in Hamma (ed.), 85–106.

—— (1996c), "Liminal perils. Early Roman receptions of Greek medicine", in Ragep and Ragep (eds.), 369–418.

—— (1997a), "Galen and the 'Second Sophistic'", in Sorabji (ed.), 33–54.

—— (1997b), "Teleology and mechanism: Aristotelian biology and early Hellenistic medicine", in Kullmann and Föllinger (eds.), 183–208.

—— (1998), "Dynamis: the Hippocratics and Plato", in Boudouris (ed.), 262–279.

—— (1999a), "Celsus as historian?", in Van der Eijk (ed.), 251–294.

—— (1999b), "Rupture and continuity: Hellenistic reflections on the history of medicine", in Van der Eijk (ed.), 143–187.

—— (2000), "Body, soul, and nerves: Epicurus, Herophilus, Erasistratus, the Stoics, and Galen", in Wright and Potter (eds.), 79–116.

Stein, M. (1997), "La thériaque chez Galien: sa préparation et son usage thérapeutique", in Debru (ed.), 199–209.

Stratton G.M. (1917), *Theophrastus and the Greek Physiological Psychology before Aristotle*, London.

Striker, G. (1974), "κριτήριον τῆς ἀληθείας", Nachrichten der Akademie der Wissenschaft zu Göttingen, Phil.-hist. klasse, 2, 48–100 (reprinted: idem, (1996), *Essays on Hellenistic Epistemology and Ethics*, Cambridge, 22–76).

Stok, F. (1993), "La scuola medica Empirica a Roma. Problemi storici e prospettive di ricerca", *ANRW* II. 37.1, 600–645.

Strohmaier, G. (1970), "Dura mater, pia mater. Die Geschichte zweier anatomischer Termini", *Medizinhistorisches Journal*, 5, 201–216.
—— (1998), "Reception and tradition: medicine in the Byzantine and Arab world", in Grmek (ed.), 139–169.
Sudhoff, K. (1913), "Die Lehre von den Hirnventrikeln in textlicher und graphischer Tradition des Alterums und Mittelalters", *AGM*, 7, 149–205.
Sullivan, S.D. (1981), "The function of θυμός in Hesiod and the Greek lyric Poets", *Glotta*, LIX, 147–155.
Swain, S. (1996), *Hellenism and Empire*, Oxford.
Tanner, R.G. (1985), "Early Greek origins of the idea of pneuma: notes towards a theory", in Dockrill and Tanner (eds.), 49–53.
Tarrant, H.A.S. (1985a), "Pneuma-related concepts in Platonism", in Dockrill and Tanner (eds.), 55–60.
—— (1985b), *Scepticism or Platonism? The Philosophy of the Fourth Academy*, Cambridge.
—— (1998), "The proximity of philosophy and medicine in the age of Galen", in Boudouris (ed.), 231–240.
Telfer, W. (1955), *Cyril of Jerusalem and Nemesius of Emesa*, London.
Temkin, O. (1935), "Celsus' 'On medicine' and the ancient medical sects", *BHM*, 3, 249–264.
—— (1951), "On Galen's pneumatology", *Gesnerus*, 8, 180–189 (reprinted: (1977), 154–161).
—— (1956), *Soranus' Gynecology*, Baltimore.
—— (1960), "Nutrition from classical antiquity to the Baroque", in Galdston (ed.), 78–97.
—— (1961), "A Galenic model for quantitative physiological reasoning?", *BHM*, 35, 470–475 (reprinted: (1977), 162–166).
—— (1962), "Byzantine medicine: tradition and empiricism", *Dumbarton Oaks Papers*, 16, 97–115 (reprinted: (1977), 202–222).
—— (1971), *The Falling Sickness: A History of Epilepsy from the Greeks to the Beginnings of Modern Neurology*, Second Edition, Baltimore.
—— (1973), *Galenism. Rise and Decline of a Medical Philosophy*, Cornell.
—— (1977), *The Double Face of Janus and Other Essays in the History of Medicine*, Baltimore.
—— (1991), *Hippocrates in a World of Pagans and Christians*, Baltimore.
Temkin, O.; Temkin, C.L. (eds.) (1967), *Ancient Medicine. Selected Papers of Ludwig Edelstein*, Baltimore.
Thompson, A.R. (1952), "Homer as a surgical anatomist", *Proceedings of the Royal Society of Medicine*, 45, 765–767.
Thompson, C.J.S. (1938), "The evolution and development of surgical instruments. IV. The trepan", *British Journal of Surgery*, 25, 726–734.
Thomssen, H.; Probst, C. (1994), "De Medizin von Rufus von Ephesos", *ANRW* II. 37.2, 1254–1292.
Tieleman, T. (1991), "Diogenes of Babylon and Stoic embryology: Ps. Plutarch, Plac. V. 15. 4. reconsidered", *Mnemosyne*, 44, 106–125.
—— (1995a), "Dialectic and science: Galen, Herophilus and Aristotle on phenomena", in Van der Eijk, Horstmanshoff, Schrijvers (eds.), 487–495.
—— (1995b), Galen's concept of dialectic and the Middle Platonist background: the argument of *PHP* book II", in Vegetti and Gastaldi (eds.), 49–65.
—— (1995c), Review of: Kollesch and Nickel (eds.) (1993), *Mnemosyne*, 48, 600–607.
—— (1996a), *Galen and Chrysippus. Argument and Refutation in the De Placitis Books II and III*, Leiden.
—— (1996b), "The hunt for Galen's shadow: Alexander of Aphrodisias, *De anima* 94.7–100.17 Bruns reconsidered", in Algra, Van der Horst, Runia (eds.), 265–283.

—— (1998), "Plotinus on the seat of the soul: reverberations of Galen and Alexander in *Enn.* IV, 3 [27], 23", *Phronesis*, 43 (4), 306–325.
Tilney, F.; Warren, L.F. (1919), *The Morphology and Evolutionary Significance of the Pineal Body*, Philadelphia.
Todd, R.B. (1976), *Alexander of Aphrodisias on Stoic Physics*, Leiden.
—— (1984), "Philosophy and medicine in John Philoponus' commentary on Aristotle's *De anima*", *Dumbarton Oaks Papers*, 38, 103–110.
—— (1995), "Peripatetic epistemology before Alexander of Aphrodisias: the case of Alexander of Damascus", *Eranos*, 93, 122–8.
Toledo-Pereyra, L.H. (1973), "Galen's contribution to surgery", *J.Hist.Med.*, 28, 357–375.
Torkildsen, A. (1934), "The gross anatomy of the lateral ventricles", *Journal of Anatomy*, 68, 480–491.
Touwaide, A. (1994), "Galen et la toxicologie", *ANRW* II. 37.2, 1887–1986.
Toynbee, J.M.C. (1973), *Animals in Roman Life and Art*, London.
Tracy, T.J. (1976), "Plato, Galen, and the center of consciousness", *Illinois Classical Studies*, I, 43–52.
Triolo, V. (1966), "An interpretative analysis of Galenic renal physiology", *Clio Medica*, 1, 113–128.
Tzedakis Y.; Martlew, H. (eds.) (forthcoming), *Minoans and Mycenaeans: flavours of their time—the scientific evidence*, Oxford.
Ullmann, M. (1994), "Die arabische Überlieferung den Schriften des Rufus von Ephesos", *ANRW* II. 37.2, 1293–1349.
Ulrich, F. (1919), *Die anatomische und vivisektorische Technik des Galenos*, Werdau i Sa.
Vallance, J. (1990), *The Lost Theory of Asclepiades of Bithynia*, Oxford.
Vegetti, M. (1979), *Il coltello e lo stilo*, Milan.
—— (1980), "Anatomia e classificazione degli animali nella biologia antica", in Grmek (ed.), 469–483.
—— (1981), "Modelli di medicina in Galeno", in Nutton (ed.), 47–63.
—— (1993a), "I nervi dell' anima", in Kollesch und Nickel (eds.), 63–77.
—— (1993b), "Tra il sapere e la pratica: la medicina Ellenistica", in Algrimi et al., (eds.), 73–120.
—— (1994), "L'immagine del medico e lo statuto epistemologica della medicina in Galeno", *ANRW* II. 37.2, 1672–1717.
—— (1998), "Between knowledge and practice: Hellenistic medicine", in Grmek (ed.), 72–103.
—— (1999a), "Historiographical strategies in Galen's physiology (*De usu partium, De naturalibus facultatibus*)", in Van der Eijk (ed.), 383–395.
—— (1999b), "Tradition and truth. Forms of philosophical-scientific historiography in Galen's *De placitis*", in Van der Eijk (ed.), 333–357.
Vegetti, M.; Gastaldi, S. (eds.) (1995), *Studi Di Storia Della Medicina Antica E Medievale*, Pavia.
Verbeke, G. (1945), *L' évolution de la doctrine du pneuma du stoïcisme à St. Augustin, Etude philosophique*, Louvain.
—— (1978), "Doctrine du pneuma et entéléchisme chez Aristote", in Lloyd and Owen (eds.), 191–214.
Wallace-Hadrill, A. (1990), "Pliny the Elder and man's unnatural history", *Greece & Rome*, 37, 80–96.
Walsh, J. (1926), "Galen's discovery and promulgation of the function of the recurrent laryngeal nerve", *Annals of Medical History*, 8, 176–184.
Wear, A. (1975), "The history of Galenism" Review of: Temkin (1973), *History of Science*, xiii, 122–129.
—— (1981), "Galen in the Renaissance", in Nutton (ed.), 229–262.
—— (ed.) (1992), *Medicine in Society. Historical Essays*, Cambridge.

Wear, A.; French, R.K.; Lonie, I.M. (eds.) (1985), *The Medical Renaissance of the Sixteenth Century*, Cambridge.
Weeks, K.R. (1970), *The Anatomical Knowledge of the Ancient Egyptians and the Representation of the Human Figure in Egyptian Art*, Ph.D. Diss., Yale.
Wellmann, M. (1985), *Die pneumatische Schule. Bis auf Archigenes* (Philologische Untersuchungen, XIV), Berlin.
—— (1901), *Fragmentsammlung den griechischen Ärzte I: Dei Fragmente der sikelischen Ärzte Akron, Philiston und des Diokles von Karystos*, Berlin.
—— (1907), "Erasistratos", *RE*, 6.1, cols. 343–50.
Welskopf, E.C. (ed.) (1965), *Neue Beiträge zur Geschichte der Alten Welt, II: Römisches Reich*, Berlin.
Weyl, H. (1952), *Symmetry*, Princeton.
White, K.D. (1970), *Roman Farming*, London.
—— (1975), *Farm Equipment of the Roman World*, Cambridge.
Wiberg, J. (1914), "The anatomy of the brain in the works of Galen and 'Ali 'Abbâs, a comparative historical-anatomical study", *Janus*, 19, 17–32; 84–104.
Wiesner, J. (ed.) (1987), *Aristoteles I Werk und Wirkung, Paul Moraux gewidmet II*, Berlin.
Wilkie, J. (1937), *The Dissection and Study of the Sheep's Brain*, Oxford.
Williams, P.L. et al. (eds.) (1989), *Gray's Anatomy*, Thirty Seventh Edition, Edinburgh.
Wilson, J.R. (1980), "ΚΑΙΡΟΣ as 'due measure'", *Glotta*, LVIII, 177–204.
Wilson, L.G. (1959), "Erasistratus, Galen and the pneuma", *BHM*, 33, 293–314.
Withington, E.T. (1928), *Hippocrates*, Vol. III, Cambridge Mass.—London.
De Witt, F. (1959), "An historical study on theories of the placenta to 1900", *J.Hist.Med.*, 14, 360–374.
Wittern, R.; Pellegrin, P. (eds.) (1996), *Hippokratische Medizin und antike Philosophie. Medizin Der Antiken I*, Hildesheim.
Woolf, G. (1994), "Becoming Roman, staying Greek: culture, identity and the civilizing process in the Roman east". *PCPS*, 116–143.
Woollam, D.H.M. (1957), "The historical significance of the cerebrospinal fluid", *Med.Hist.*, 1, 91–114.
—— (1958), "Concepts of the brain and its functions in classical antiquity", in Poynter (ed.), 5–18.
Wright, J.P.; Potter, P. (eds.) (2000), *Psyche and Soma: physicians and metaphysicians on the mind-body problem from Antiquity to Enlightenment*, Oxford.
Ziman, J. (1978), *Reliable knowledge. An exploration of the grounds for belief in science*, Cambridge.

INDEX OF SOURCES AND LIST OF EDITIONS

GENERAL INDEX

STUDIES IN ANCIENT MEDICINE

1. F. Kudlien and Richard J. Durling (eds.). *Galen's Method of Healing.* Proceedings of the 2nd International Galen Symposium. 1991 ISBN 90 04 09272 2
2. Hippocrates. *Pseudepigraphic Writings.* Letters — Embassy — Speech from the Altar — Decree. Edited and translated by Wesley D. Smith. 1990. ISBN 90 04 09290 0
3. Robert I. Curtis. *Garum and Salsamenta.* Production and Commerce in Materia Medica. 1991. ISBN 90 04 09423 7
4. Jody Rubin Pinault. *Hippocratic Lives and Legends.* 1992. ISBN 90 04 09574 8
5. Richard J. Durling. *A Dictionary of Medical Terms in Galen.* 1993. ISBN 90 04 09754 6
6. Willem F. Daems. *Nomina simplicium medicinarum ex Synonymariis Medii Aevi collecta.* Semantische Untersuchungen zum Fachwortschatz hoch- und spätmittelalterlicher Drogenkunde. 1993. ISBN 90 04 09672 8
7. Irene and Walter Jacob (eds.). *The Healing Past.* Pharmaceuticals in the Biblical and Rabbinic World. 1993. ISBN 90 04 09643 4
8. Marie-Hélène Marganne. *L'ophtalmologie dans l'Égypte gréco-romaine d'après les papyrus littéraires grecs.* 1994. ISBN 90 04 09907 7
9. Samuel S. Kottek. *Medicine and Hygiene in the Works of Flavius Josephus.* 1994. ISBN 90 04 09941 7
10. Charles Burnett and Danielle Jacquart (eds.). *Constantine the African and ʿAlī ibn al-ʿAbbās al-Maǧūsī.* The *Pantegni* and Related Texts. 1994. ISBN 90 04 10014 8
11. J.N. Adams. *Pelagonius and Latin Veterinary Terminology in the Roman Empire.* 1995. ISBN 90 04 10281 7
12. Ivan Garofalo (ed.). *Anonymi medici De morbis acutis et chroniis.* Translated into English by Brian Fuchs. 1997. ISBN 90 04 10227 2
13. Armelle Debru. *Le corps respirant.* La pensée physiologique chez Galien. 1996. ISBN 90 04 10436 4
14. Guigonis de Caulhiaco (Guy de Chauliac). *Inventarium sive Chirurgia Magna.* 2 volumes. Vol.I: Text. Edited by Michael R. McVaugh; Vol.II: Commentary. Prepared by Michael R. McVaugh & ††Margaret S. Ogden. 1997. ISBN 90 04 10706 1 *(I)*; ISBN 90 04 10784 3 *(II)*; ISBN 90 04 10785 1 *(Set)*
15. Mark Grant. *Dieting for an Emperor.* A Translation of Books 1 and 4 of Oribasius' *Medical Compilations* with an Introduction and Commentary. 1997. ISBN 90 04 10790 8
16. Armelle Debru (ed.). *Galen on Pharmacology.* Philosophy, History and Medicine. Proceedings of the Vth International Galen Colloquium, Lille, 16-18 March 1995. 1997. ISBN 90 04 10403 8

17. MARIE-HÉLÈNE MARGANNE. *La chirurgie dans l'Égypte gréco-romaine d'après les papyrus littéraires grecs.* 1998. ISBN 90 04 11134 4
18. KLAUS-DIETRICH FISCHER, DIETHARD NICKEL & PAUL POTTER (eds.). *Text and Tradition.* Studies in Ancient Medicine and its Transmission. Presented to Jutta Kollesch. 1998. ISBN 90 04 11052 6
19. KEITH DICKSON. *Stephanus the Philosopher and Physician.* Commentary on Galen's *Therapeutics to Glaucon.* 1998. ISBN 90 04 10935 8
20. PHILIPP J. VAN DER EIJK (ed.). *Ancient Histories of Medicine.* Essays in Medical Doxography and Historiography in Classical Antiquity. 1999. ISBN 90 04 10555 7
21. CHRISTINE F. SALAZAR. *The Treatment of War Wounds in Graeco-Roman Antiquity.* 2000. ISBN 90 04 11479 3
22. PHILIP J. VAN DER EIJK. *Diocles of Carystus.* A Collection of Fragments with Translation and Commentary. Volume I: Text and Translation. 2000. ISBN 90 04 10265 5, ISBN 90 04 12013 0 (*Set*)
23. PHILIP J. VAN DER EIJK. *Diocles of Carystus.* A Collection of Fragments with Translation and Commentary. Volume II: Commentary. 2001. ISBN 90 04 12012 2, ISBN 90 04 12013 0 (*Set*)
24. TECUSAN, M. *The Fragments of the Methodists, Volume One.* Methodism outside Soranus. 2003. ISBN 90 04 12451 9
25. JULIE LASKARIS. *The Art is Long.* On the Sacred Disease and the Scientific Tradition. 2002. ISBN 90 04 12152 8
26. JULIUS ROCCA. *Galen on the Brain.* Anatomical Knowledge and Physiological Speculation in the Second Century AD. 2003. ISBN 90 04 12512 4

www.ingramcontent.com/pod-product-compliance
Lightning Source LLC
Chambersburg PA
CBHW071326230926
51849CB00002B/323
* 9 7 8 9 0 0 4 1 2 5 1 2 4 *